MW01517973

Computer-Aided Mechanical Design and Analysis

THIRD EDITION

Computer-Aided Mechanical Design and Analysis

THIRD EDITION

V RAMAMURTI
(F.N.A, F.N.A.E.)

Department of Applied Mechanics
Indian Institute of Technology
Madras

McGraw-Hill
New York   San Francisco   Washington, D.C.   Auckland   Bogotá
Caracas   Lisbon   London   Madrid   Mexico City   Milan
Montreal   New Delhi   San Juan   Singapore
Sydney   Tokyo   Toronto

McGraw-Hill

A Division of The McGraw-Hill Companies

1 2 3 4 5 6 7 8 9 0   DOC/DOC   9 0 3 2 1 0 9 8

ISBN 0-07-060036-8

This book was previously published © 1996, 1992, 1987 by Tata McGraw-Hill Publishing Company Limited.

Printed and bound by R. R. Donnelley & Sons Company.

McGraw-Hill books are available at special quantity discounts to use as premiums and sales promotions, or for use in corporate training programs. For more information, please write to the Director of Special Sales, McGraw-Hill, 11 West 19th Street, New York, NY 10011. Or contact your local bookstore.

 This book is printed on recycled, acid-free paper containing a minimum of 50% recycled, de-inked fiber.

Respectfully Dedicated to the
Memory of
My Beloved Parents

BHAGIRATHI AND N VISWANATHAN

Preface to the Third Edition

This edition is brought out after a lapse of four years. During this period, awareness for teaching this subject to the senior level undergraduates and to the postgraduates in mechanical engineering has been realized by most of the technical institutions. Besides, most manufacturing industries are equipped with a few PCs in their design offices. This facilitates designing all their products in a scientific manner using in-house computing facility. This, in turn, has necessitated the training of their design personnel in computer aided design. Keeping these in mind, these contents have been modified. The chapter on cyclic symmetry has been substantially improved. The size of the book has gone by about 20%.

I would like to thank my following students who have helped me in bringing out this edition: M. Hanumantha Rao, P. Laxmi Prasad, Mahesh M. Bhat, P. Malathi, R. Nandakumar, D. Subramani and Sumanta Neogy. I express my gratitude to Messrs K.C.P. Ltd., Madras with whom I had more than 20 years of interaction. Origin of some of the photographs included and case studies cited, can be traced to the association I had with this industry.

V RAMAMURTI

Preface to the First Edition

The primary aim of writing this book is to indicate the enormous amount of numerical work involved in solving large-size problems of interest to practising engineers. It is imperative to get to know the most efficient algorithm to solve these problems. Throughout the course of the book, emphasis has been on the core and time needed to solve any given problem by different methods. It is presumed that the reader has the basic knowledge of strength of materials, theory of machines and computer programming.

The study has been restricted to the behaviour in the linear range in order to reduce the size of the book. Transient and steady-state vibration problems as well as static problems have been considered. Use of finite difference and finite element methods of formulation has been indicated. A number of computer programs (classroom-tested) have been given. Example problems have been worked out and an adequate number of additional exercises have been included. The last chapter is on "case studies" using the subject matter covered in the five chapters. The study of cyclic symmetric objects has been given importance in chapter 4.

The material reported in this book will be useful to practising engineers in industries having digital computers and to college seniors and research students in the field of machine design.

At the outset I would like to thank wholeheartedly the authorities of the Indian Institute of Technology, Madras for the congenial atmosphere provided for writing this book. I gratefully acknowledge the enormous amount of work put in by my students Mr. M. Ananda Rao, Mr. S. Ramadurai, Dr. P. Balasubramanyam, Mr. V. Krishnan, Mr. V. Om Prakash, Mr. K. Ramesh, Mr. N.G.V. Sai Diwakar, Mr. P. Seshu and Mr. T. Sthanu Subramanyam in writing the computer programs and running sample problems. A major portion of Chapter 4 is the PhD work of Mr. P. Balasubramanyam. The enthusiasm shown by him in preparing this chapter is appreciated. The support given by the Computer Centre at IIT Madras in preparing the computer listings is acknowledged. The information shown in Tables 1.1 to 1.5 has been collected from the *Handbook of Finite Element Systems* edited by C.A. Brebbia. I am grateful to him.

Appreciation is also due to my wife Raji and children Rama and Ramesh without whose cooperation and help it would have been impossible for me to complete this assignment.

V RAMAMURTI

Contents

1   Introduction

The common features of practical problems a design engineer faces today are the problems of complexity and diversity. Let us consider a few examples: a pressure vessel designed to withstand high pressures and thermal gradients; a turbo rotor carrying several stages of blades and disks; a multi-cylinder I.C. Engine crank-shaft; and a machine tool structure and gear wheel subjected to moving loads. All are critical machine components. Some are important because of their sheer size, e.g. a multi-layered pressure vessel for an internal pressure of 200 atm and 6 m diameter involves 400 tonnes of steel just in material alone. There has, hence, been an increasing demand on the designer to optimise his designs—primarily to reduce the wastage of material. The days of empirical design, wherein experience was the best knowledge book, were thus numbered and there emerged a clear need for a rational approach towards optimal design. Soon, it has been recognised that the first step after the initial, tentative design has been completed, is to know how well it works, after all, only then can an improvement be thought of. The "code" based designer immediately turned to the analyst seeking a solution. That, in essence, is the break-away point in the history of engineering design. The analysts could no longer continue to work with idealised uniform beams, plates and shells. They were required to offer reasonably accurate solutions to the practical problems of a design engineer whose structural geometry was complicated, loading approximately estimated and boundary conditions not found written on the structure. Classical mechanics soon evolved a series of approximate methods of structural analysis based on the principle of minimization of the total potential. The classical methods formulated a differential equation governing a class of problems, a solution for which is then sought. This approach necessarily demanded a certain regularity from the geometry of structure, material properties, nature of loading and deformation and, of course, elegant boundary conditions. The energy methods sought to satisfy the principle of minimum total potential by a displacement function assumed a priorii: The usual choices had been the trignometric or algebraic series. Though approximate theory yielded valuable results of practical significance, for complex structures the assumption on deflection curve was really demanding. Let us take a simple example—that of a pre-twisted, cantilever beam. This is a valid representation of a helicopter rotor blade or steam/gas turbine or compressor blading. There is simply no function readily available for approximating the deflection curve taking into account the angle of the pre-twist. In came the weighted residual formulation, culminating in the establishment of the finite element method as the tool for analysing a wide range of practical problems. Mention must also be made of another age-old powerful method—the finite difference method which seeks to solve the difference form of the governing differential equation, though its main drawback is the requirement on the existence of governing differential equations. Still whenever such a formulation is available, its performance is comparable only to the best of solution schemes.

Thus, with time, engineering design and analysis evolved to such an extent that they can no more be separated except in trivial cases. This has been possible essentially because of two developments: the integral formulation of structural mechanics problems (e.g. finite element method) and the advent of high-speed digital computers. Today, the designer-cum-analyst can ask for the stress distribution in or the natural frequencies of practically any type of structure. But this depends on his ability to solve the problem at hand in the most efficient way possible. The algorithms of equation solvers have been constantly improved to a stage of nearing optimum efficiency. But that is going too far too fast. Let us first briefly review as to how exactly such a physical entity as a structure in front of our eyes is reduced to a set of simultaneous algebraic equations.

The complexity of structures and complicated loading such as static, steady state, dynamic, transient and random are no more matters of concern. When using the finite element method, their behaviour, in all generality, can be characterised by just three quantities—stiffness, mass and damping. We now seek to review the formulation of these characteristics by the two approaches—the finite element and finite difference.

By and large, the skill of the designer revolves around his ability to decide on the best finite element model for his analysis. Programs incorporating these features are available in many standard packages like ADINA, ANSYS, APPLESAP, ASKA, MARC, MSC NASTRAN, NISA, SAP7 and SESAM [1.1]. Modified versions of these are being marketed all over the world by many software companies. One may wonder why there should be any necessity at all to know as to how these ready-made packages work. It is natural to think that as long as one knows how to input the parameters of interest and how to interpret the results, it should be more than adequate. This is not at all true. If one is aiming to buy these programs, the following are some of the problems he encounters: the right choice of package, the ability of the computer at his disposal to handle the large-size program procured and the Himalayan blunder of buying unwanted programs which the organisation is unlikely to use at any point of time. Hence, it is essential to know what these programs contain, the basic methods used in solving these complicated problems, the kind of infrastructure these programs need and how to modify these programs should such an eventuality arise.

Most of the problems analysed in this book fall under the category of statically indeterminate structures. Given the forces and moments acting on the unit under consideration, one cannot expect the thickness or the breadth to be computed by specifying the working stress. On the contrary, assuming the dimensions of the unit, doing the analysis for the given system of forces is straightforward. If the resultant stresses and deformations are within limits, the design is acceptable; otherwise the dimensions have to be modified and the analysis redone till the acceptable limit on stresses and displacements is obtained. The foregoing approach emphasizes analysis but uses it for checking the adequacy of the design. In other words, design and analysis are becoming inseparable for most of the practical problems. It is not fair to term this approach as purely analysis. An explanation regarding the justification of the title of the book is also warranted at this juncture. Classical books on machine element design deal with design only and not with machine drawing. Likewise, a book on computer aided design must deal with the design procedures for fixing the sizes of complicated machine elements. They should not deal with computer aided graphics. They should not also deal with the hardware associated with a computer. Besides, it was felt that it is not desirable to deal with computer programs for doing the analysis of machine elements like springs, bolts, nuts and shafts since they are too elementary.

The most commonly used finite elements in mechanical engineering will now be presented. The development of these elements has been discussed in detail in other books and papers [1.2 to 1.8]. An excellent treatment of this method to problems in the area of machine dynamics is reported by Krämer [1.9].

► 1.1 FINITE ELEMENT METHOD

The large-size problems handled by modern digital computers connected with the static and dynamic linear analysis of complicated machines or structures are generally of the form

$$[M]\{\ddot{u}\} + [C]\{\dot{u}\} + [K]\{u\} = \{f(t)\} \tag{1.1}$$

where $[M]$ is the global mass matrix, $[C]$ the global damping matrix and $[K]$ the global stiffness matrix. $\{f(t)\}$ is the forcing function vector in time, $\{u\}$ is the resultant displacement vector, $\{\dot{u}\}$ and $\{\ddot{u}\}$ represent its velocity and acceleration vectors respectively. Generally, $[M]$, $[C]$ and $[K]$ are banded. Depending upon the nature of these coefficients, the problems are classified as static, dynamic, linear or non-linear. The following are some of the specific classifications:

(i) When $[C] = 0$, $[M] = 0$, $[K]$, and $\{f(t)\}$ are constants, the result is a static linear problem.

(ii) When $[M]$ and $[C]$ are absent, $[K]$ is a function of $\{u\}$, and $\{f(t)\}$ a constant, the result is a non-linear static problem.

(iii) If $\{f(t)\}$ and $[C]$ are absent, and $[M]$ and $[K]$ are constants, one gets an eigenvalue problem.

(iv) If $[M]$, $[C]$ and $[K]$ are constants and $\{f(t)\}$ is a periodic forcing function, the result is a multi-degree of freedom steady state vibration problem.

(v) If $[M]$, $[C]$ and $[K]$ are constants and $\{f(t)\}$ is a transient function of time, the result is a transient vibration problem.

Considerable amount of effort has gone into the solution of such problems. The use of the finite element or finite difference method for analysing the varieties of problems leads us ultimately to Eq. (1.1).

1.1.1 Finite Element Procedure

The structure is idealised by just subdividing the original object into an assembly of discrete elements such that the resulting structure will simulate the original one. The elements are connected to each other at points known as nodes. After making a reasonable assumption on the behaviour of an element, the kinetic energy T and strain energy U are calculated as functions of nodal point displacements.

If the structure is composed of N elements, we can write

$$T = \sum_{i=1}^{N} T_i$$

$$U = \sum_{i=1}^{N} U_i \qquad (1.2)$$

The application of Lagrange's equation then results in the governing equation of type (1.1) when damping is ignored.

1.1.2 Discretisation

Discretising the structure requires experience and complete understanding of the behaviour of the structure. The structure can behave like a beam, truss, plate, shell, etc. Having chosen one of the models, one can compute the following:

$$\{f\} = [N]\{\delta\} \qquad (1.3)$$

$\{f\}$ is the displacement vector at an arbitrary point inside the element, $\{\delta\}$ the nodal displacement vector and $[N]$ the shape function. Likewise, components of strain in an arbitrary point can also be written as

$$\{\varepsilon\} = [B]\{\delta\} \qquad (1.4)$$

where $[B]$ is the strain displacement vector.
The stress component $\{\sigma\}$ can be expressed as

$$\{\sigma\} = [D]\{\varepsilon\} \qquad (1.5)$$

where $[D]$ is the elasticity matrix.
The strain energy U for the element can be written as

$$U = \frac{1}{2} \int_{vol} \varepsilon^T \sigma d\,vol \qquad (1.6)$$

The velocity vector of an arbitrary point can be written as

$$\{\dot{f}\} = [N]\{\dot{\delta}\} \qquad (1.7)$$

where $\{\dot{\delta}\}$ is the time derivative of $\{\delta\}$.
    Hence the kinetic energy T can be written as

$$T = \frac{1}{2} \int_{vol} \rho\{\dot{\delta}\}^T [N]^T [N]\{\dot{\delta}\}\, dvol \qquad (1.8)$$

where ρ is the density of the material.
Equation (1.6) can be recast as

$$U = \frac{1}{2}\{\delta\}^T [K_e]\{\delta\} \qquad (1.9)$$

$$[K_e] = \int_{vol} [B]^T [D][B]\, dvol \qquad (1.10)$$

Here $[K_e]$ is known as the element stiffness matrix.

Similarly Eq. (1.8) can be recast as

$$T = \frac{1}{2}\{\dot{\delta}\}^T [M_e]\{\dot{\delta}\} \tag{1.11}$$

where

$$[M_e] = \int_{\text{vol}} \rho [N]^T [N] \, d\text{vol} \tag{1.12}$$

$[M_e]$ is known as the mass matrix.

➤ **1.2 TYPICAL ELEMENTS**

A number of finite elements have been developed to solve complicated problems. They are dealt with exclusively in books on finite element method. In this section five typical elements are discussed which can handle most of the problems in mechanical design.

1.2.1 Beam Element (Bending in One Plane)

A typical beam element of length l with three degrees of freedom at either of its nodes is shown in Fig. 1.1. Any point within the element, we write

$$u = a_1 + a_2 x$$
$$v = a_3 + a_4 x + a_5 x^2 + a_6 x^3 \tag{1.13}$$
$$\theta = dv/dx = a_4 + 2a_5 x + 3a_6 x^2$$

As is clear, a linear variation has been assumed for the longitudinal displacement u and a cubic polynomial for the bending displacement v. Substituting the nodal coordinate

at $x = 0$, $\qquad u_1 = a_1 \quad v_1 = a_3 \quad \theta_1 = a_4$

at $x = 1$, $\qquad u_2 = a_1 + a_2$

$$v_2 = a_3 + a_4 l + a_5 l + a_6 l^3 \tag{1.14}$$
$$\theta_2 = a_4 + 2a_5 l + 3a_6 l$$

Back-substitution of a_i in Eq. (1.13) yields

$$\{f\} = [N]\{\delta\}^e$$

where $\{f\}$ is the displacement field $\begin{Bmatrix} u \\ v \end{Bmatrix}$ at any point within the element, $\{\delta\}^e$ the nodal displacement vector and N the interpolation functions. They can be expressed in the following form.

The nodal displacement vector for this beam is

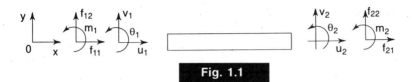

Fig. 1.1

2-D Beam element

$$
\{\delta\}^e = \begin{Bmatrix} u_1 \\ v_1 \\ \theta_1 \\ u_2 \\ v_2 \\ \theta_2 \end{Bmatrix} \qquad (1.15)
$$

One can write the displacement field inside the element as

$$
\{f\} = \begin{Bmatrix} u \\ v \end{Bmatrix} = \begin{bmatrix} a_1 & 0 & 0 & a_2 & 0 & 0 \\ 0 & a_3 & a_4 & 0 & a_5 & a_6 \end{bmatrix} \begin{Bmatrix} u_1 \\ v_1 \\ \theta_1 \\ u_2 \\ v_2 \\ \theta_2 \end{Bmatrix} \qquad (1.16)
$$

The constants a_1 to a_6 can be evaluated from the boundary conditions at $x = 0$ and $x = l$.

$$
a_1 = (l - x)/l
$$

$$
a_2 = x/l
$$

$$
a_3 = 1 - (3x^2/l^2) + (2x^3/l^3) \qquad (1.17)
$$

$$
a_4 = x - (2x^2/l) + (x^3/l^2)
$$

$$
a_5 = (3x^2/l^2) - (2x^3/l^3)
$$

$$
a_6 = - (x^2/l) + (x^3/l^2)
$$

For this problem, the longitudinal strain

$$
\varepsilon = (du/dx) - y(d^2v/dx^2) \qquad (1.18)
$$

Following the procedure underlined in Eq. (1.10), the stiffness matrix can be expressed as [1.10]

$$[K_e] = \begin{bmatrix} AE/l & 0 & 0 & -AE/l & 0 & 0 \\ & 12EI/l^3 & 6EI/l^2 & 0 & -12EI/l^3 & 6EI/l^2 \\ & & 4EI/l & 0 & -6EI/l^2 & 2EI/l \\ & & & AE/l & 0 & 0 \\ & & & & 12EI/l^3 & -6EI/l^2 \\ \text{Symmetric} & & & & & 4EI/l \end{bmatrix}$$

(1.19)

The mass matrix can also be expressed in a similar fashion using Eq. (1.12). This can be written as [1.10].

$$[M]_e = \frac{\rho Al}{420} \begin{bmatrix} 140 & 0 & 0 & 70 & 0 & 0 \\ 0 & 156 & 22l & 0 & 54 & -13l \\ 0 & 22l & 4l^2 & 0 & 13l & -3l^2 \\ 70 & 0 & 0 & 140 & 0 & 0 \\ 0 & 54 & 13l & 0 & 156 & -22l \\ 0 & -13l & -3l^2 & 0 & -22l^2 & 4l^2 \end{bmatrix}$$

(1.20)

If the mass of the beam is assumed to be lumped at the two ends (one half at each end), the mass matrix will assume the following simple form:

$$[M]_e = \frac{\rho lA}{2} \begin{bmatrix} 1 & & & & & \\ & 1 & & & & \\ & & 0 & & & \\ & & & 1 & & \\ & & & & 1 & \\ & & & & & 0 \end{bmatrix}$$

(1.21)

1.2.1.1   Typical Applications

The behaviour of superstructures made of frames to support boilers, motors and other industrial equipment deforming predominantly in one plane, the behaviour of rotating uniform and non-uniform shafts, portal and semi-portal frames of electric overhead travelling cranes, fall under this category.

1.2.2 3D Beam Element (Bending in Two Planes Mutually at Right Angles to Each Other)

This finds very wide application in industry [1.11]. The previous case is a subset of problems of this type. A typical beam element having axial deformation u in x direction and lateral deflection, due to bending in y and z directions, is shown in Fig. 1.2. Extending the arguments of the previous case to this type (using the right-handed co-ordinate system), one can write the stiffness matrix as shown in Eq. (1.22) on p. 10.

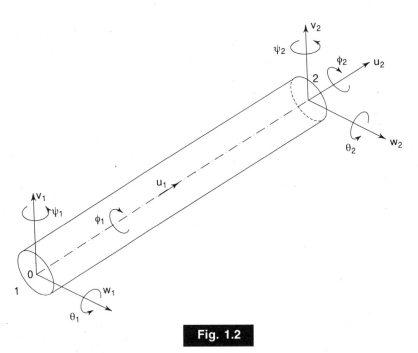

Fig. 1.2

3D Beam element in local coordinate system

Lumped mass matrix will be of the size (12×12) in nature, similar to Eq. (1.21) with 1 in the first three and 1 in the 7th to the 9th diagonal elements. The nodal displacement vector is of the form

$$\{u_1, \ v_1, \ w_1, \ \phi_1, \ \psi_1, \ \theta_1, \ u_2, \ v_2, \ w_2, \ \phi_2, \ \psi_2, \ \theta_2\}^T$$

It may be observed that the first, second, sixth, seventh, eighth and twelfth degrees of freedom correspond to the 2D beam element (Eq. (1.15)).

$$[K] = \begin{bmatrix}
\frac{EA}{l} & 0 & 0 & 0 & 0 & 0 & -\frac{EA}{l} & 0 & 0 & 0 & 0 & 0 \\[4pt]
 & \frac{12EI_z}{l^3} & 0 & 0 & 0 & \frac{6EI_z}{l^2} & 0 & -\frac{12EI_z}{l^3} & 0 & 0 & 0 & \frac{6EI_z}{l^2} \\[4pt]
 & & \frac{12EI_y}{l^3} & 0 & -\frac{6EI_y}{l^2} & 0 & 0 & 0 & -\frac{12EI_y}{l^3} & 0 & -\frac{6EI_y}{l^2} & 0 \\[4pt]
 & & & \frac{GJ}{l} & 0 & 0 & 0 & 0 & 0 & -\frac{GJ}{l} & 0 & 0 \\[4pt]
 & & & & \frac{4EI_y}{l} & 0 & 0 & 0 & \frac{6EI_y}{l^2} & 0 & \frac{2EI_y}{l} & 0 \\[4pt]
 & & & & & \frac{4EI_z}{l} & 0 & -\frac{6EI_z}{l^2} & 0 & 0 & 0 & \frac{2EI_z}{l} \\[4pt]
 & & & & & & \frac{AE}{l} & 0 & 0 & 0 & 0 & 0 \\[4pt]
 & \text{Symmetric} & & & & & & \frac{12EI_z}{l^3} & 0 & 0 & 0 & -\frac{6EI_z}{l^2} \\[4pt]
 & & & & & & & & \frac{12EI_y}{l^3} & 0 & \frac{6EI_y}{l^2} & 0 \\[4pt]
 & & & & & & & & & \frac{GJ}{l} & 0 & 0 \\[4pt]
 & & & & & & & & & & \frac{4EI_y}{l} & 0 \\[4pt]
 & & & & & & & & & & & \frac{4EI_z}{l}
\end{bmatrix} \qquad (1.22)$$

Typical Applications

These elements are used to represent the chassis of automobiles, railway bogie frames, pipe lines, 3-dimensional frame structures supporting various equipment in industrial plants.

1.2.3 Plate and Shell Element

A triangular plate and shell element with 6 d.o.f. per node is shown in Fig. 1.3. This can be used in all plate-bending problems as well as for approximating a shell as an assemblage of flat plate elements. Since the in-plane and bending behaviours are uncoupled for an isotropic plate, the individual stiffness matrices can be separately obtained and then combined to give the element stiffness matrix, using natural co-ordinates [1.2], assuming a linear variation for in-plane displacements u, v and a cubic variation for the bending displacement w.

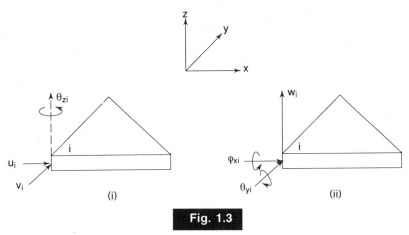

Fig. 1.3

A flat triangular element subjected to (i) In-plane displacement (ii) Bending displacement

1.2.3.1 Shape Functions

The polynomials for the in-plane displacements u and v are linear in the area co-ordinates L_1, L_2 and L_3 while for the bending displacement w, the polynomial is cubic. For a three-noded triangular element, a linear polynomial in L_1, L_2 and L_3 is satisfactory for the in-plane displacements u and v, since there will be three displacements and three generalised co-ordinates for each u and v, i.e.

$$u = \alpha_1 L_1 + \alpha_2 L_2 + \alpha_3 L_3 \tag{1.23}$$

$$v = \alpha_4 L_1 + \alpha_5 L_2 + \alpha_6 L_3 \qquad (1.24)$$

For the bending problem, each node has three degrees of freedom, viz. the displacement w and the rotations θ_x and θ_y and hence, the element requires a polynomial having at least nine generalised co-ordinates. w is expressed in the form of a cubic polynomial [1.2], [1.12].

$$w = \beta_1 L_1 + \beta_2 L_2 + \beta_3 L_3 + \beta_4 \left(L_2^2 L_1 + \frac{1}{2} L_1 L_2 L_3 \right) + \ldots +$$

$$\beta_9 \left(L_1^2 L_3 + \frac{1}{2} L_1 L_2 L_3 \right) \qquad (1.25)$$

The area co-ordinates L_1, L_2 and L_3 expressed as

$$L_1 = A_1/A; \qquad L_2 = A_2/A: \qquad L_3 = A_3/A \qquad (1.26)$$

with

$$2A_i = 2A_{jk} + b_i x + c_i y \qquad (1.27)$$

varying i, j and k cyclically.

A is the area of the element and A_{12}, A_{23}, A_{13} are the areas of the triangles defined by the typical point M. A_{jk} is the area of the triangle for which the nodes j, k and the origin of the Cartesian co-ordinate system are the vertices.

Further, let

$$c_1 = x_3 - x_2; \qquad c_2 = x_1 - x_3; \qquad c_3 = x_2 - x_1 \quad (1.28)$$

$$- b_1 = y_3 - y_2; \qquad - b_2 = y_1 - y_3; \qquad - b_3 = y_2 - y_1 \quad (1.29)$$

The in-plane nodal displacement vector is defined by

$$\{q_1\} = [u_1 \quad v_1 \quad u_2 \quad v_2 \quad u_3 \quad v_3]^T \qquad (1.30)$$

and the bending nodal displacement vector is defined by

$$\{q_2\} = [w_1, \theta_{x_1}, \theta_{y_1}, w_2, \theta_{x_2}, \theta_{y_2}, w_3, \theta_{x_3}, \theta_{y_3}]^T \qquad (1.31)$$

where

$$\theta_{x_i} = (\partial w / \partial y)_i \qquad (1.32)$$

and

$$\theta_{y_i} = (- \partial w / \partial x)_i \qquad (1.33)$$

Evaluating $i = (1, 2, 3)$, the constants α_1 to α_6 and β_1 to β_9, we get

$$\begin{Bmatrix} u \\ v \end{Bmatrix} = [N_P]\{q_1\} \qquad (1.34)$$

where $[N_P]$ and $[N_b]$ are the in-plane and bending shape functions.

$$[N_P] = \begin{bmatrix} L_1 & 0 & L_2 & 0 & L_3 & 0 \\ 0 & L_1 & 0 & L_2 & 0 & L_3 \end{bmatrix} \qquad (1.35)$$

and

$$\{w\} = [N_b]\{q_2\} \qquad (1.36)$$

$$[N_b] = [N_{b_1}, N_{b_2}, ..., N_{b_9}] \qquad (1.37)$$

$$N_{b_1} = L_1 + L_1^2 L_2 - L_2^2 L_1 + L_1^2 L_3 - L_3^2 L_1 \qquad (1.38)$$

$$N_{b_2} = b_2 \left(L_1^2 L_3 + \frac{1}{2} L_1 L_2 L_3 \right) - b_3 \left(L_1^2 L_2 + \frac{1}{2} L_1 L_2 L_3 \right) \quad (1.39)$$

$$N_{b_3} = c_2 \left(L_1^2 L_3 + \frac{1}{2} L_1 L_2 L_3 \right) - c_3 \left(L_1^2 L_2 + \frac{1}{2} L_1 L_2 L_3 \right) \quad (1.40)$$

The other shape functions for nodes 2 and 3 can be written down by a cyclic permutation of the suffixes 1, 2, 3.

Derivation of Stiffness Matrix

The stiffness matrix consists of two parts, viz. the in-plane stiffness and bending stiffness matrices.
We have the in-plane strain,

$$\{\varepsilon_p\} = \left\{ \begin{array}{c} \varepsilon_x \\ \varepsilon_y \\ \varepsilon_{xy} \end{array} \right\} = \left\{ \begin{array}{c} \partial u / \partial x \\ \partial v / \partial y \\ (\partial u / \partial y) + (\partial v / \partial x) \end{array} \right\}$$

$$= \left\{ \begin{array}{c} \sum_{i=1}^{3} (\partial u / \partial L_i)(\partial L_i / \partial x) \\ \sum_{i=1}^{3} (\partial v / \partial L_i)(\partial L_i / \partial y) \\ \sum_{i=1}^{3} (\partial u / \partial L_i)(\partial L_i / \partial y) + (\partial v / \partial L_i)(\partial L_i / \partial x) \end{array} \right\}$$

$$= \frac{1}{2A} \begin{bmatrix} b_1 & 0 & b_2 & 0 & b_3 & 0 \\ 0 & c_1 & 0 & c_2 & 0 & c_3 \\ c_1 & b_1 & c_2 & b_2 & c_3 & b_3 \end{bmatrix} \left\{ \begin{array}{c} u_1 \\ v_1 \\ u_2 \\ v_2 \\ u_3 \\ v_3 \end{array} \right\} \qquad (1.41)$$

Hence $\qquad\qquad \{\varepsilon_p\} = [B_p]\{q_1\} \qquad (1.42)$

In-plane stress vector

$$\{\sigma_P\} = \begin{Bmatrix} \sigma_x \\ \sigma_y \\ \sigma_{xy} \end{Bmatrix} = [D_P]\{\varepsilon_P\}$$

$[D_p]$ in-plane elasticity matrix is given in Appendix 3. With this, the in-plane strain energy P_1 is given by

$$P_1 = \frac{1}{2} \iint\limits_A \{\varepsilon_P\}^T \{\sigma_P\} \, dA$$

$$= \frac{1}{2} \iint\limits_A \{\varepsilon_P\}^T [D_P]\{\varepsilon_P\} \, dA \qquad (1.43)$$

$$= \frac{1}{2} \{q_1\}^T [k_P]\{q_1\} \qquad (1.44)$$

where $[k_P]$ is the element in-plane stiffness matrix and is given by

$$[k_P] = \iint\limits_A [B_P]^T [D_P][B_P] \, dA \qquad (1.45)$$

With the shape functions for bending displacement, the strain vector $\{\varepsilon_f\}$ is expressed as

$$\{\varepsilon_f\} = \begin{bmatrix} -\dfrac{\partial^2 w}{\partial x^2} \\[2mm] -\dfrac{\partial^2 w}{\partial y^2} \\[2mm] -2\dfrac{\partial^2 w}{\partial x \partial y} \end{bmatrix} = \begin{bmatrix} -\dfrac{\partial^2 Nb_1}{\partial x^2} & -\dfrac{\partial^2 Nb_2}{\partial x^2} & \cdots & -\dfrac{\partial^2 Nb_9}{\partial x^2} \\[2mm] -\dfrac{\partial^2 Nb_1}{\partial y^2} & -\dfrac{\partial^2 Nb_2}{\partial y^2} & \cdots & -\dfrac{\partial^2 Nb_9}{\partial y^2} \\[2mm] -2\dfrac{\partial^2 Nb_1}{\partial x \partial y} & -2\dfrac{\partial^2 Nb_2}{\partial x \partial y} & \cdots & -\dfrac{\partial^2 Nb_9}{\partial x \partial y} \end{bmatrix} \{q_2\} \quad (1.46)$$

$$\{\varepsilon_f\} = [S_1][S_2]\{q_2\} \qquad (1.47)$$

where

$$[S_1] = \frac{-1}{4A^2} \begin{bmatrix} b_1 & b_2 & b_3 & 2b_1b_2 & 2b_2b_3 & 2b_3b_1 \\ c_1 & c_2 & c_3 & 2c_1c_2 & 2c_2c_3 & 2c_3c_1 \\ 2b_1c_1 & 2b_2c_2 & 2b_3c_3 & (2b_1c_2 + & (2b_2c_3 + & (2b_2c_3 + \\ & & & 2b_2c_1) & 2b_3c_2) & 2b_3c_1) \end{bmatrix}$$

$$(1.48)$$

and

$$[S_2] = \begin{bmatrix} \dfrac{\partial^2 Nb_1}{\partial L_1^2} & \dfrac{\partial^2 Nb_2}{\partial L_1^2} & \cdots & \dfrac{\partial^2 Nb_9}{\partial L_1^2} \\[2mm] \dfrac{\partial^2 Nb_1}{\partial L_2^2} & \dfrac{\partial^2 Nb_2}{\partial L_2^2} & \cdots & \dfrac{\partial^2 Nb_9}{\partial L_2^2} \\[2mm] \dfrac{\partial^2 Nb_1}{\partial L_3^2} & \dfrac{\partial^2 Nb_2}{\partial L_3^2} & \cdots & \dfrac{\partial^2 Nb_9}{\partial L_3^2} \\[2mm] \dfrac{\partial^2 Nb_1}{\partial L_1 \partial L_2} & \dfrac{\partial^2 Nb_2}{\partial L_1 \partial L_2} & \cdots & \dfrac{\partial^2 Nb_9}{\partial L_1 \partial L_2} \\[2mm] \dfrac{\partial^2 Nb_1}{\partial L_2 \partial L_3} & \dfrac{\partial^2 Nb_2}{\partial L_2 \partial L_3} & \cdots & \dfrac{\partial^2 Nb_9}{\partial L_2 \partial L_3} \\[2mm] \dfrac{\partial^2 Nb_1}{\partial L_3 \partial L_1} & \dfrac{\partial^2 Nb_2}{\partial L_3 \partial L_1} & \cdots & \dfrac{\partial^2 Nb_9}{\partial L_3 \partial L_1} \end{bmatrix} \qquad (1.49)$$

Stress vector is given by

$$\{\sigma_f\} = [D_f]\{\varepsilon_f\}$$

$[D_f]$ elasticity matrix in plate flexure is given in Appendix 3.

But

$$P_2 = \frac{1}{2} \iint_A \{\varepsilon_f\}^T [D_f]\{\varepsilon_f\}\, dA \qquad (1.50)$$

$$= \frac{1}{2}\{q_2\}^T [k_f]\{q_2\} \qquad (1.51)$$

where $[k_f]$ is the element bending stiffness matrix and is given by

$$[k_f] = \iint_A [B_f]^T [D_f][B_f]\, dA \qquad (1.52)$$

The integration is performed by using the numerical three-point Gaussian integration over the triangular area (Appendix I).

The minimisation of total energy potentials P_1 and P_2 leads to two elastic stiffness matrices $[k_P]$ and $[k_f]$ associated with the nodal displacement vectors $\{q_1\}$ and $\{q_2\}$. It is important to note that rotation θ_z does not enter as a parameter into the definition of deformations in either $\{q_1\}$ or $\{q_2\}$. However, it is convenient for the assembly process to take this rotation into account and associate with it a fictitious couple M_z. The fact that it does not enter into the minimisation procedure can be accounted for by inserting an appropriate number of zeros in the stiffness matrix.

Redefining the combined nodal displacements as

$$\{q\} = [\delta_i, \quad \delta_j, \quad \delta_k]^T \tag{1.53}$$

$$\{\delta_i\} = [u_i, \ v_i, \ w_i, \ \theta_{xi}, \ \theta_{yi}, \ \theta_{zi}]^T \tag{1.54}$$

The elastic stiffness matrix is now made up from the following submatrices

$$[k_E] = \begin{bmatrix} k_{ii} & k_{ij} & k_{lm} \\ k_{ji} & k_{jj} & k_{jm} \\ k_{mi} & k_{mj} & k_{mm} \end{bmatrix} \tag{1.55}$$

with each submatrix in the form

$$\begin{bmatrix} k_P & & 0 \\ & k_f & \\ 0 & & k_{\theta_z} \end{bmatrix} \tag{1.56}$$

where k_{θ_z} is an arbitrary fictitious stiffness coefficient.

The element mass matrix

$$[m_E] = \int_{\text{vol}} \rho [N]^T [N] \, d(\text{vol}) \tag{1.57}$$

where the displacement vector within the element is

$$\begin{Bmatrix} u \\ v \\ w \end{Bmatrix} = \begin{matrix} [N] \\ (3,18) \end{matrix} \begin{matrix} \{q\} \\ (18,1) \end{matrix} \tag{1.58}$$

Typical Applications

These elements are used in analysing housings of machine tools, gear boxes, automobile radiator fan blades, fabricated gear wheels, bearing pedestals, storage vessels with or without partitions, bladed discs, pressure vessels, bus bodies, and rail coaches. This element is an ideal choice for more than 50% of machine elements. Use of this element comes in handy for optimising the weight of housing of hydraulic presses, press brakes, guillotine shears and eccentric presses.

1.2.4   Axi-symmetric Element

Axi-symmetric Loading

Many problems in engineering have loading rotationally symmetric on objects which are also rotationally symmetric. They will be experiencing deformations u and v only in the radial and axial directions. In its simplest form, these can be expressed as linear variations in r (radial co-ordinate) and in z (axial co-ordinate) (Fig. 1.4). If a triangular element is chosen as the finite element, the displacement vector can be expressed as

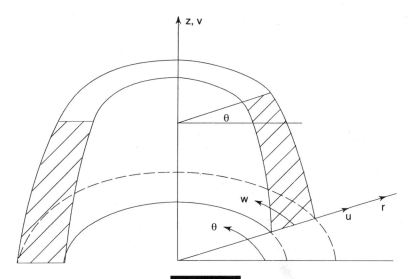

Fig. 1.4

An axi-symmetric solid, co-ordinate and displacements

$$\delta = \left\{ \begin{array}{c} u \\ v \end{array} \right\} \tag{1.59}$$

$$u = L_1 u_1 + L_2 u_2 + L_3 u_3$$
$$v = L_1 v_1 + L_2 v_2 + L_3 v_3 \tag{1.60}$$

Asymmetric Loading (Semi-analytic Approach)

The same shape function can be extended to general loading on a rotationally symmetric structure by expressing these displacements as trignometric functions in the circumferential direction thus

$$u = \Sigma \, u^n \, \cos \, n\theta$$
$$v = \Sigma \, v^n \, \cos \, n\theta \tag{1.61}$$
$$w = \Sigma \, w^n \, \sin \, n\theta$$

w is the circumferential displacement present in the case of an asymmetric problem. (For the most general type, the second set comprising of sine series for u and v and cosine series for w should also be considered.) The nodal displacement vector consisting of the three displacements at each node can be expressed as [1.2] and [1.13].

$$\{\delta^n\}^e = \{u_1^n, u_1^n, v_1^n, w_1^n, u_2^n, v_2^n, w_2^n, u_3^n, v_3^n, w_3^n\}^T \qquad (1.62)$$

The displacement function can be genaralised in the following form:

$$u^n = L_1(u_1 - u_3) + L_2(u_2 - u_3) + u_3$$
$$v^n = L_1(v_1 - v_3) + L_2(v_2 - v_3) + v_3 \qquad (1.63)$$
$$w^n = L_1(w_1 - w_3) + L_2(w_2 - w_3) + w_3$$

To evaluate the stiffness matrix of an element, we need to calculate the strain displacement transformation matrix. The element strains are obtained in terms of derivatives of the element displacements with respect to the local co-ordinates. But the element strains are defined in terms of natural co-ordinates. Therefore, we need to relate r and z derivatives to u, v and w. Relating r and z to the natural co-ordinates with linear shape functions in the natural co-ordinates, the co-ordinate interpolations are

$$r = L_1(r_1 - r_3) + L_2(r_2 - r_3) + r_3$$
$$z = L_1(z_1 - z_3) + L_2(z_2 - z_3) + z_3 \qquad (1.64)$$

$$\begin{bmatrix} \partial/\partial L_1 \\ \partial/\partial L_2 \end{bmatrix} = \begin{bmatrix} (\partial r/\partial L_1) \cdot (\partial z/\partial L_1) \\ (\partial r/\partial L_2) \cdot (\partial z/\partial L_2) \end{bmatrix} \begin{bmatrix} \partial/\partial r \\ \partial/\partial z \end{bmatrix} \qquad (1.65)$$

$$= \begin{bmatrix} (r_1 - r_3) & (z_1 - z_3) \\ (r_2 - r_3) & (z_2 - z_3) \end{bmatrix} \begin{bmatrix} (\partial/\partial r) \\ (\partial/\partial z) \end{bmatrix} \qquad (1.66)$$

$$= [J] \begin{bmatrix} (\partial/\partial r) \\ (\partial/\partial z) \end{bmatrix} \qquad (1.67)$$

where $[J]$ is the Jacobian operator relating the natural co-ordinate derivatives to the local co-ordinate derivatives. The Jacobian operator can be easily found using Eq. (1.66).

Now, it is required to obtain $\partial/\partial r$, $\partial/\partial z$, etc. to get the strain displacement relations. They can be calculated in the following way:

$$\left\{\begin{array}{c} (\partial/\partial r) \\ (\partial/\partial z) \end{array}\right\} = [J]^{-1}\left\{\begin{array}{c} (\partial/\partial L_1) \\ (\partial/\partial L_2) \end{array}\right\} = \left[\begin{array}{cc} a_{11} & a_{12} \\ a_{21} & a_{22} \end{array}\right]\left\{\begin{array}{c} (\partial/\partial L_1) \\ (\partial/\partial L_2) \end{array}\right\} \quad (1.68)$$

where a_{11}, a_{12}, a_{21}, a_{22} are elements of the inverse Jacobian matrix

$$\left\{\begin{array}{c} (\partial u/\partial r) \\ (\partial u/\partial z) \end{array}\right\} = [J]^{-1}\left\{\begin{array}{c} (\partial u/\partial L_1) \\ (\partial u/\partial L_2) \end{array}\right\} = [J]^{-1}\left\{\begin{array}{c} u_1 - u_3 \\ u_2 - u_3 \end{array}\right\} \quad (1.69)$$

Equation (1.69) can be written as

$$\left\{\begin{array}{c} \dfrac{\partial u}{\partial r} \\[2mm] \dfrac{\partial u}{\partial z} \end{array}\right\} = [J]^{-1}\left[\begin{array}{ccccccccc} 1 & 0 & 0 & 0 & 0 & 0 & -1 & 0 & 0 \\ 0 & 0 & 0 & 1 & 0 & 0 & -1 & 0 & 0 \end{array}\right]\left\{\begin{array}{c} u_1 \\ v_1 \\ w_1 \\ u_2 \\ v_2 \\ w_2 \\ u_3 \\ v_3 \\ w_3 \end{array}\right\} \quad (1.70)$$

Substituting for $[J]^{-1}$ from Eq. (1.68) in Eq. (1.70), we get

$$\left\{\begin{array}{c} (\partial u/\partial r) \\ (\partial u/\partial z) \end{array}\right\} = \left[\begin{array}{ccccccccc} a_{11} & 0 & 0 & a_{12} & 0 & 0 & (-a_{11}-a_{12}) & 0 & 0 \\ a_{22} & 0 & 0 & a_{22} & 0 & 0 & (-a_{21}-a_{22}) & 0 & 0 \end{array}\right][q] \quad (1.71)$$

where $\{q\}$ is nodal displacement vector defined as

$$\{u_1,\ v_1 \cdot w_1,\ u_2,\ v_2 \cdot w_2,\ u_3,\ v_3 \cdot w_3\ \}^T$$

Similarly, strains with respect to v and w displacements can be written as

$$\left\{\begin{array}{c} (\partial v^n/\partial r) \\ (\partial v^n/\partial z) \end{array}\right\} = \left[\begin{array}{ccccccccc} 0 & a_{11} & 0 & 0 & a_{12} & 0 & 0 & (-a_{11}-a_{12}) & 0 \\ 0 & a_{21} & 0 & 0 & a_{22} & 0 & 0 & (-a_{21}-a_{22}) & 0 \end{array}\right]\{q\} \quad (1.72)$$

$$\left\{\begin{array}{c} (\partial w^n/\partial r) \\ (\partial w^n/\partial z) \end{array}\right\} = \left[\begin{array}{ccccccccc} 0 & 0 & a_{11} & 0 & 0 & a_{12} & 0 & 0 & (-a_{11}-a_{12}) \\ 0 & 0 & a_{21} & 0 & 0 & a_{22} & 0 & 0 & (-a_{21}-a_{22}) \end{array}\right]\{q\} \quad (1.73)$$

The three-dimensional expressions for strain in cylindrical co-ordinates can be written as

$$[\varepsilon^n]^e = \begin{Bmatrix} \varepsilon_r^n \\ \varepsilon_z^n \\ \varepsilon_\theta^n \\ \gamma_{rz} \\ \gamma_{r\theta} \\ \gamma_{z\theta} \end{Bmatrix} = \begin{Bmatrix} \dfrac{\partial u}{\partial r} \\[2mm] \dfrac{\partial v}{\partial z} \\[2mm] \dfrac{u}{r} + \dfrac{1}{r}\dfrac{\partial w}{\partial \theta} \\[2mm] \dfrac{\partial u}{\partial z} + \dfrac{\partial v}{\partial r} \\[2mm] \dfrac{1}{r}\dfrac{\partial u}{\partial \theta} + \dfrac{\partial w}{\partial r} - \dfrac{w}{r} \\[2mm] \dfrac{1}{r}\dfrac{\partial v}{\partial \theta} + \dfrac{\partial w}{\partial z} \end{Bmatrix} = \begin{Bmatrix} \dfrac{\partial u^n}{\partial r} \\[2mm] \dfrac{\partial v^n}{\partial z} \\[2mm] \dfrac{u^n}{r} + n\dfrac{w^n}{r} \\[2mm] \dfrac{\partial u^n}{\partial z} + \dfrac{\partial v^n}{\partial r} \\[2mm] -n\dfrac{u^n}{r} + \dfrac{\partial w^n}{\partial r} - \dfrac{w^n}{r} \\[2mm] -n\dfrac{v^n}{r} + \dfrac{\partial w^n}{\partial z} \end{Bmatrix} \qquad (1.74)$$

In the above matrices, the first four rows are multiplied by $\cos n\theta$ and the last two rows by $\sin n\theta$, to get $[\varepsilon_n]$ as given on p. 21, Eq. (1.75).

Substituting Eq. (1.71) to (1.73) in Eq. (1.74), the relation between strain and displacements can be written in the matrix form as

$$\{\varepsilon_n\} = [B]\{q\} \qquad (1.76)$$

where $[B]$ is the strain displacement matrix
 $[q]$ the nodal displacement vector

Now, the element stiffness matrix corresponding to the local degrees of freedom is obtained from the variational principles. It is given by

$$[k]_e = \int_v [B]^T [D][B]\, dV \qquad (1.77)$$

where $[k]$ is the element stiffness matrix
 $[D]$ is the material properties matrix, as given below:

$$[D] = \begin{bmatrix} E_1 & E_2 & E_2 & 0 & 0 & 0 \\ E_2 & E_1 & E_2 & 0 & 0 & 0 \\ E_2 & E_2 & E_1 & 0 & 0 & 0 \\ 0 & 0 & 0 & E_3 & 0 & 0 \\ 0 & 0 & 0 & 0 & E_3 & 0 \\ 0 & 0 & 0 & 0 & 0 & E_3 \end{bmatrix} \qquad (1.78)$$

where

$$E_1 = \frac{E(1-\mu)}{(1+\mu)(1-2\mu)}$$

$$\{\varepsilon^{n}\}=
\begin{bmatrix}
a_{11} & 0 & 0 & a_{12} & 0 & 0 & (-a_{11}-a_{12}) & 0 & 0 \\[4pt]
0 & a_{21} & 0 & 0 & a_{22} & 0 & 0 & (-a_{21}-a_{22}) & 0 \\[4pt]
\dfrac{L_{1}}{r} & 0 & \dfrac{nL_{1}}{r} & \dfrac{L_{2}}{r} & 0 & \dfrac{nL_{2}}{r} & \dfrac{L_{3}}{r} & 0 & \dfrac{nL_{3}}{r} \\[8pt]
a_{21} & a_{11} & 0 & a_{22} & a_{12} & 0 & (-a_{21}-a_{22}) & (-a_{11}-a_{12}) & 0 \\[4pt]
\dfrac{nL_{1}}{r} & \left(a_{11}-\dfrac{L_{1}}{r}\right) & \dfrac{-nL_{1}}{r} & \dfrac{nL_{2}}{r} & \left(a_{12}-\dfrac{L_{2}}{r}\right) & \dfrac{-nL_{2}}{r} & \dfrac{nL_{3}}{r} & \left(-a_{11}-a_{12}-\dfrac{L_{3}}{r}\right) & \dfrac{-L_{3}}{r} \\[8pt]
0 & \dfrac{nL_{1}}{r} & a_{21} & 0 & \dfrac{nL_{2}}{r} & a_{22} & 0 & \dfrac{nL_{3}}{r} & (-a_{21}-a_{22})
\end{bmatrix}
\begin{Bmatrix}
u_{1} \\ v_{1} \\ w_{1} \\ u_{2} \\ v_{2} \\ w_{2} \\ u_{3} \\ v_{3} \\ w_{3}
\end{Bmatrix}$$

$$(1.75)$$

$$E_3 = \frac{E_\mu}{(1 + \mu)(1 - 2\mu)} \qquad (1.79)$$

$$E_3 = \frac{E}{2(1 + \mu)}$$

Note that in Eq. (1.77), the elements of B are functions of the natural co-ordinates L_1, L_2, L_3 and r. We can write dV in Eq. (1.77) per unit radian as

$$dV = r \det J \, dL_1 \, dL_2 \qquad (1.80)$$

where r is the centroidal radius, det J is the determinant of the Jacobian operator. Combining Eq. (1.77) and (1.80)

$$[k]_e = \int B^T \, DBr \det J \, dL_1 \, dL_2 \qquad (1.81)$$

An explicit evaluation of the integral in Eq. (1.81) is generally not possible and numerical integration is to be used. Now, stiffness matrix can be written

$$[k]_e = \int_v F \, dL_1 \, dL_2 \qquad (1.82)$$

where $F = B^T DBr \det J$ and the integration is performed in the natural co-ordinate system of the element as explained in Appendix I. The elements of F depend on L_1, L_2 and L_3.

Typical Applications

Rotating equipment like kilns, kiln tyres, tumbling mills, cement mills, pressure vessels, cylindrical or conical containers are typical examples.

1.2.5   3D Isoparametric Brick Element

Simple 3D element is an eight-noded brick element. The eight nodes and local axes of this isoparametric element are given in Fig. 1.5. It is advantageous to use isoparametric elements because we do not have to bother about the transformation of the element stiffness matrix from the local to global co-ordinates and also these can take curved boundaries into account.

Out of the available standard shape functions, Serendipity poly-nomial is chosen, as it contains most of the lower order terms in Pascal's triangle, as compared to polynomials like the ones belonging to the Lagrange family. This Serendipity polynomial, in general, is given by [1.2].

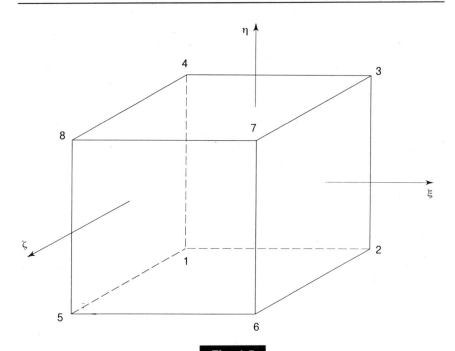

Fig. 1.5

3D Brick element

$$N_i = 1/8(1 + \xi\xi_i)(1 + \eta\eta_i)(1 + \zeta\zeta_i) \qquad [1.83]$$

The eight shape functions got by substituting

$$\xi_i = \pm 1, \quad \eta_i = \pm 1, \quad \zeta_i = \pm 1$$

Therefore

$$N_1 = 1/8(1 - \xi)(1 - \eta)(1 - \zeta)$$

$$N_2 = 1/8(1 + \xi)(1 - \eta)(1 - \zeta)$$

$$N_3 = 1/8(1 + \xi)(1 + \eta)(1 - \zeta)$$

$$N_4 = 1/8(1 - \xi)(1 + \eta)(1 - \zeta) \qquad (1.84)$$

$$N_5 = 1/8(1 - \xi)(1 - \eta)(1 + \zeta)$$

$$N_6 = 1/8(1 + \xi)(1 - \eta)(1 + \zeta)$$

$$N_7 = 1/8(1 + \xi)(1 + \eta)(1 + \zeta)$$

$$N_8 = 1/8(1 - \xi)(1 + \eta)(1 + \zeta)$$

The above shape functions were obtained by assuming a linear variation of displacements along each axis. Only displacements will be continuous in the neighbouring elements.

The displacement at any point within the element is given by

$$
\begin{bmatrix} u \\ v \\ w \end{bmatrix}_{3\times1} = \begin{bmatrix} u_1 & u_2 & u_3 & u_4 & u_5 & u_6 & u_7 & u_8 \\ v_1 & v_2 & v_3 & v_4 & v_5 & v_6 & v_7 & v_8 \\ w_1 & w_2 & w_3 & w_4 & w_5 & w_6 & w_7 & w_8 \end{bmatrix}_{3\times8} \begin{bmatrix} N_1 \\ N_2 \\ N_3 \\ N_4 \\ N_5 \\ N_6 \\ N_7 \\ N_8 \end{bmatrix}_{8\times1} \tag{1.85}
$$

where N_1 to N_8 are the shape functions and u_i, v_i, w_i are the nodal displacement of the ith node. To take into account the curved boundaries, the co-ordinates of any point within the element are given by

$$
\begin{bmatrix} x \\ y \\ z \end{bmatrix}_{3\times1} = \begin{bmatrix} x_1 & x_2 & x_3 & x_4 & x_5 & x_6 & x_7 & x_8 \\ y_1 & y_2 & y_3 & y_4 & y_5 & y_6 & y_7 & y_8 \\ z_1 & z_2 & z_3 & z_4 & z_5 & z_6 & z_7 & z_8 \end{bmatrix}_{3\times8} \begin{bmatrix} N_1 \\ N_2 \\ N_3 \\ N_4 \\ N_5 \\ N_6 \\ N_7 \\ N_8 \end{bmatrix}_{8\times1} \tag{1.86}
$$

where again N_1 to N_8 are the shape functions and x_i, y_i, z_i are the nodal co-ordinates of the ith node. All the six components of stress and strain will be present. Hence, $\{\sigma\}$ and $\{\varepsilon\}$ are of size 6×1. $\{\varepsilon\}$ is given by

$$
\{\varepsilon\} = \begin{Bmatrix} \varepsilon_x \\ \varepsilon_y \\ \varepsilon_z \\ \gamma_{xy} \\ \gamma_{yz} \\ \gamma_{xz} \end{Bmatrix} = \begin{Bmatrix} \partial u/\partial x \\ \partial v/\partial y \\ \partial w/\partial z \\ (\partial u/\partial y) + (\partial v/\partial x) \\ (\partial v/\partial z) + (\partial w/\partial y) \\ (\partial u/\partial z) + (\partial w/\partial x) \end{Bmatrix} \tag{1.87}
$$

Derivatives of displacements u, v and w with respect to x, y, z are required to formulate the strain matrix. So, let us first get the derivative of u with respect to x, y and z. Using Eq. (1.85), we can write

$$
\left\{ \begin{array}{c} \dfrac{\partial u}{\partial x} \\[2mm] \dfrac{\partial u}{\partial y} \\[2mm] \dfrac{\partial u}{\partial z} \end{array} \right\}_{3\times1}
=
\begin{bmatrix}
\dfrac{\partial N_1}{\partial x} & \dfrac{\partial N_2}{\partial x} & \dfrac{\partial N_3}{\partial x} & \dfrac{\partial N_4}{\partial x} & \dfrac{\partial N_5}{\partial x} & \dfrac{\partial N_6}{\partial x} & \dfrac{\partial N_7}{\partial x} & \dfrac{\partial N_8}{\partial x} \\[2mm]
\dfrac{\partial N_1}{\partial y} & \dfrac{\partial N_2}{\partial y} & \dfrac{\partial N_3}{\partial y} & \dfrac{\partial N_4}{\partial y} & \dfrac{\partial N_5}{\partial y} & \dfrac{\partial N_6}{\partial y} & \dfrac{\partial N_7}{\partial y} & \dfrac{\partial N_8}{\partial y} \\[2mm]
\dfrac{\partial N_1}{\partial z} & \dfrac{\partial N_2}{\partial z} & \dfrac{\partial N_3}{\partial z} & \dfrac{\partial N_4}{\partial z} & \dfrac{\partial N_5}{\partial z} & \dfrac{\partial N_6}{\partial z} & \dfrac{\partial N_7}{\partial z} & \dfrac{\partial N_8}{\partial z}
\end{bmatrix}_{3\times8}
\left\{ \begin{array}{c} u_1 \\ u_2 \\ u_3 \\ u_4 \\ u_5 \\ u_6 \\ u_7 \\ u_8 \end{array} \right\}_{8\times1}
$$

(1.88)

But $\partial N_i/\partial x_n$ is not known because we do not know N_i as a function of x.

Since

$$
\left\{ \begin{array}{c} \dfrac{\partial u}{\partial \xi} \\[2mm] \dfrac{\partial u}{\partial \eta} \\[2mm] \dfrac{\partial u}{\partial \zeta} \end{array} \right\}_{3\times1}
=
\begin{bmatrix}
\dfrac{\partial x}{\partial \xi} & \dfrac{\partial y}{\partial \xi} & \dfrac{\partial z}{\partial \xi} \\[2mm]
\dfrac{\partial x}{\partial \eta} & \dfrac{\partial y}{\partial \eta} & \dfrac{\partial z}{\partial \eta} \\[2mm]
\dfrac{\partial x}{\partial \zeta} & \dfrac{\partial y}{\partial \zeta} & \dfrac{\partial z}{\partial \zeta}
\end{bmatrix}_{3\times3}
\left\{ \begin{array}{c} \dfrac{\partial u}{\partial x} \\[2mm] \dfrac{\partial u}{\partial y} \\[2mm] \dfrac{\partial u}{\partial z} \end{array} \right\}_{3\times1}
= [AJ] \left\{ \begin{array}{c} \dfrac{\partial u}{\partial x} \\[2mm] \dfrac{\partial u}{\partial y} \\[2mm] \dfrac{\partial u}{\partial z} \end{array} \right\}
$$

(1.89)

We have

$$
\left\{ \begin{array}{c} \dfrac{\partial u}{\partial x} \\[2mm] \dfrac{\partial u}{\partial y} \\[2mm] \dfrac{\partial u}{\partial z} \end{array} \right\}
= [AJ]^{-1} \left\{ \begin{array}{c} \dfrac{\partial u}{\partial \varepsilon} \\[2mm] \dfrac{\partial u}{\partial \eta} \\[2mm] \dfrac{\partial u}{\partial \zeta} \end{array} \right\}
$$

(1.90)

where $[AJ]$ is the Jacobian matrix. We can get u's derivative w.r.t. ξ, η, ζ and from (1.86) and as we know x, y, z as functions of ε, η, ζ. Therefore, $[AJ]$ can be calculated as follows. From Eq. (1.86), we can write

$$
\begin{Bmatrix} \dfrac{\partial x}{\partial \xi} \\[6pt] \dfrac{\partial x}{\partial \eta} \\[6pt] \dfrac{\partial x}{\partial \zeta} \end{Bmatrix}_{3\times 1}
=
\begin{bmatrix}
\dfrac{\partial N_1}{\partial \xi} & \dfrac{\partial N_2}{\partial \xi} & \dfrac{\partial N_3}{\partial \xi} & \dfrac{\partial N_4}{\partial \xi} & \dfrac{\partial N_5}{\partial \xi} & \dfrac{\partial N_6}{\partial \xi} & \dfrac{\partial N_7}{\partial \xi} & \dfrac{\partial N_8}{\partial \xi} \\[6pt]
\dfrac{\partial N_1}{\partial \eta} & \dfrac{\partial N_2}{\partial \eta} & \dfrac{\partial N_3}{\partial \eta} & \dfrac{\partial N_4}{\partial \eta} & \dfrac{\partial N_5}{\partial \eta} & \dfrac{\partial N_6}{\partial \eta} & \dfrac{\partial N_7}{\partial \eta} & \dfrac{\partial N_8}{\partial \eta} \\[6pt]
\dfrac{\partial N_1}{\partial \zeta} & \dfrac{\partial N_2}{\partial \zeta} & \dfrac{\partial N_3}{\partial \zeta} & \dfrac{\partial N_4}{\partial \zeta} & \dfrac{\partial N_5}{\partial \zeta} & \dfrac{\partial N_6}{\partial \zeta} & \dfrac{\partial N_7}{\partial \zeta} & \dfrac{\partial N_8}{\partial \zeta}
\end{bmatrix}_{3\times 8}
\begin{Bmatrix} x_1 \\ x_2 \\ x_3 \\ x_4 \\ x_5 \\ x_6 \\ x_7 \\ x_8 \end{Bmatrix}_{8\times 1}
$$

$$
\text{(1.91)}
$$

$$
= [EN]_{3\times 8}\{X\}_{8\times 1} \tag{1.92}
$$

The elements of $[EN]$ matrix are given by

$$
EN(1,\,1) = \frac{\partial N_1}{\partial \xi} = -\frac{1}{8}(1-\eta)(1-\zeta)
$$

$$
EN(1,\,2) = \frac{\partial N_2}{\partial \xi} = \frac{1}{8}(1-\eta)(1-\zeta)
$$

$$
EN(1,\,3) = \frac{\partial N_3}{\partial \xi} = \frac{1}{8}(1+\eta)(1-\zeta)
$$

$$
EN(1,\,4) = \frac{\partial N_4}{\partial \xi} = -\frac{1}{8}(1+\eta)(1-\zeta) \tag{1.93}
$$

$$
EN(1,\,5) = \frac{\partial N_5}{\partial \xi} = -\frac{1}{8}(1-\eta)(1+\zeta)
$$

$$
EN(1,\,6) = \frac{\partial N_6}{\partial \xi} = \frac{1}{8}(1-\eta)(1+\zeta)
$$

$$
EN(1,\,7) = \frac{\partial N_7}{\partial \xi} = \frac{1}{8}(1+\eta)(1+\zeta)
$$

$$
EN(1,\,8) = \frac{\partial N_8}{\partial \xi} = -\frac{1}{8}(1+\eta)(1+\zeta)
$$

Similarly, we can get the other elements of $[EN]$.

$$[AJ]_{3\times3} = \begin{bmatrix} \dfrac{\partial N_1}{\partial \xi} & \dfrac{\partial N_2}{\partial \xi} & \cdots & \dfrac{\partial N_8}{\partial \xi} \\[2ex] \dfrac{\partial N_1}{\partial \eta} & \dfrac{\partial N_2}{\partial \eta} & \cdots & \dfrac{\partial N_8}{\partial \eta} \\[2ex] \dfrac{\partial N_1}{\partial \zeta} & \dfrac{\partial N_2}{\partial \zeta} & \cdots & \dfrac{\partial N_8}{\partial \zeta} \end{bmatrix}_{3\times8} \begin{Bmatrix} x_1 & y_1 & z_1 \\ x_2 & y_2 & z_2 \\ x_3 & y_3 & z_3 \\ x_4 & y_4 & z_4 \\ x_5 & y_5 & z_5 \\ x_6 & y_6 & z_6 \\ x_7 & y_7 & z_7 \\ x_8 & y_8 & z_8 \end{Bmatrix}_{8\times3} \qquad [1.94)$$

$$= [EN]_{3\times8}\,[XE]_{8\times3} \tag{1.95}$$

Now, in order to calculate u's derivative with respect to the x, y and z co-ordinates, we need its derivative with respect to ξ, η, ζ. As we have already expressed u as a function of ξ, η, ζ in (1.85), we can get u's derivative with respect to ξ, η, ζ as follows:

$$\begin{Bmatrix} \dfrac{\partial u}{\partial \xi} \\[2ex] \dfrac{\partial u}{\partial \eta} \\[2ex] \dfrac{\partial u}{\partial \zeta} \end{Bmatrix} = \begin{bmatrix} \dfrac{\partial N_1}{\partial \xi} & \dfrac{\partial N_2}{\partial \xi} & \dfrac{\partial N_3}{\partial \xi} & \cdots & \dfrac{\partial N_8}{\partial \xi} \\[2ex] \dfrac{\partial N_1}{\partial \eta} & \dfrac{\partial N_2}{\partial \eta} & \dfrac{\partial N_3}{\partial \eta} & \cdots & \dfrac{\partial N_8}{\partial \eta} \\[2ex] \dfrac{\partial N_1}{\partial \zeta} & \dfrac{\partial N_2}{\partial \zeta} & \dfrac{\partial N_2}{\partial \zeta} & \cdots & \dfrac{\partial N_8}{\partial \zeta} \end{bmatrix} \begin{Bmatrix} u_1 \\ u_2 \\ u_3 \\ u_4 \\ u_5 \\ u_6 \\ u_7 \\ u_8 \end{Bmatrix} \tag{1.96}$$

$$= [EN]_{3\times8}\{u\}$$

Now from Eq. (1.81), we can get the derivative of u with respect to x, y, z as

$$\begin{Bmatrix} \dfrac{\partial u}{\partial x} \\[2mm] \dfrac{\partial u}{\partial y} \\[2mm] \dfrac{\partial u}{\partial z} \end{Bmatrix} = \begin{Bmatrix} J_{11} & J_{12} & J_{13} \\ J_{21} & J_{22} & J_{23} \\ J_{31} & J_{32} & J_{33} \end{Bmatrix} \begin{bmatrix} \dfrac{\partial N_1}{\partial \xi} & \dfrac{\partial N_2}{\partial \xi} & \cdots & \dfrac{\partial N_8}{\partial \xi} \\[2mm] \dfrac{\partial N_1}{\partial \eta} & \dfrac{\partial N_2}{\partial \eta} & \cdots & \dfrac{\partial N_8}{\partial \eta} \\[2mm] \dfrac{\partial N_1}{\partial \zeta} & \dfrac{\partial N_2}{\partial \zeta} & \cdots & \dfrac{\partial N_8}{\partial \zeta} \end{bmatrix} \begin{Bmatrix} u_1 \\ u_2 \\ u_3 \\ u_4 \\ u_5 \\ u_6 \\ u_7 \\ u_8 \end{Bmatrix} \quad (1.97)$$

$$= [AJ]^{-1}[EN]\{u\}$$

where J_{ij} is the ith row jth column of $[AJ]^{-1}$.

Similarly, we can get the derivatives of v, and w with respect to x, y and z. The strain vector can be easily obtained in the following manner using Eq. (1.87).

$$\{\varepsilon\} = \begin{bmatrix} J_{11} & 0 & 0 & J_{12} & 0 & 0 & J_{13} & 0 & 0 \\ 0 & J_{21} & 0 & 0 & J_{22} & 0 & 0 & J_{23} & 0 \\ 0 & 0 & J_{31} & 0 & 0 & J_{32} & 0 & 0 & J_{33} \\ J_{21} & J_{11} & 0 & J_{22} & J_{12} & 0 & J_{23} & J_{13} & 0 \\ 0 & J_{31} & J_{21} & 0 & J_{32} & J_{22} & 0 & J_{33} & J_{23} \\ J_{31} & 0 & J_{11} & J_{32} & 0 & J_{12} & J_{33} & 0 & J_{13} \end{bmatrix}_{6\times 9} \begin{Bmatrix} \partial u/\partial \xi \\ \partial v/\partial \xi \\ \partial w/\partial \xi \\ \partial u/\partial \eta \\ \partial v/\partial \eta \\ \partial w/\partial \eta \\ \partial u/\partial \zeta \\ \partial v/\partial \zeta \\ \partial w/\partial \zeta \end{Bmatrix}_{9\times 1}$$

$$(1.98)$$

Hence $\{\varepsilon\}$ is given by Eq. (1.99) given on p. 29.
Simplifying

$$\{\varepsilon\} = [BA]_{6\times 9}[AB]_{9\times 24}\{U\} \quad (1.100)$$

or

$$= [B]_{6\times 24}\{U\} \quad (1.101)$$

Having got the element strain vector, let us formulate the element stiffness matrix.

$$
\{\varepsilon\} = [BA]_{6\times 9}
\underbrace{
\begin{bmatrix}
\dfrac{\partial N_1}{\partial \xi} & 0 & 0 & \dfrac{\partial N_2}{\partial \xi} & 0 & 0 & \dfrac{\partial N_1}{\partial \xi} & 0 & 0 & \cdots & \dfrac{\partial N_8}{\partial \xi} & 0 & 0 \\[2mm]
0 & \dfrac{\partial N_1}{\partial \eta} & 0 & 0 & \dfrac{\partial N_2}{\partial \eta} & 0 & 0 & \dfrac{\partial N_1}{\partial \eta} & 0 & \cdots & 0 & \dfrac{\partial N_8}{\partial \eta} & 0 \\[2mm]
0 & 0 & \dfrac{\partial N_1}{\partial \zeta} & 0 & 0 & \dfrac{\partial N_2}{\partial \zeta} & 0 & 0 & \dfrac{\partial N_1}{\partial \zeta} & \cdots & 0 & 0 & \dfrac{\partial N_8}{\partial \zeta} \\[2mm]
\dfrac{\partial N_1}{\partial \eta} & 0 & 0 & \dfrac{\partial N_2}{\partial \eta} & 0 & 0 & \dfrac{\partial N_1}{\partial \eta} & 0 & 0 & \cdots & \dfrac{\partial N_8}{\partial \eta} & 0 & 0 \\[2mm]
0 & \dfrac{\partial N_1}{\partial \zeta} & 0 & 0 & \dfrac{\partial N_2}{\partial \zeta} & 0 & 0 & \dfrac{\partial N_1}{\partial \zeta} & 0 & \cdots & 0 & \dfrac{\partial N_8}{\partial \zeta} & 0 \\[2mm]
0 & 0 & \dfrac{\partial N_1}{\partial \xi} & 0 & 0 & \dfrac{\partial N_2}{\partial \xi} & 0 & 0 & \dfrac{\partial N_8}{\partial \xi} & \cdots & 0 & 0 & \dfrac{\partial N_8}{\partial \xi}
\end{bmatrix}
}_{9\times 24}
\underbrace{
\begin{Bmatrix}
u_1 \\ v_1 \\ w_1 \\ u_2 \\ v_2 \\ w_2 \\ u_3 \\ v_3 \\ w_3 \\ \vdots \\ u_8 \\ v_8 \\ w_8
\end{Bmatrix}
}_{24\times 1}
\tag{1.99}
$$

It is given by

$$[K] = \int [B]^T [D][B]d(\text{vol}) = \int [B]^T [D][B][AJ]\,d\xi\,d\eta\,d\zeta \qquad (1.102)$$

where $[B]$ is the strain-displacement matrix. Integral in Eq. (1.102) is numerically evaluated as explained in Appendix I and $[D]$ is the elasticity matrix.

The stress vector $\{\sigma\}$ of the element is given by

$$\{\sigma\}_{6\times1} = [D]_{6\times6}\{\varepsilon\} \qquad (1.103)$$

$[D]$, the elasticity matrix, is given by

$$[D] = \frac{E(1-\mu)}{(1+\mu)(1-2\mu)}
\begin{array}{l}
\text{Sym-} \\
\text{metric}
\end{array}
\begin{bmatrix}
1 & \dfrac{\mu}{1-\mu} & \dfrac{\mu}{1-\mu} & 0 & 0 & 0 \\[2mm]
 & 1 & \dfrac{\mu}{1-\mu} & 0 & 0 & 0 \\[2mm]
 & & 1 & 0 & 0 & 0 \\[2mm]
 & & & \dfrac{1-2\mu}{2(1-\mu)} & 0 & 0 \\[2mm]
 & & & & \dfrac{1-2\mu}{2(1-\mu)} & 0 \\[2mm]
 & & & & & \dfrac{1-2\mu}{2(1-\mu)}
\end{bmatrix}$$

$$(1.104)$$

1.2.5.1   Non-conforming Elements for Stress Analysis

Bilinear elements are attractive because they are simple and have only corner nodes. Unfortunately, they are too stiff in bending. These elements can be improved by adding the missing modes as internal degrees of freedom [1.14].

We now add to the dependent variables the incompatible modes in Eq. (1.85). The expression now becomes

$$u = \sum_{i=1}^{8} N_i u_i + \sum_{i=1}^{3} P_i a_i \qquad (1.105)$$

(similarly for v and w)
where

$$\left.
\begin{aligned}
P_1 &= (1 - \xi^2) \\
P_2 &= (1 - \eta^2) \\
P_3 &= (1 - \zeta^2)
\end{aligned}
\right\} \qquad (1.106)$$

However, for the element geometry, we continue to use the shape functions of basic nodes only.

Writing the strain matrix

$$[\varepsilon] = \sum_{i=1}^{8} [B_i]\{u_i\} + \sum_{i=1}^{3} G_i\{a_i\} \tag{1.107}$$

where

$$\{u_i\}^T = [u_1, u_2, ..., u_8, v_1, v_2, ..., v_8, w_1, w_2, ..., w_8] \tag{1.108}$$

$$\{a_i\}^T = [a_1, a_2, a_3, a_4, a_5, a_6, a_7, a_8, a_9]$$

$$\left.\begin{array}{l} [B_i] = [BA][AB] \\ [G_i] = [BA][AG] \end{array}\right\} \tag{1.109}$$

where

$$[AG] = \begin{bmatrix}
P_1\xi & 0 & 0 & P_2\xi & 0 & 0 & P_3\xi & 0 & 0 \\
0 & P_1\xi & 0 & 0 & P_2\xi & 0 & 0 & P_3\xi & 0 \\
0 & 0 & P_1\xi & 0 & 0 & P_2\xi & 0 & 0 & P_3\xi \\
P_1\eta & 0 & 0 & P_2\eta & 0 & 0 & P_3\eta & 0 & 0 \\
0 & P_1\eta & 0 & 0 & P_2\eta & 0 & 0 & P_3\eta & 0 \\
0 & 0 & P_1\eta & 0 & 0 & P_2\eta & 0 & 0 & P_3\eta \\
P_1\zeta & 0 & 0 & P_2\zeta & 0 & 0 & P_3\zeta & 0 & 0 \\
0 & P_1\zeta & 0 & 0 & P_2\zeta & 0 & 0 & P_3\zeta & 0 \\
0 & 0 & P_1\zeta & 0 & 0 & P_2\zeta & 0 & 0 & P_3\zeta
\end{bmatrix} \tag{1.110}$$

We now compute the element stiffness matrix and load matrix and obtain [1.14]

$$\begin{bmatrix} K_{uu} & K_{ua} \\ K_{ua} & K_{aa} \end{bmatrix} \begin{Bmatrix} u_0 \\ a_0 \end{Bmatrix} = \begin{Bmatrix} F_u \\ 0 \end{Bmatrix} \tag{1.111}$$

Incompatible elements often yield results of high quality. Interelement gaps and overlaps tend to soften a structure. Softening counters the inherent over stiffness of assumed displacement approximation. A good balance of the two effects leads to good results with a loose mesh [1.14]. After formulation of stiffness, matrix condensation removes the a_0. Thus for the 3D brick element, a_1 through a_9 are eliminated, leaving 24×24 condensed $[K]$ matrix. The solid element described pass the

patch test only if they are parallelepipeds. A modified integration
scheme corrects this failure and the element is known as QM6 element.
The contribution of nodeless degrees of freedom a_1, to the consistent
element load vector $\{Fu\}$ is

$$\{Fua\} = \int_{V_e} [G]^T [D][B] \; dv \; \{u_0\}$$

$$= \int_{V_e} [G]^T \{\sigma_0\} \; dv \; \{u_0\} \tag{1.112}$$

where $\{u_0\}$ corresponds to set of nodal displacements corresponding
to rigid body motions or constant strain conditions. Imagine that
instead of representing the initial stress, $\{\sigma_0\}$ represents element
stresses produced by nodal displacements $\{u_0\}$ on the element boundary.
The basic isoparametric element with neither a_1 nor $\{Fua\}$ present,
is able to pass the patch test. In other words, when $\{\sigma_0\}$ is constant
and produced in the essential or basic dof $\{u_0\}$, certain nodal loads
associated with $\{\sigma_0\}$ are applied by an element to its nodes. These
loads should not be disturbed if incompatible modes are added.
Accordingly, in a patch test no additional nodal loads should be
associated with a_i. This means that $\{Fua\}$ must vanish when $\{\sigma_0\}$ is
constant. When $\{\sigma_0\}$ is constant $\{Fua\}$ will be zero if

$$\iint_{V_e} [G] \; dV = 0$$

i.e.
$$\int_{-1}^{1} \int_{-1}^{1} \int_{-1}^{1} [G]^T J \, d\zeta \, d\eta \, d\tau = 0 \tag{1.113}$$

where J is the Jacobian determinant. For parallelepipeds, J is constant
and $[G]$ contains first powers of ζ, η and τ so that the above equation
is automatically satisfied. However for the general shapes, J and $[G]$
are more complicated and the above equation is not satisfied and the
patch test is failed. But the above equation is automatically satisfied
by forming $[G]$ and integrating using the constant values of $[J_0]^{-1}$
and J_0 where $[J_0]$ and J_0 are the Jacobian matrix and its determinant
at $\zeta = \eta = \tau = 0$, instead of using the correct $[J]^{-1}$ and J at the Gauss
Quadrature points.

Typical Applications

Single and multi-point cutting tools, spiral bevel and hypoid gear teeth, and objects of comparable length, breadth and depth are typical examples.

It may be realised that the applications mentioned for all the five element choices cover components in machine design which have no rigorous analytical design procedure existing at present based on purely scientific reasoning.

When a series of elements are involved, the overall stiffness and mass matrices can be obtained by combining the matrices of the elements. The logic of this assembly is explained later in Example 2.11.

➤ 1.3 PRE- AND POST-PROCESSING

Enormous amount of material has been pouring in on the subject "Computer Aided Design" in the recent past. It either deals with Computer Aided Graphics (CAG) or Computer Aided Analysis (CAA). Material under CAG usually deals with making component drawings from assembly drawings, layout drawings, enlarging, reducing or rotating a given three-dimensional object, preparing developments of components or interpretation of solids. Under computer aided analysis can be grouped finite element modelling of the given object, identifying the input forces and estimating the resulting response (deformations) and subsequently determining the stress levels at various locations of the components. If the component stresses are within limits, the design is considered safe. Otherwise, dimensions have to be altered, the iteration process repeated till the members are optimally designed, keeping the manufacturing constraints in mind. Even this analysis needs quite a lot of graphics, by way of discretising the object, identifying the nodal points of the finite element discretisation and reconstructing the object with the discretised information making sure that the input geometry is free of mistakes. [1.15, 1.16, 1.17]. This is known as pre-processing.

Once the deformations are obtained, constructing the deformed shape of the object and identifying the zones of equal intensity of stress on the object under load are also exercises in graphics. Drawing the animated view of the deformed object under dynamic load at discrete time intervals is an interesting exercise in drafting. These can be grouped under post-processing. Both pre- and post-processing come under CAG.

➤ 1.4 FINITE DIFFERENCE METHOD

There are occasions when the behaviour of the object under load can be mathematically formulated. If this leads to a governing differential equation whose closed form solution is not easily available, approximate methods of solution must be employed. Initial value (for transient vibration) and boundary value (for static problems) problems involving either ordinary or partial differential equations may be solved by such methods. The derivatives of functions appearing in the differential equation are approximated by Taylor series expansion of the unknown function.

Taylor series expansion of f_j is given by

$$f_j = f_{i\pm1} = f(x_i \pm h) = f_i \pm hf'_i + \frac{h^2}{\lfloor 2}f''_i \pm \frac{h^3}{\lfloor 3}f'''_i + \dots \quad (1.114)$$

where $(\ )'$ stands for df/dx and h is the spacing.

From Eq. (1.114) it is seen that

$$f'_i = \frac{f_{i+1} - f_{i-1}}{2h} \qquad (1.115)$$

with an error of the order of $\frac{h^2}{6}f'''$.

Similarly,

$$f''_i = \frac{1}{h^2}(f_{i+1} - 2f_i + f_{i-1}) \qquad (1.116)$$

with an error of the order $\frac{h^2}{12}f_i^{\text{IV}}$.

When this is extended to partial differential equations in two variables x and y, the same argument can be extended. If h is the constant spacing in x as well as in y directions, and if (i, j) denotes any location, one can write

$$\left\{\frac{\partial^2 f}{\partial y^2}\right\}_{\text{at } i,j} = \frac{1}{h^2}(f_{i,j+1} - 2f_{i,j} + f_{i,j-1}) \qquad (1.117)$$

with an error of the order of $\frac{h^2}{12} \cdot \frac{\partial^4 f}{\partial x^4}$

Similarly,

$$\left\{\frac{\partial^2 f}{\partial y^2}\right\}_{\text{at } i,j} = \frac{1}{h^2}(f_{i,j+1} - 2f_{i,j} + f_{i,j-1}) \qquad (1.118)$$

with an error of the order of $\frac{h^2}{12} \cdot \frac{\partial^4 f}{\partial x^4}$.

All other combinations of interest have been discussed in Ref. [1.19].

1.4.1 Beam Problem

The governing differential equation can be written in the following form

$$EI \frac{d^2W}{dX^2} = -M_X$$

$$\frac{d^2M_X}{dX^2} = P \tag{1.119}$$

where EI = Flexural rigidity of the beam
  W = Lateral deflection
  X = Axial location
  P = Intensity of loading

Non-dimensionalisation of the variables can be done as follows:

$$X = a\xi \text{ where } a \text{ is a reference length}$$

$$W = aw$$

$$M_X = \frac{EI}{a} \cdot (m_\xi)$$

ξ, w and m_ξ are now dimensionless

$$\frac{d}{dX} = \frac{d}{a \, d\xi}, \qquad \frac{d^2}{dX^2} = \frac{1}{a^2} \frac{d^2}{d\xi^2}$$

Equations (1.119) get recast in the following form

$$\left. \begin{array}{c} w'' + m_\xi = 0 \\[2mm] m_\xi'' = \dfrac{Pa^3}{EI} \end{array} \right\} \tag{1.120}$$

where $(\ )' = d/d\xi$

If w and m_ξ are treated as dependent variables and ξ as the independent variable,

$$w'' = \frac{w_{i+1} - 2w_i + w_{i-1}}{h^2}$$

and

$$m_\xi'' = \frac{(m_\xi)_{i+1} - 2(m_\xi)_i + (m_\xi)_{i-1}}{h^2} \tag{1.121}$$

where h is the spacing.

If L is the total length of the beam and if there are totally n intervals, h can be expressed as

$$h = \frac{1}{a} \cdot \frac{L}{n} \tag{1.122}$$

a can be so chosen as to make h much smaller than 1. This will lead to an error of order h^2.

The governing differential equation at the various locations on the beam together with the boundary conditions will result in a set of simultaneous equations when the differential equations are recast as difference equations.

1.4.2   Shell Problem

For a general shell as shown in Fig. 1.6 the governing differential equation of equilibrium can be written as [1.20].

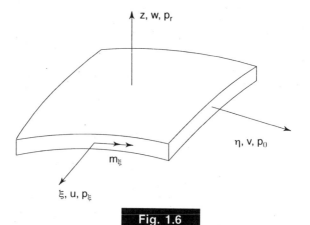

Fig. 1.6

Element of a shell

$$a_1 u'' + a_2 u + a_3 u^{\cdot} + a_4 v'^{\cdot} + a_5 v^{\cdot} + a_6 w'$$
$$+ a_7 w + a_8 m'_\xi + a_9 m_\xi = -p_\xi$$
$$a_{10} u'^{\cdot} + a_{11} u^{\cdot} + a_{12} v'' + a_{13} v' + a_{14} v'^{\cdots}$$
$$+ a_{15} w''' \cdot + a_{16} w' + a_{17} w'^{\cdot} + a_{18} m'_\xi = -p_\theta$$
$$a_{19} u' + a_{20} u + a_{21} v'''\cdot + a_{22} v'^{\cdot} + a_{23} v^{\cdot}$$
$$+ a_{24} w'''^{\cdots} + a_{25} w'^{\cdots} + a_{26} w^{\cdots}$$

$$+ a_{27} m''_\xi + a_{28} m'_\xi + a_{29} m^{\cdots}_\xi = -p_r$$
$$a_{30} u' + a_{31} u + a_{32} v^{\cdot} + a_{33} w'' + a_{34} w' + a_{35} w^{\cdots} + a_{36} m_\xi = 0 \tag{1.123}$$

Here u, v, w are the three non-dimensional displacements and m_ξ is the non-dimensional bending moment.

$$(\ )' = \frac{\partial}{\partial \xi}, (\ )\dot{} = \frac{\partial}{\partial \eta} \tag{1.124}$$

where ξ and η are non-dimensional x and y co-ordinates a_1 to a are coefficients. These equations can also be recast as difference equations using equtions of the type (1.115) and (1.116). These together with the boundary conditions will again lead to a set of simultaneous equations. When the two curvatures of the shell are zero and when u and v are absent, they will lead to equations of plates. In these simplified equations, if differentiation with respect to y (or η) is also ignored, they will reduce to the beam equations (1.120) discussed earlier (only coefficients (a_{27}, a_{33} and a_{36} will remain). To summarise, Eqs (1.120) are the most general governing differential equations for beams, plates and shells.

Shells of revolution assume special importance in the field of mechanical engineering since units like pressure vessels, boilers, tube mills, kiln shells and many process equipment are shells of revolution. In all these cases, the displacements and stress resultants can be expressed as trignometric functions in the circumferential direction thereby eliminating the differentiation in y (or η) direction. The governing partial differential equations reduce to ordinary differential equations in x (or ξ) only. These have been derived in Ref. [1.19] and used in [1.20]. They are of the following form:

$$a_1 u'' + a_2 u' + a_3 u + a_4 v' + a_5 v + a_6 w' + a_7 w + a_8 m_\xi' + a_9 m_\xi = -p_\xi$$

$$a_{10} u' + a_{11} u + a_{12} v'' + a_{13} v' + a_{14} v$$
$$+ a_{15} w'' + a_{16} w' + a_{17} w + a_{18} m_\xi = -p_\theta$$

$$a_{19} u' + a_{20} u + a_{21} v'' + a_{22} v' + a_{23} v + a_{24} w''$$
$$+ a_{25} w' + a_{26} w + a_{27} m_\xi'' + a_{28} m_\xi' + a_{29} m_\xi = -p_r$$

$$a_{30} u' + a_{31} u + a_{32} v + a_{33} w'' + a_{34} w' + a_{35} w + a_{36} m_\xi = 0 \tag{1.125}$$

where a_1 to a_{36} are coefficients which may vary along the meridian of the shell. These are possible when the symmetric loads are expressed in the following form:

$$p_\xi = \Sigma \, (p_\xi)_n \, \cos n\theta$$

$$p_\theta = \Sigma \, (p_\theta)_n \, \sin n\theta \tag{1.126}$$

$$p_r = \Sigma \, (p_r)_n \, \cos n\theta$$

for anti-symmetric loads, these are of the form

$$p_\xi = \Sigma \, (p_\xi)_n \, \sin \, n\theta$$
$$p_\theta = \Sigma \, (p_\theta)_n \, \cos \, n\theta \qquad (1.127)$$
$$p_r = \Sigma \, (p_r)_n \, \sin \, n\theta$$

In Eq. (1.126), $n = 0$ corresponds to axi-symmetric problem.

In Chapter 2 it will be shown that this formulation leads to simultaneous equations of a very narrow band width. These equations could be further simplified for circular plates subjected to asymmetric or axi-symmetric loads.

The same concept can be extended to solids of revolution subjected to general loading. The axi-symmetric elastic wave propagation of uniform bars using finite difference method has been presented in Ref. [1.21]. The governing differential equation takes the following form

$$\left(u'' + \frac{1}{r}u'^\cdot - \frac{u}{r^2} + w'^\cdot \right) + \left(\frac{V_S}{V_C} \right)^2 (u^{\cdot\cdot} - w'^\cdot)$$

$$= \frac{1}{V_C^2} \frac{\partial^2 u}{\partial t^2}$$

$$\left(w^{\cdot\cdot} + \frac{1}{r}u^\cdot + u'^\cdot \right) + \left(\frac{V_S}{V_C} \right)^2 \left(w'' - u'^\cdot - \frac{1}{r}u'^\cdot + \frac{1}{r}w' \right)$$

$$= \frac{1}{V_C^2} \frac{\partial^2 w}{\partial t^2} \qquad (1.128)$$

$$(\,)' = \frac{\partial}{\partial r}, \, (\,)^\cdot = \frac{\partial}{\partial z} \qquad (1.129)$$

u and w are radial and axial displacements, V_C and V_S are compressional and shear speeds of waves, r and z are radial and axial coordinates, $\partial^2 u/\partial t^2$ and $\partial^2 w/\partial t^2$ are accelerations. Extending this to asymmetric problem does not pose any difficulty.

Collection of nineteen papers on "Advances in finite element and finite difference methods" has been edited by Noor [1.22].

➤ 1.5 CONCLUDING REMARKS

There are two very important points to be noted here. Firstly, as any practising engineer in this area knows, the size of the matrices $[M]$, $[C]$ and $[K]$ will very often be around 1000 or more. Hence, it is certainly required to take advantage of factors like the system symmetry, bandedness, efficient algorithm besides, in the field of dynamic analysis, iterative solution procedure.

The second point concerns the mistaken faith in the results of these numerical methods (e.g. FEM), specially in the light of ever-proliferating packages. It is foolish to think that once the computer prints a few numbers as outputs, the results are reliable. The completely numerical nature of FEM makes it possible to have a physical insight into the problem. Hence, the user must be doubly concerned to check the validity of the results.

The aim of writing this book is to enable the reader to grasp the logical growth of the subject matter over a period of time. At the end, the reader must be in a position to know what is inside all these software packages given as "black boxes". Keeping this aim in view, we must necessarily go through the mathematics of each algorithm to gain a clear understanding. It is all simple matrix manipulation. So while reading the subsequent chapters, let us always keep our objective in mind—we want to understand each solution scheme and compare its performance relative to others so that for a given problem, we can choose the best.

REFERENCES

1.1 Brebbia, C.A. (Ed.), *A Handbook of Finite Element Systems*, C.M.I. Publications, Southampton, 1981.

1.2 Zienkiewicz, O.C., *The Finite Element Method in Engineering Sciences*, 2nd edn, McGraw-Hill, London, 1971.

1.3 Desai, C.S. and J.F. Abel, *Introduction to the Finite Element Method: A Numerical Method for Engineering Analysis*, Van Nostrand Reinhold, New York, 1972.

1.4 Rao, S.S., *The Finite Element Method in Engineering*, Pergamon Press, New York, 1982.

1.5 Gallagher, R.H., *Finite Element Analysis Fundamentals*, Prentice-Hall, Englewood Cliffs, N. J., 1975.

1.6 Argyris, J.H., Continua and discontinua, *Proc. Conf. Matrix Methods in Struct. Mech. Air Force Inst. Tech.*, Wright Patternson Base, Ohio, 1965.

1.7 Krishnamurthy, C.S., *Finite Element Analysis—Theory and Programming*, Tata McGraw-Hill, New Delhi, 1987.

1.8 Reddy, J.N., *Introduction to Finite Element Methods*, McGraw-Hill, New York, 1984.

1.9 Krämer, E., *Maschinendynamik*, Springer Verlag, Berlin, 1984.

1.10 Lalanne, M.P., Berthier and J.D. Hagopian, *Mechanical Vibrations for Engineers*, John Wiley & Sons, New York, 1983.

1.11 Przemieniecki, J.A., *Theory of Matrix Structural Analysis*, McGraw-Hill, New York, 1968.

1.12 Ramamurti, V. and Sreenivasamurthy S., Coriolis effect on rotating cantilever plate, *J. Strain Analysis*, vol. 16(3), pp 97–106, 1981.

1.13 Ramamurti, V. and L.S. Gupta, Design of Rotary Kiln Tyres, *Zement Kalk Crips*, vol. 31 (11), pp 614–618, 1978.

1.14 Cook, R.D., Malkus D.S. and Plesha, M.E., *Concepts and Applications of Finite Element Analysis*, John Wiley & Sons, New York, 1989.

1.15 Schroeder, W.J. and M.S., Shepard, Automatic mesh generation and the Delaunay triangulation, *Int. J. Num. Methods in Engg.*, vol. 26, pp 2503–2515, 1988.

1.16 Hole, K., Finite element mesh generation methods—A review and classification, *Computer Aided Design*, vol. 20, no. 1, pp 27–38, 1988.

1.17 Srinivasan, V., L.R. Nackman, Tung Jung Mu and S.N. Meshkat, Automatic mesh generation using the symmetric axis transformation of polygonal domains, *IBM Research Report RC16132*, 9/27/90.

1.18 Salvadori, M.G. and M.L. Baron, *Numerical Methods in Engineering*, Prentice-Hall, Englewood Cliffs, New Jersey, 1961.

1.19 Budiansky, B. and P.P. Radkowski, Numerical analysis of unsymmetrical bending of shells of revolution, *AIAAJ* 1 1833–1842, 1963.

1.20 Ramamurti, V. and R.S. Alwar, Stress analysis of tube mills, *J. Strain Analysis*, vol. 8, 200–208, 1973.

1.21 Alterman, Z. and F.C. Karal, Propagation of elastic waves in semi-finite cylindrical rod, using finite difference methods, *J. Sound and Vib.*, vol. 13, no. 2, pp 115–145, 1970.

1.22 Noor, A.K. (Ed.), Advances in computational, structural and solid mechanics, *Computers and Stuctures*, vol. 27, no. 1, 1987.

EXERCISE

1.1 A shaft AD 450 mm long has a diameter of 20 mm for the first one third AB of its length and has a diameter of 25 mm for the remaining two third BD. It is supported on self aligning ball bearings at A and C (at a distance of 300 mm from A) and is held by double row deep groove ball bearings at D. The shaft carries a load of 100 N at B. Calculate the maximum bending stress under the load B. (Ans. 6.1 N/mm^2) (Hint: Treat the shaft as hinged at A and C and fixed at D. Consider two degrees of freedom each at A, B, C and D out of which the deflections at A, C and D are zero, besides the slope at D. Use the stiffness matrix given by Eq. 1.119)

Static Problems

Before proceeding to study the behaviour of static problems, it is useful to recapitulate some basic definitions.

➤ 2.1 DEFINITIONS

2.1.1 Lower and Upper Triangular Matrix

The lower triangular matrix, [L], which is square, has all its elements above the principal diagonal equal to zero. The upper triangular matrix [U], which is square, has all its elements below the principal diagonal equal to zero.

2.1.2 Symmetric Matrix

The elements of the symmetric matrix are symmetrical about the principal diagonal.

Example 2.1

$$\begin{bmatrix} 10 & 4 & 5 \\ 4 & 20 & 6 \\ 5 & 6 & 30 \end{bmatrix}$$

2.1.3 Skew Symmetric Matrix

If in a matrix $[A], A_{ij(i \neq j)} = -A_{ji}$, this matrix is said to be skew symmetric.

Example 2.2

$$\begin{bmatrix} 10 & 4 & -5 \\ -4 & 20 & -6 \\ 5 & 6 & 30 \end{bmatrix}$$

2.1.4   Banded Matrix

A banded matrix is one whose non-zero elements form a definite band around the principal diagonal.

Example 2.3

$$[A] = \begin{bmatrix} a_{11} & a_{12} & a_{13} & 0 & 0 \\ a_{21} & a_{22} & a_{23} & a_{24} & 0 \\ a_{31} & a_{32} & a_{33} & a_{34} & a_{35} \\ 0 & a_{42} & a_{43} & a_{44} & a_{45} \\ 0 & 0 & a_{53} & a_{54} & a_{55} \end{bmatrix}$$

There are only two elements on either side of the principal diagonal which are non-zero. This is a banded matrix having a maximum bandwidth (number of non-zero elements along any row) of 5. The semi-bandwidth of this matrix is 3. This denotes the maximum number of non-zero elements in any row on one side of the principal diagonal including the principal diagonal. Let us assume b to denote the semi-bandwidth. Hence the bandwidth according to this definition is $(2b - 1)$. Some authors use carpetwidth to denote bandwidth. Henceforth let us use b to denote semi-bandwidth.

2.1.5   Tridiagonal Matrix

A banded matrix of specific interest is a banded matrix of semi-bandwidth 2 (carpetwidth 3). This is known as a tridiagonal matrix. (A diagonal matrix has non-zero elements only in its principal diagonal.)

➤ 2.2   GENERAL APPROACH

The present-day computer handles problems of the order of 10,000 unknowns or more. It is absolutely essential to optimise the solution algorithm to result in the most efficient procedure. For example, the structural analyses of nuclear power plant design may consume 180 engineering man months and require 30 different mathematical models [2.1]. If inappropriate methods of solutions are employed, the cost of analysis may go up several fold. It is normally argued that in most engineering problems the cost of productive computer runs is usually very small as compared to the cost of man hours required for pre- and post-processing (input preparation and output compilation). But it is also true that one productive run is preceded by 10 or more non-productive runs. Especially when the size of the problem is large, an efficient solution algorithm really contributes to considerable saving in running cost. With this in view, we will now deal with different approaches available to solve large size problems.

The general nature of problems falling under the category of static loading will be of the type

$$[K]\{u\} = \{f\} \tag{2.1}$$

where $[K]$ is the stiffness matrix of size $(n \times n)$, $\{u\}$ is the displacement matrix of size $(n \times 1)$ and $\{f\}$ is the force vector of size $(n \times 1)$. $[K]$ and $\{f\}$ will be known $\{u\}$ will have to be evaluated.

Equations (2.1) are a set of n simultaneous equations to be solved. The well established approach is to solve these equations by Cramer's rule: n determinants are evaluated by replacing one column at a time (first to the nth) by the R.H.S. force vector. Lastly the determinants of the coefficients on the L.H.S. (Δ) is evaluated. Then $u_i(i = 1, n)$ is given by

$$u_i = \frac{\Delta_i}{\Delta} \tag{2.2}$$

This is a very general approach. The number of multiplications involved in evaluating a determinant of size n is $n^3/3$. For $(n + 1)$ determinants encountered here, the total number of multiplications is $[(n + 1)n^3]/3$. The time taken for addition or subtraction in a computer is usually ignored as compared to multiplication or division. When n is very large, then the number of multiplications is $n^4/3$. If, for example, each multiplication takes approximately 50×10^{-6} s in a digital computer, the total time taken

$$= 50 \times 10^{-6}\,\frac{n^4}{3}\,\text{s}$$

When $n = 100$, the time taken is around 26 min.

➤ 2.3 MATRIX INVERSION

Matrix inversion followed by multiplication is a very effective method of solving simultaneous equations.

Since $\qquad\qquad\qquad [K]\{u\} = \{f\}$

we have $\qquad\qquad\qquad \{u\} = [K]^{-1}\{f\} \tag{2.3}$

It will be shown that the number of multiplications involved in inverting a matrix of size $(n \times n)$ is n^3. Hence if $\{u\}$ is computed by

Eq. (2.3), the total number of multiplications will be equal to $n^3 + n^2$, since $[K]^{-1}\{f\}$ involves n^2 multiplications besides n^3 multiplications in inversion. When $n = 100$, total number of multiplications = $(100)^3 + (100)^2 \approx 100^3$.

Time taken = $10^6 \times 50 \times 10^{-6} = 50$ s (using the unit time taken in the previous example).

This is $n/3$ times more efficient than Cramer's rule.

This time taken should be compared with the time taken by a hand calculator. To quote from Ref. [2.2], "Thus a system of 10 equations can be easily solved in approximately 10 h." Whereas a hand calculator takes 10 h for solving a 10×10 problem, computer takes about 50 s for solving a 100×100 problem.

2.3.1   Inversion by Partitioning

If $[K]$ is a square matrix of size $n \times n$ the inversion is achieved in stages by first inverting a matrix of size (1×1), using this information (2×2) and so on till one gets $(n \times n)$ [2.3].
For a typical partition, let $[K_i]$ be given by

$$[K_i] = \begin{bmatrix} A_i & B_i \\ C_i & D_i \end{bmatrix} \tag{2.4}$$

$[K_i]$ is the truncated left hand square matrix of size $(m \times m)$, m being less than n. $[A_i]$ will have a size $(m - 1, m - 1)$, B_i of size $(m - 1, 1)$, C_i of size $(1, m - 1)$ and D_i of size $(1, 1)$.
Let the inverted matrix of $[K_i]$ be

$$[K_i]^{-1} = \begin{bmatrix} W_i & X_i \\ Y_i & Z_i \end{bmatrix} \tag{2.5}$$

where W_i, X_i, Y_i and Z_i are of size $(m - 1, m - 1)$, $(m - 1, 1)$, $(1, m - 1)$ and $(1, 1)$ respectively.

Since $[K_i] [K_i]^{-1} = [I]$, we have

$$\begin{bmatrix} A_i & B_i \\ C_i & D_i \end{bmatrix} \begin{bmatrix} W_i & X_i \\ Y_i & Z_i \end{bmatrix} = [I] \tag{2.6}$$

Hence
$$[A][W] + [B][Y] = [I] \tag{2.7}$$

$$[A][X] + [B][Z] = [0] \tag{2.8}$$

$$[C][W] + [D][Y] = [0] \tag{2.9}$$

$$[C][X] + [D][Z] = [I] \tag{2.10}$$

dropping subscripts.

Alternately,

$$\begin{bmatrix} \mathsf{'} & X \\ Y & Z \end{bmatrix} \begin{bmatrix} A & B \\ C & D \end{bmatrix} = [I]$$

Expanding,

$$[W][A] + [X][C] = [I] \tag{2.11}$$

$$[W][B] + [X][D] = [0] \tag{2.12}$$

$$[Y][A] + [Z][C] = [0] \tag{2.13}$$

$$[Y][B] + [Z][D] = [I] \tag{2.14}$$

Premultiplying Eq. (2.8) by $[C][A]^{-1}$

$$[C][X] + [C][A]^{-1}[B][Z] = [0]$$

Subtracting this from Eq. (2.10), we have

$$\left[[D] - [C][A]^{-1}[B]\right][Z] = [I]$$

Hence

$$[Z] = \left[[D] - [C][A]^{-1}[B]\right]^{-1} \tag{2.15}$$

From Eq. (2.13), we have

$$[Y][A] = -[Z][C]$$

Hence

$$[Y] = -[Z][C][A]^{-1} \tag{2.16}$$

From Eq. (2.8), we have

$$[A][X] = -[B][Z]$$

Hence

$$[X] = -[A]^{-1}[B][Z] \tag{2.17}$$

From Eq. (2.7),

$$[W] + [A]^{-1}[B][Y] = [A]^{-1}$$

Hence

$$[W] = [A]^{-1} - [A]^{-1}[B][Y] \tag{2.18}$$

Equations (2.15) to (2.18) are the recurrence relationships for computing $[Z]$, $[Y]$, $[X]$, and $[W]$.

Assuming $[A^{-1}]$ for a problem of size $(m-1, m-1)$ is known, $[B]$, $[C]$, $[D]$ are inputs of size $(m-1, 1)$, $(1, m-1)$ and $(1, 1)$ corresponding to the mth row and mth column, Eqs (2.15) to (2.18) compute $[Z]$ of size $(1, 1)$, $[Y]$ of size $(1, m-1)$, $[X]$ of size $(m-1, 1)$ and $[W]$ of $(m-1, m-1)$.

Now values $[W]$, $[X]$, $[Y]$ and $[Z]$ combined give us $[K]^{-1}$ of size

(m, m) according to Eq. (2.5). This process is continued till one gets $[K]^{-1}$ of size $(n \times n)$. It is needless to point out that for $m = 1$, $[A^{-1}]$ is just $1/A$.

Total Number of Multiplications

When $m = 1$, the number of multiplications = 1.

When $m = 2$, $[Z]$, $[Y]$, $[X]$ and $[W]$ involve 3, 1, 2 and 1 multiplications respectively.

For values of $m = 1$ and $m = 2$, the total number of multiplications $= 1 + 7 = 8$.

When $m = 3$, the number of multiplications involved in Z, Y, X, and W are 7, 2, 6, and 4 respectively.

Hence for values up to $m = 3$, the number of multiplications = 8 + 19 = 27. It is easy to prove for the mth cycle that the number of multiplications involved in Z, Y, X, and W are $m^2 + m^1 + 1$, m, $m^2 + m^1$, m^2 respectively ($= 3m^2 + 3m + 1$).

Hence the total number of multiplications up to $(m + 1) = m^3 + (3m^2 + 3m + 1) = (m + 1)^3$.

Core Needed

For the mth cycle core needed for Z, Y, X and W are $(1, 1)$, $(1, m - 1)$, $(m - 1, 1)$, $(m - 1, m - 1)$. The program for this inversion is given as Prog. 2.1.

Example 2.4
Find the inverse of the following matrix:

$$[A] = \begin{bmatrix} 5 & -4 & +1 & 0 \\ -4 & 6 & -4 & 1 \\ 1 & -4 & 6 & -4 \\ 0 & 1 & -4 & 5 \end{bmatrix}$$

I Stage

$$[A_1] = [5]$$

$$[A_1]^{-1} = [1/5]$$

II Stage

$$[A_2] = \begin{bmatrix} 5 & -4 \\ -4 & 6 \end{bmatrix}$$

$$[A] = [5] \quad [B] = [-4]$$

$$[C] = [-4] \quad [D] = [6]$$

$$[A^{-1}] = 1/5 \quad [A^{-1}][B] = -4/5$$

$$[C][A^{-1}][B] = 16/5$$

$$[Z] = ([D] - [C][A]^{-1}[B])^{-1} = 5/14$$

$$[Y] = -[Z][C][A]^{-1} = 2/7$$

$$[X] = -[Z][A]^{-1}[B] = 2/7$$

$$[W] = [A]^{-1} - [A]^{-1}[B][Y] = 3/7$$

Hence
$$[A_2]^{-1} = \begin{bmatrix} 3/7 & 2/7 \\ 2/7 & 5/14 \end{bmatrix}$$

III Stage

$$[B] = \begin{bmatrix} 1 \\ -4 \end{bmatrix} \quad [C] = [1 - 4] \quad [D] = [6]$$

$$[A]^{-1}[B] = \begin{bmatrix} 3/7 & 2/7 \\ 2/7 & 5/14 \end{bmatrix} \begin{bmatrix} 1 \\ -4 \end{bmatrix} = \begin{bmatrix} -5/7 \\ -8/7 \end{bmatrix}$$

$$[C][A]^{-1}[B] = [1, -4] \begin{bmatrix} -5/7 \\ -8/7 \end{bmatrix} = 27/7$$

$$[Z] = \left(6 - \frac{27}{7}\right)^{-1} = 7/15$$

$$[Y] = -[Z][C][A]^{-1} = [1/3, 8/15]$$

$$[X] = [-A]^{-1}[B][Z] = \begin{bmatrix} 1/3 \\ 8/15 \end{bmatrix}$$

$$[W] = [-A]^{-1} - [A]^{-1}[B][Y] = \begin{bmatrix} 2/3 & 2/3 \\ 2/3 & 29/30 \end{bmatrix}$$

$$[A_3^{-1}] = \begin{bmatrix} 2/3 & 2/3 & 1/3 \\ 2/3 & 29/30 & 8/15 \\ 1/3 & 8/15 & 7/15 \end{bmatrix}$$

IV Stage

$$[B] = \begin{bmatrix} 0 \\ 1 \\ -4 \end{bmatrix} \qquad [C] = [0, 1, -4] \qquad [D] = [5]$$

$\therefore$ $\qquad$ $[Z] = 6/5$ $\qquad$ $[Y] = [4/5, 7/5, 8/5]$

$$[X] = \begin{bmatrix} 4/5 \\ 7/5 \\ 8/5 \end{bmatrix} \qquad [W] = \begin{bmatrix} 18/15 & 24/15 & 21/15 \\ 24/15 & 78/30 & 36/15 \\ 21/15 & 36/15 & 39/15 \end{bmatrix}$$

$\therefore$ $\qquad$ $$[A]_v^{-1} = \begin{bmatrix} 6/5 & 8/5 & 7/5 & 4/5 \\ 8/5 & 13/5 & 12/5 & 7/5 \\ 7/5 & 12/5 & 13/5 & 8/5 \\ 4/5 & 7/5 & 8/5 & 6/5 \end{bmatrix}$$

2.3.2 Inversion by Gauss Jordan Method

Let the matrix be [A] and its inverse [B]. Then since $[A] \cdot [B] = [I]$, we have

$$\begin{bmatrix} a_{11} & a_{12} & \dots & a_{1n} \\ a_{21} & a_{22} & \dots & a_{2n} \\ \dots & \dots & \dots & \dots \\ a_{n1} & a_{n2} & \dots & a_{nn} \end{bmatrix} \begin{bmatrix} b_{11} & b_{12} & \dots & b_{1n} \\ b_{21} & \dots & \dots & \dots \\ \dots & \dots & \dots & \dots \\ b_{n1} & b_{n2} & \dots & b_{nn} \end{bmatrix}$$

$$= \begin{bmatrix} 1 & & & \\ & 1 & & \\ & & 1 & \\ & & & 1 \end{bmatrix} \qquad (2.19)$$

Expanding the matrix multiplication [2.4],

$$a_{11}b_{11} + a_{12}b_{21} + \dots + a_{1n}b_{n1} = 1$$
$$a_{21}b_{11} + a_{22}b_{21} + \dots + a_{2n}b_{n1} = 0 \qquad (2.20)$$
$$\dots \qquad \dots \qquad \dots \qquad \dots$$
$$a_{n1}b_{11} + a_{n2}b_{21} + \dots + a_{nn}b_{n1} = 0$$

$$a_{11}b_{12} + a_{12}b_{22} + \ldots + a_{1n}\overline{b_{n2}} = 0 \qquad (2.21)$$
$$a_{21}b_{12} + b_{22}b_{22} + \ldots + a_{2n}b_{n2} = 1$$
$$\ldots \quad \ldots \quad \ldots \quad \ldots$$
$$a_{n1}b_{12} + a_{n2}b_{22} + \ldots + a_{nn}b_{n2} = 0$$

The n sets of equations contained in Eqs (2.20) can be used to evaluate the unknown coefficients $b_{11}, b_{21} \ldots b_{n1}$. Similarly, Eqs (2.21) are used to evaluate $b_{12}, b_{22} \ldots b_{n2}$ and this process can be continued till one obtains the coefficients $b_{1n}, b_{2n} \ldots b_{nn}$. This procedure can easily be explained by working out an example.

Example 2.5
Compute the inverse of

$$[A] = \begin{bmatrix} 4.00 & 2.00 & 1.00 & 1.00 \\ 2.00 & 10.00 & 2.00 & 1.00 \\ 1.00 & 2.00 & 4.00 & 2.00 \\ 1.00 & 2.00 & 4.00 & 8.00 \end{bmatrix}$$

Equations (2.19) can be written in the following form:

	[A]				[B]			
4.00	2.00	1.00	1.00	1.00	0.00	0.00	0.00	(1)
2.00	10.00	2.00	1.00	0.00	1.00	0.00	0.00	(2)
1.00	2.00	4.00	2.00	0.00	0.00	1.00	0.00	(3)
1.00	2.00	4.00	8.00	0.00	0.00	0.00	1.00	(4)

I Stage of reduction

(1)	4.00	2.00	1.00	1.00	1.00	0.00	0.00	0.00	(1)
$(2) - \dfrac{(1)}{2}$	0.00	9.00	1.50	0.50	-0.50	1.00	0.00	0.00	(6)
$(3) - \dfrac{(1)}{4}$	0.00	1.50	3.75	1.75	-0.25	0.00	1.00	0.00	(7)
$(4) - \dfrac{(1)}{4}$	0.00	1.50	3.75	7.75	-0.25	0.00	0.00	1.00	(8)

II Stage of reduction

(1)/4	1.00	0.50	0.25	0.25	0.25	0.00	0.00	0.00	(5)
	0.00	9.00	1.50	0.50	-0.50	1.00	0.00	0.00	(6)
	0.00	1.50	3.75	1.75	-0.25	0.00	1.00	0.00	(7)
	0.00	1.50	3.75	7.75	-0.25	0.00	0.00	1.00	(8)

III Stage of reduction

(5) $-\dfrac{(6)}{18}$ 1.00   0.00   0.17   0.22   0.28 $-$ 0.06   0.00   0.00   (9)

(6)   0.00   9.00   1.50   0.50 $-$ 0.50   1.00   0.00   0.00   (6)

(7) $-\dfrac{(6)}{6}$ 0.00   0.00   3.50   1.67 $-$ 0.17 $-$ 0.16   1.00   0.00 (11)

(8) $-\dfrac{(6)}{6}$ 0.00   0.00   3.50   7.67 $-$ 0.17 $-$ 0.16   0.00   1.0 (12)

IV Stage of reduction

       1.00   0.00   0.17   0.22   0.28 $-$ 0.06   0.00   0.00   (9)

$\dfrac{(6)}{6}$ 0.00   1.00   0.17   0.06 $-$ 0.06   0.11   0.00   0.00 (10)

       0.00   0.00   3.50   1.67 $-$ 0.17 $-$ 0.16   1.00   0.00 (11)

       0.00   0.00   3.50   7.67 $-$ 0.17 $-$ 0.16   0.00   1.00 (12)

V Stage of reduction

[9] $-\dfrac{(11)}{20}$ 1.00 0.00 0.00   0.14   0.29 $-$ 0.05 $-$ 0.05 0.00 (13)

[10] $-\dfrac{(11)}{20}$ 0.00 1.00 0.00 $-$ 0.02 $-$ 0.05   0.12 $-$ 0.05 0.00 (14)

       0.00 0.00 3.50   1.67 $-$ 0.17 $-$ 0.16   1.00 0.00 (11)

[12] $-$ [11] 0.00 0.00 0.00   6.00 $-$ 0.00   0.00 $-$ 1.00 1.00 (13)

VI Stage of reduction

       1.00 0.00 0.00   0.14   0.29 $-$0.05 $-$0.05 $-$0.00 (13)

[10]/3.50 0.00 1.00 0.00 $-$0.02 $-$0.05   0.12 $-$0.05   0.00 (14)

       0.00 0.00 1.00   0.48 $-$0.05 $-$0.05   0.28   0.00 (15)

       0.00 0.00 0.00   6.00   0.00   0.00 $-$1.00   1.00 (16)

VII Stage of reduction

[13] $-\dfrac{(16)}{42}$ 1.00 0.00 0.00 0.00   0.29 $-$ 0.05 $-$ 0.02 $-$ 0.02 (17)

[14] $-\dfrac{(16)}{300}$ 0.00 1.00 0.00 0.00 $-$ 0.05   0.12 $-$ 0.05 0.003 (18)

[15] $- \dfrac{(16)}{12.5}$ 0.00 0.00 1.00 0.00 $-$ 0.05 $-$ 0.05 0.37 $-$ 0.08 (19)

0.00 0.00 0.00 6.00 0.00 0.00 $-$ 1.00 1.00 (16)

VIII Stage of reduction

1.00	0.00	0.00	0.00	0.29	−0.05	−0.02	−0.02 (17)
0.00	1.00	0.00	0.00	−0.05	0.12	−0.05	0.003 (18)
0.00	0.00	1.00	0.00	−0.05	−0.05	0.37	−0.08 (19)

$\dfrac{(16)}{6}$ 0.00 0.00 0.00 1.00 0.00 0.00 −0.17 0.17 (20)

Hence the inverted matrix

$$[B] = \begin{bmatrix} 0.29 & -0.05 & -0.02 & -0.02 \\ -0.05 & 0.12 & -0.05 & 0.003 \\ -0.05 & -0.05 & 0.37 & -0.08 \\ 0.00 & 0.00 & -0.17 & 0.17 \end{bmatrix}$$

Total Number of Multiplications

At every stage of simplifications every row of $[B]$ involves n divisions and n rows involve $(n \times n) = n^2$ divisions. Since there are n such stages the total number of multiplications (or divisions) is equal to n^3.

Since the modified elements of $[B]$ can be accommodated in their original locations, the core needed is only $(n \times n)$. The listing of this program (along with row, column interchange, if needed) is given in Prog. 2.2.

The modifications at every stage can be summarised as follows:

(i) $a'_{kj} = \dfrac{a_{kj}}{a_{kk}}$ for $j = 1, 2, \dots n$ (except $j \neq k$)

(ii) $a'_{kk} = \dfrac{1}{a_{kk}}$

(iii) $a'_{ij} = a_{ij} - a'_{kj} a_{ik}$ for $\begin{cases} i = 1, 2, \dots n \text{ (except } i \neq k) \\ j = 1, 2, \dots \text{ (for each } i \text{ except } j \neq k) \end{cases}$

(iv) $a'_{ik} = 0 - a_{ik} a'_{kk}$ for $i = 1, 2, \dots n$ (except $i \neq k$)   (2.22)

➤ **2.4   GAUSSIAN ELIMINATION METHOD**

One of the well-known techniques used for solving a set of simultaneous equations is the Gaussian elimination technique.

Let
$$a_{11}x_1 + a_{12}x_2 + \ldots + a_{1n}x_n = b_1$$
$$a_{21}x_1 + a_{22}x_2 + \ldots + a_{2n}x_n = b_2 \qquad (2.23)$$
$$\ldots \qquad \ldots \qquad \ldots$$
$$a_{n1}x_1 + a_{n2}x_2 + \ldots + a_{nn}x_n = b_n$$

Modifying the second equation by adding $-a_{21}/a_{11}$ of the first equation, we have

$$(0) + \left(a_{22} - \frac{a_{21}}{a_{11}}a_{12}\right)x_2 + \ldots = b_2 - b_1\frac{a_{21}}{a_{11}}$$

Similarly, the third and the rest can be modified by multiplying them by adding $-a_{31}/a_{11}$ of the first, $-a_{41}/a_{11}$ of the first, and so on. The net resultant would be

$$a_{11}x_1 + a_{12}x_2 \ldots = b_1$$
$$0 + a'_{22}x_2 + \ldots = b'_2$$
$$0 + a'_{32}x_2 + \ldots = b'_3 \qquad (2.24)$$
$$\ldots \qquad \ldots \qquad \ldots$$
$$0 + a'_{n2}x_2 + \ldots = b'_n$$

Proceeding along similar lines for the elimination of the second column, third column, and so on, one can get the following resultant set of equations.

$$a_{11}x_1 + a_{12}x_2 \ldots = b_1$$
$$a'_{22}x_2 + \ldots = b'_2$$
$$a''_{33}x_3 + \ldots = b'_3 \qquad (2.25)$$
$$\ldots \qquad \ldots \qquad \ldots$$
$$a'''_{nn}x_n = b'''_n$$

Values of x_n, x_{n-1}, ..., x_1 can be obtained from Eqs (2.25) by back substitution.

Example 2.6

$$4x_1 + 2x_2 + x_3 + x_4 = 10 \qquad (1)$$
$$2x_1 + 10x_2 + 2x_3 + x_4 = 20 \qquad (2)$$
$$x_1 + 2x_2 + 4x_3 + 2x_4 = 30 \qquad (3)$$
$$x_1 + 2x_2 + 4x_3 + 8x_4 = 40 \qquad (4)$$
$$4x_1 + 2x_2 + x_3 + x_4 = 10 \qquad (1)$$

$(2) - \dfrac{(1)}{2}$
$$0x_1 + 9_{x2} + 1.5x_3 + 5x_4 = 15 \qquad (5)$$

$(3) - \dfrac{(1)}{4}$ $\qquad 0x_1 + 1.5x_2 + 3.75x_3 + 1.75x_4 = 27.5$ $\qquad (6)$

$(4) - \dfrac{(1)}{4}$ $\qquad 0x_1 + 1.5x_2 + 3.75x_3 + 7.75x_4 = 37.5$ $\qquad (7)$

$$4x_1 + 2x_2 + x_3 + x_4 = 10 \qquad (1)$$
$$9x_2 + 1.5x_3 + 5x_4 = 15 \qquad (5)$$

$(6) - \dfrac{(5)}{6}$ $\qquad 0x_2 + 3.5x_3 + 1.67x_4 = 25$ $\qquad (8)$

$(7) - \dfrac{(5)}{6}$ $\qquad 0x_2 + 3.5x_3 + 7.67x_4 = 35$ $\qquad (9)$

$$4x_1 + 2x_2 + x_3 + x_4 = 10 \qquad (1)$$
$$9x_2 + 1.5x_3 + 5x_4 = 15 \qquad (5)$$
$$3.5x_3 + 1.67x_4 = 25 \qquad (8)$$

$(9)-(8)$ $\qquad 6x_4 = 10$ $\qquad (10)$

$\therefore \qquad x_4 = 1.66$

From (8), $\qquad x_3 = \dfrac{25 - (1.66) \times (1.67)}{3.5} = 6.35$

From (5), $\qquad x_2 = 0.52$

From (1), $\qquad x_1 = 0.24$

Total Number of Multiplications

For the L.H.S. it is

$$n(n - 1) + (n - 1)(n - 2) + \ldots + 2.1$$

For the R.H.S. it is

$$(n - 1) + (n - 2) + \ldots + 1$$

For back substitution it is

$$1 + 2 + \ldots (n - 1)$$

The addition of these three when n is large

$$\frac{n^3}{3} + n^2 \approx \frac{n^3}{3}$$

To improve the accuracy on the computer, the largest element is found and brought to the leading or pivot position before elimination of elements is done in any column.

Let us consider the previous example.

Example 2.7

$$\begin{bmatrix} 4 & 2 & 1 & 1 \\ 2 & 10 & 2 & 1 \\ 1 & 2 & 4 & 2 \\ 1 & 2 & 4 & 8 \end{bmatrix} \begin{Bmatrix} x_1 \\ x_2 \\ x_3 \\ x_4 \end{Bmatrix} = \begin{Bmatrix} 10 \\ 20 \\ 30 \\ 40 \end{Bmatrix}$$

Since 10 is the largest element, let us make it the first row, first column by interchanging the first two rows and first two columns.

$$\begin{bmatrix} 10 & 2 & 2 & 1 \\ 2 & 4 & 1 & 1 \\ 2 & 1 & 4 & 2 \\ 2 & 1 & 4 & 8 \end{bmatrix} \begin{Bmatrix} x_2 \\ x_1 \\ x_3 \\ x_4 \end{Bmatrix} = \begin{Bmatrix} 20 \\ 10 \\ 30 \\ 40 \end{Bmatrix}$$

Eliminating the first column of the second, third and fourth rows, we have

$$\begin{bmatrix} 10 & 2 & 2 & 1 \\ 0 & 3.6 & 0.6 & 0.8 \\ 0 & 0.6 & 3.6 & 1.8 \\ 0 & 0.6 & 3.6 & 7.8 \end{bmatrix} \begin{Bmatrix} x_2 \\ x_1 \\ x_3 \\ x_4 \end{Bmatrix} = \begin{Bmatrix} 20 \\ 6 \\ 26 \\ 36 \end{Bmatrix}$$

Since 7.8 is the largest element, let us interchange the second row and fourth row, and second row and second column.

$$\begin{bmatrix} 10 & 1 & 2 & 2 \\ 0 & 7.8 & 3.6 & 0.6 \\ 0 & 1.8 & 3.6 & 0.6 \\ 0 & 0.8 & 0.6 & 3.6 \end{bmatrix} \begin{Bmatrix} x_2 \\ x_4 \\ x_3 \\ x_1 \end{Bmatrix} = \begin{Bmatrix} 20 \\ 36 \\ 26 \\ 6 \end{Bmatrix}$$

Eliminating the second column of third and fourth rows, we have

$$\begin{bmatrix} 10 & 1 & 2 & 2 \\ 0 & 7.8 & 3.6 & 0.6 \\ 0 & 0 & 2.77 & 0.46 \\ 0 & 0 & 0.24 & 3.54 \end{bmatrix} \begin{Bmatrix} x_2 \\ x_4 \\ x_3 \\ x_1 \end{Bmatrix} = \begin{Bmatrix} 20 \\ 36 \\ 17.7 \\ 240 \end{Bmatrix}$$

Interchanging third and fourth rows and hence third and fourth columns, we have

$$\begin{bmatrix} 10 & 1 & 2 & 2 \\ 0 & 7.8 & 0.6 & 3.6 \\ 0 & 0 & 3.54 & 0.24 \\ 0 & 0 & 0.46 & 2.77 \end{bmatrix} \begin{Bmatrix} x_2 \\ x_4 \\ x_1 \\ x_3 \end{Bmatrix} = \begin{Bmatrix} 20 \\ 36 \\ 2.40 \\ 17.7 \end{Bmatrix}$$

Eliminating the third column of the fourth row

$$\begin{bmatrix} 10 & 1 & 2 & 2 \\ 0 & 7.8 & 0.6 & 3.6 \\ 0 & 0 & 3.54 & 0.24 \\ 0 & 0 & 0 & 2.74 \end{bmatrix} \begin{Bmatrix} x_2 \\ x_4 \\ x_1 \\ x_3 \end{Bmatrix} = \begin{Bmatrix} 20 \\ 36 \\ 2.40 \\ 17.40 \end{Bmatrix}$$

$\therefore$
$$x_3 = 6.34$$
$$x_1 = 0.24$$
$$x_4 = 1.66$$
$$x_2 = 0.52$$

➤ 2.5 CROUT'S PROCEDURE

Let us consider the simultaneous equations

$$[K]\{\delta\} = \{f\} \qquad (2.26)$$

In this approach $[K]$ is written as [2.6]

$$[K] = [L] \cdot [U] \qquad (2.27)$$

where $[L]$ and $[U]$ are the lower and upper triangular matrices of the same size as $[K]$. The principal diagonal elements of $[U]$ are unity.

Expanding Eqs (2.27),

$$\begin{bmatrix} k_{11} & k_{12} & k_{13} & \cdots \\ k_{21} & k_{22} & k_{23} & \cdots \\ k_{31} & k_{32} & k_{33} & \cdots \\ \cdots & \cdots & \cdots & \cdots \end{bmatrix}$$

$$= \begin{bmatrix} l_{11} & & & \\ l_{21} & l_{22} & & \\ l_{31} & l_{32} & l_{33} & \cdots \\ \cdots & \cdots & \cdots & \cdots \end{bmatrix} \begin{bmatrix} 1 & u_{12} & u_{13} & \cdots \\ & 1 & u_{23} & \cdots \\ & & 1 & \cdots \\ & & & \cdots \end{bmatrix} \qquad (2.28)$$

$$k_{11} = l_{11}, \quad k_{21} = l_{21}, \quad k_{31} = l_{31} \tag{2.29}$$

$$k_{12} = l_{11}u_{12}$$

$$k_{13} = l_{11}u_{13}$$

$$\therefore \quad u_{12} = k_{12}/l_{11}$$

$$u_{13} = k_{13}/l_{11} \tag{2.30}$$

Next
$$l_{22} = k_{22} - l_{21}u_{12}; \quad l_{32} = k_{32} - l_{31}u_{12}$$

$$u_{23} = (k_{23} - l_{21}u_{13})/l_{22} \tag{2.31}$$

$$\cdots \quad \cdots \quad \cdots \quad \cdots$$

The order of computation is first column, first row, second column, second row, and so on.

Example 2.8

$$4x_1 + 2x_2 + x_3 + x_4 = 10$$

$$2x_1 + 10x_2 + 2x_3 + x_4 = 20$$

$$x_1 + 2x_2 + 4x_3 + 2x_4 = 30$$

$$x_1 + 2x_2 + 4x_3 + 8x_4 = 40$$

$$\begin{bmatrix} 4 & 2 & 1 & 1 & 10 \\ 2 & 10 & 2 & 1 & 20 \\ 1 & 2 & 4 & 2 & 30 \\ 1 & 2 & 4 & 8 & 40 \end{bmatrix}$$

L.H.S   R.H.S.

Simplifying as per Eqs (2.29) to (2.31), we have

$$\begin{bmatrix} 4 & 0.5 & 0.25 & 0.25 & 2.5 \\ 2 & 9 & 0.17 & 0.06 & 1.66 \\ 1 & 1.5 & 3.495 & 0.47 & 7.14 \\ 1 & 1.5 & 3.495 & 6.00 & 1.66 \end{bmatrix}$$

2 Stage
4 Stage
6 Stage
8 Stage

1   3   5   7th Stage

Besides Eqs (2.29) to (2.31), one can write

$$[L][U]\{\delta\} = \{f\}$$

$$[U]\{\delta\} = [L]^{-1}\{f\}$$

Let
$$[L]^{-1}\{f\} = \{y\} \tag{2.32}$$

Then
$$[U]\{\delta\} = \{y\} \qquad (2.33)$$

$$\begin{bmatrix} 1 & u_{12} & u_{13} & \\ & 1 & \cdots & \\ & & 1 & \end{bmatrix} \begin{Bmatrix} \delta_1 \\ \delta_2 \\ \vdots \\ \delta_n \end{Bmatrix} = \begin{Bmatrix} y_1 \\ y_2 \\ \vdots \\ y_n \end{Bmatrix}$$

$$\therefore \qquad \delta_n = y_n$$

$$\delta_{n-1} = y_{n-1} - u_{n-1,n}y_n, \text{ and so on}$$

From Eq. (2.32), we have

$$f_1 = l_{11}y_1$$

$$f_2 = l_{21}y_1 + l_{22}y_2$$

and so on

Now R.H.S. of the equations has been modified to $\{y\}$ as shown.
Now

$$\delta_4 = y_4 = 1.66$$

$$\delta_3 = y_3 - u_{34}\delta_4$$

$$= 7.14 - (0.47 \times 1.66)$$

$$= 6.35$$

$$\delta_2 = y_2 - u_{23}\delta_3 - u_{24}\delta_4$$

$$= 0.52$$

$$\delta_1 = y_1 - u_{12}\delta_2 - u_{13}\delta_3 - u_{14}\delta_4$$

$$= 0.24$$

Steps 1 to 8 of the example can be explained in the following way.

Step I: Keep column 1 unchanged.

Step II: In the first row change all the elements other than the first dividing them by k_{11}. Even the right hand force (augmented matrix) is divided by k_{11}. This corresponds to computing y_i.

Step III: Modify the second to fourth elements of column 2 as follows:

$$k_{22} = k_{22} - k_{21}k_{12}$$

$$k_{32} = k_{32} - k_{31}k_{12}$$

$$k_{42} = k_{42} - k_{41}k_{12}$$

Step IV: Modify the third, fourth elements of the second row and f_2 as

$$k_{23} = k_{23} - k_{21} \cdot k_{13}/k_{22}$$

$$k_{24} = k_{24} - k_{21} \cdot k_{14}/k_{22}$$

$$f_2 = f_2 - k_{21} \cdot f_1/k_{22}$$

Steps V to VIII: Modify the third row, third column, fourth row and fourth column in the same manner.

This programme is suitable for real unsymmetric matrix $[K]$. The number of multiplications involved is $(n^3/3) + n^2$ when n is large.

➤ 2.6   CHOLESKY'S METHOD

This is the most efficient method for solving simultaneous equations when $[K]$ is symmetric. The procedure is similar to Crout's procedure. $[K]$ is decomposed and written either as $[L][L]^T$ or $[L][D][L]^T$. Let us discuss both the schemes here.

$$[K] = [L][L]^T \tag{2.34}$$

It can also be written as:

$$[K] = [U]^T [U]$$

Rewriting Eq. (2.24), we have:

$$\begin{bmatrix} k_{11} & & \\ k_{21} & k_{22} & \text{Sym} \\ k_{31} & k_{32} & k_{33} \end{bmatrix} = \begin{bmatrix} l_{11} & & 0 \\ l_{21} & l_{22} & \\ l_{31} & l_{32} & l_{33} \end{bmatrix} \begin{bmatrix} l_{11} & l_{21} & l_{31} \\ & l_{22} & l_{32} \\ 0 & & l_{23} \end{bmatrix}$$

Expanding, we have

$$k_{11} = l_{11}^2$$

$$k_{12} = l_{11}l_{21} \tag{2.35}$$

$$k_{13} = l_{11}l_{31}$$

$$\ldots \qquad \ldots$$

$$k_{21} = l_{11}l_{21}$$

$$k_{22} = l_{21}^2 + l_{22}^2 \tag{2.36}$$

$$k_{23} = l_{21}l_{31} + l_{22}l_{32}$$

Hence

$$l_{11} = \sqrt{k_{11}}$$
$$l_{21} = k_{12}/l_{11}$$
$$l_{31} = k_{13}/l_{11} \qquad (2.37)$$

$$\cdots \quad \cdots$$

$$l_{22} = \sqrt{k_{22} - l_{21}^2}$$
$$l_{32} = (k_{23} - l_{21}l_{31})/l_{22}$$

This is a fairly straightforward procedure. But there are occasions when computation of l_{ii} may become difficult when the quantity inside the square root becomes negative. To get over this problem either L can be declared complex or $[K]$ can be expressed as

$$[K] = [L][D][L]^T \qquad (2.38)$$

where

$$[L] = \begin{bmatrix} 1 & & & 0 \\ l_{21} & 1 & & \\ l_{31} & l_{32} & 1 & \\ l_{41} & l_{42} & l_{43} & 1 \end{bmatrix}$$

$$[D] = \begin{bmatrix} d_{11} & & & 0 \\ & d_{22} & & \\ & & d_{33} & \\ 0 & & & d_{44} \end{bmatrix} \qquad (2.39)$$

For any typical row, for example the fourth row, we have

$$g_{41} = k_{41}$$
$$g_{42} = k_{42} - l_{21}g_{41}$$
$$g_{43} = k_{43} - l_{31}g_{41} - l_{32}g_{42} \qquad (2.40)$$
$$l_{41} = g_{41}/d_{11}$$
$$l_{42} = g_{42}/d_{22}$$
$$l_{43} = g_{43}/d_{33}$$

Finally,

$$d_{44} = k_{44} - l_{41}g_{41} - l_{42}g_{42} - l_{43}g_{43}$$

Elements of the principal diagonal of $[D]$ avoid evaluating square roots as in Eqs (2.37).

It may be observed that the determinant of [K] is merely the product of the elements of the principal diagonal of [D].
Solving the simultaneous equations.

$$[K]\{\delta\} = \{f\}$$

when
$$[K] = [L][L]^T$$

then
$$[L][L]^T\{\delta\} = \{f\} \tag{2.41}$$

Premultiplying both sides by $[L]^{-1}$, we have

$$[L]^T\{\delta\} = [L]^{-1}\{f\}$$

Let
$$[L]^{-1}\{f\} = \{y\}$$

or
$$\{f\} = [L]\{y\} \tag{2.42}$$

Since $\{f\}$ and $[L]$ are known, $\{y\}$ can be computed by forward substitution

$$\begin{Bmatrix} f_1 \\ f_2 \\ \cdots \\ f_n \end{Bmatrix} = \begin{bmatrix} l_{11} & & & \\ l_{22} & l_{22} & & \\ \cdots & & & \\ \cdots & \cdots & l_{nn} \end{bmatrix} \begin{Bmatrix} y_1 \\ y_2 \\ \vdots \\ y_n \end{Bmatrix}$$

$$f = l_{11}y_1$$

$$f_2 = l_{21}y_1 + l_{22}y_2$$

and so on

Having computed $\{y\}$, $\{\delta\}$ can be computed by backward substitution in the equation

$$[L]^T\{\delta\} = \{y\} \tag{2.43}$$

or
$$\begin{bmatrix} l_{11} & l_{21} & l_{31} & \cdot \\ & l_{22} & l_{32} & \cdot \\ & & l_{33} & \cdot \\ \cdots & \cdots & \cdots & \cdots \end{bmatrix} \begin{Bmatrix} \delta_1 \\ \delta_2 \\ \vdots \\ \delta_n \end{Bmatrix} = \begin{Bmatrix} y_1 \\ y_2 \\ \vdots \\ y_n \end{Bmatrix}$$

In the case of [K] being expressed as $[L][D][L]^T$, the same procedure basically holds. The steps involved in solving the simultaneous equations are the following:

$$[K]\{\delta\} = \{f\}$$

$$[L][D][L]^T\{\delta\} = \{f\}$$

$$[L]^{-1}\{f\} = \{y\} \tag{2.44}$$

$$[D]^{-1}\{y\} = \{x\} \qquad (2.45)$$

$$[L]^{T}\{\delta\} = \{x\} \qquad (2.46)$$

Equations (2.44) are solved by forward substitution, (2.45) by straightforward multiplication and (2.46) by backward substitution.

Total Number of Multiplication Involved

In Cholesky's factorisation, it is $n^3/6$ since it is roughly one-half of the labour involved in Gaussian elimination due to symmetry. In Eqs (2.42) it is $n^2/2$ and Eqs (2.43) it is $n^2/2$. Hence the total number of multiplications are $n^3/6 + n^2$. When the given stiffness matrix is banded with semi-bandwidth (including diagonal) b, the number of multiplications = $nb^2/2 + 2nb$.

Core Needed

If the stiffness matrix is banded, the core needed for it is $n \times b$. This can be made use of to accommodate the lower triangular matrix also. Besides $\{f\}$ and $\{y\}$ have also to be stored. Locations of $\{f\}$ can be used for $\{\delta\}$. Hence the total core = $(n \times b) + 2n$.

➤ **2.7 POTTERS' METHOD** _____

This is a method of substructuring [2.7]. The basic approach is to split the problem into a number of smaller problems. Let the given problem be

$$[K]\{\delta\} = \{f\} \quad (2.26)$$

Let this be split in the following form

$$\begin{bmatrix} B_1 & A_1 & & & \cdots & \\ C_2 & B_2 & A_2 & & \cdots & \\ & C_3 & B_3 & A_3 & & \\ \cdots & \cdots & \cdots & \cdots & \cdots & \\ & & & C_n & B_n \end{bmatrix} \begin{Bmatrix} z_1 \\ z_2 \\ \vdots \\ z_n \end{Bmatrix} = \begin{Bmatrix} g_1 \\ g_2 \\ \vdots \\ g_n \end{Bmatrix} \qquad (2.47)$$

$z_1, z_2, z_3 \ldots$ are vectors of smaller size, $g_1, g_2, g_3 \ldots$ are force vectors of smaller size such that

$$\begin{Bmatrix} z_1 \\ z_2 \\ \vdots \\ z_n \end{Bmatrix} = \{\delta\} \qquad \begin{Bmatrix} g_1 \\ g_2 \\ \vdots \\ g_n \end{Bmatrix} = \{f\} \qquad (2.48)$$

In the case of finite element method

$$[C_2] = [A_1]^T, \qquad [C_3] = [A_2]^T$$

and so on whereas in the case of the finite difference method this is not necessarily true.

$\{z_1\}$, $\{z_2\}$, $\{z_3\}$, etc. need not necessarily have the same size. After (z_1, z_2, z_3) etc. need not necessarily have the same size. Recasting Eq. (2.47), one can write

$$\begin{bmatrix} I & P_1 & & & & \\ & I & P_2 & & & \\ & & I & P_3 & & \\ & & & & I & P_{n-1} \\ & & & & & I \end{bmatrix} \begin{Bmatrix} z_1 \\ z_2 \\ \cdots \\ \cdots \\ z_{n-1} \\ z_n \end{Bmatrix} = \begin{Bmatrix} q_1 \\ q_2 \\ \cdots \\ \cdots \\ q_{n-1} \\ q_n \end{Bmatrix} \qquad (2.49)$$

Expanding

$$\{z_1\} + [P_1]\{z_2\} = \{q_1\}$$

$$\{z_2\} + [P_2]\{z_3\} = \{q_2\}$$

$$\cdots \qquad \cdots\cdots \qquad \cdots$$

$$\{z_{n-1}\} + [P_{n-1}]\{z_n\} = \{q_{n-1}\} \qquad (2.50)$$

$$\{z_n\} = \{q_n\}$$

The first equation in (2.47) gives

$$[B_1]\{z_1\} + [A_1]\{z_2\} = \{g_1\}$$

or

$$\{z_1\} + [B_1]^{-1}[A_1]\{z_2\} = [B_1]^{-1}\{g_1\} \qquad (2.51)$$

Comparing Eq. (2.51) and the first equation in (2.50), we have

$$[P_1] = [B_1]^{-1}[A_1], \quad \{q_1\} = [B_1]^{-1}\{g_1\} \qquad (2.52)$$

The second equation in (2.47) gives

$$[C_2]\{z_1\} + [B_2]\{z_2\} + [A_2]\{z_3\} = \{g_2\}$$

Substituting for $\{z_1\}$ from the first equation in (2.50), we have

$$[C_2](\{q_1\} - [P_1]\{z_2\}) + [B_2]\{z_2\} + [A_2]\{z_3\} = \{g_2\}$$

Rearranging, we have

$$([B_2] - [C_2][P_1])\{z_2\} + [A_2]\{z_3\} = \{g_2\} - [C_2]\{q_1\}$$

or we have

$$\{z_2\} + ([B_2] - [C_2][P_1])^{-1} [A_2]\{z_3\} = ([B_2] - [C_2][P_1])^{-1} (\{g_2\} - [C_2]\{q_1\})$$
(2.53)

Comparing Eq. (2.53) with the second equation in (2.50), we have

$$[P_2] = ([B_2] - [C_2][P_1])^{-1} [A_2]$$
$$\{q_2\} = ([B_2] - [C_2][P_2])^{-1} (\{g_2\} - [C_2]\{q_1\}) \qquad (2.54)$$

Equation (2.54) can now be generalised and written as

$$[P_i] = ([B_i] - [C_i][P_{i-1}])^{-1} [A_i]$$
$$\{q_i\} = ([B_i] - [C_i][P_{i-1}])^{-1} (\{g_i\} - [C_i]\{q_{i-1}\}) \qquad (2.55)$$
$$(i = 2, \ldots, n - 1)$$

For the nth substructure

$$\{q_n\} = ([B_n] - [C_n][P_{n-1}])^{-1} (\{g_n\} - [C_n]\{q_{n-1}\}) \qquad (2.56)$$

To summarise the method, the following steps are involved.

Step I: Once $[A_1]$, $[B_1]$ and $\{g_1\}$ are generated, $[P_1]$ and $\{q_1\}$ are computed (Eq. 2.52)).

Step II: With $[P_{i-1}]$ and $\{q_{i-1}\}$ known, $[P_i]$ and $\{q_i\}$ are computed for all i's $(2, \ldots, n)$ (Eq. (2.55)).

Step III: $\{q_n\}$ is computed using Eq. (2.56), (Backward substitution).

Step IV: From the last equation of (2.50), $\{z_n\}$ is equal $\{q_n\}$.

Step V: $\{z_{n-1}\}$ to $\{z_1\}$ are back calculated in the reverse order using Eq. (2.50).

Core and Time Requirements

Core The size of P's and q's can be very small as compared to the overall size of the problem. If the total number of equations for the problem is n and the problem can be split into m equal intervals, the size of P's will be $(n/m, n/m)$ and q's (n/m). For example, if $n = 1000$ and $m = 250$, the size of $[A]$, $[B]$, $[C]$ and $[P]$ will be just (4×4).

This is typical of shell equations when the finite difference method is used. The core needed is just for $[A]$, $[B]$, $[C]$, $[P]$ and $\{q\}$ and $\{z\}$ (only one of each). In any calculation, for example, when computing $[P_i]$ and $\{q_i\}$, only $[P_{i-1}]$ and $\{q_{i-1}\}$ are needed. The ones not needed can be transferred to a tape or disk. In effect this lends itself to the out-of-core solution of large problems. The individual substructures are small enough to be reduced within the central memory. In this case there will be data blocks to be transferred from and to back up storage.

Time needed   Let us assume that the intervals are equal in size (s) such that $ms = n$.

Number of multiplications involved in each $P : 3s^3$.

This can be inferred from Eq. (2.54) since $C_i P_{i-1}$, inversion and $([B_i] + [C_i] [P_{i-1}])[A_i]$ involve s^3 multiplications each. Each $\{q_i\}$ involves $2s^2$ multiplications.

Assuming there are approximately m sets of $[P\text{'s}]$, $\{q\text{'s}\}$ and $\{z\text{'s}\}$,

$$\text{Total number} = 3ms^3 + 2s^2m + s^2m = 3m(s^3 + s^2)$$

Inversion operation in Eq. (2.53) can be replaced by Gaussian elimination. Then the labour involved gets reduced from s^3 to $s^3/3$ (see Example 2.9). Then

$$\text{Total number of multiplications} = m\left(\frac{7s^3}{3} + 3s^2\right)$$

The main advantage of this approach is its ability to handle simultaneous equations whose square matrix of coefficients is unsymmetric. The intervals chosen can be of non-uniform width also. Then this serves as a tool for solving problems of variable bandwidth also. As shown in Fig. 2.1, this is the method of substructuring. Figure 2.1 shows an object being cut into number of sections. The state variable associated with each substructure can be designated as $\{z_1\}$, $\{z_2\}$, $\{z_3\}$, etc. It is easy to write the static equation of equilibrium of the whole structure in the form of Eq. (2.48).

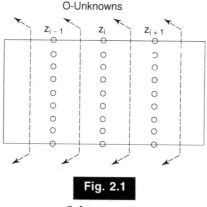

Fig. 2.1

Substructuring

The real advantage of using Potters' method as an out-of-core solution is achieved only when $[A_i]$, $[B_i]$ and $[C_i]$ are generated just when $[P_i]$, $\{q_i\}$ are calculated at the ith station.

While modelling the problem by the finite difference technique, the problem formulation itself assures the direct computation of matrices $[A_i]$, $[B_i]$ and $[C_i]$.

On the contrary, while using finite element modelling, one has to make a conscious effort to assemble the element stiffness matrices such that we assemble only the portions $[A_i]$, $[B_i]$ and $[C_i]$ of the global stiffness matrix.

The above idea would become clear by considering a simple example (Fig. 2.2). The overall stiffness matrix of the structure is shown in Fig. 2.3.

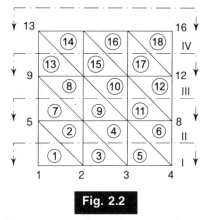

Fig. 2.2

Partitioning a finite element scheme

Full line indicates the overall stiffness matrix obtained by assembling one element after the other in the usual way. Region (a) indicates the portion of the overall stiffness matrix contributed by elements 1 to 6, (b) 7 to 12 and (c) 13 to 18.

For efficient use of Potters' algorithm only matrices $[A_i]$, $[B_i]$ and $[C_i]$ are required to be assembled every time and not the overall stiffness matrix completely.

Let us see how to assemble the matrices $[A_2]$, $[B_2]$ and $[C_2]$. On looking at Fig. 2.3, we can see that $[C_2]$ lies wholly in region (a), $[A_2]$ in region (b) but $[B_2]$ has contributions from both regions (a) and (b).

To get $[B_2]$, elements 1 to 6 as well as 7 to 12 are to be assembled such that their nodal contribution corresponding to nodes 5, 6, 7 and 8 only are taken. Thus, unlike in the usual procedure, this requires the calculation of regions (a), (b) and (c) of the overall stiffness matrix twice. This could be avoided by storing them on an auxiliary storing device such as a tape.

Once portions (a), (b) and (c) are stored on the tape, assembling

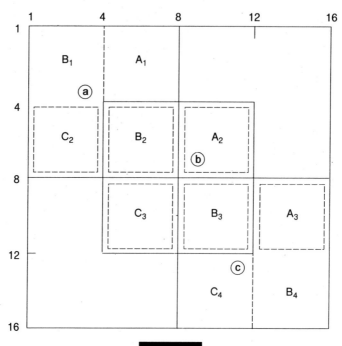

Fig. 2.3

Grouping of submatrices

them appropriately to get matrices $[A_i]$, $[B_i]$ and $[C_i]$ needs some more parameters. First, the structure is divided in the form of partitions connected in series. The nodes belonging to a partition and the elements contributing to it are also to be specified.

For example, to assemble $[B_2]$ we need to specify nodes 5, 6, 7 and 8 and elements 1 to 12. This defines partition II completely. From the partitioning concept we can say that $[C_2]$ is the effect of the earlier partition on the present one and $[A_2]$ the effect of the next partition on the present one. Program 2.8 takes all these factors into account to assemble matrices $[A_i]$, $[B_i]$ and $[C_i]$ and store them in a tape.

Example 2.9
Solve the problem given in Example 2.7 by Potters' scheme

$$
\begin{bmatrix} 4 & 2 & 1 & 1 \\ 2 & 10 & 2 & 1 \\ 1 & 2 & 4 & 2 \\ 1 & 2 & 4 & 8 \end{bmatrix} \begin{Bmatrix} x_1 \\ x_2 \\ x_3 \\ x_4 \end{Bmatrix} = \begin{Bmatrix} 10 \\ 20 \\ 30 \\ 40 \end{Bmatrix}
$$

Dividing into groups of 2×2

$$[A_1] = \begin{bmatrix} 1 & 1 \\ 2 & 1 \end{bmatrix} \qquad [B_1] = \begin{bmatrix} 4 & 2 \\ 2 & 10 \end{bmatrix} \qquad \{g_1\} = \begin{Bmatrix} 10 \\ 20 \end{Bmatrix}$$

$$[C_2] = \begin{bmatrix} 1 & 2 \\ 1 & 2 \end{bmatrix} \qquad [B_2] = \begin{bmatrix} 4 & 2 \\ 4 & 8 \end{bmatrix} \qquad \{g_2\} = \begin{Bmatrix} 30 \\ 40 \end{Bmatrix}$$

$$[B_1^{-1}] = \begin{bmatrix} \dfrac{5}{18} & -\dfrac{1}{18} \\ -\dfrac{1}{18} & \dfrac{2}{18} \end{bmatrix}$$

$$[P_1] = B_1^{-1}A_1 = \frac{1}{18}\begin{bmatrix} 3 & 4 \\ 3 & 1 \end{bmatrix}$$

$$\{q_1\} = B_1^{-1}g_1 = \frac{1}{18}\begin{Bmatrix} 30 \\ 30 \end{Bmatrix}$$

$$\therefore \qquad \{q_2\} = (B_2 - C_2 P_1)^{-1}(g_2 - C_2 q_1)$$

$$= \begin{Bmatrix} 6.325 \\ 1666 \end{Bmatrix}$$

$$\therefore \qquad \{z_2\} = \{q_2\} = \begin{Bmatrix} 6.325 \\ 1.166 \end{Bmatrix}$$

But $\quad \{z_1\} = -[P_1]\{z_2\} + \{q_1\}$.

$$= -\frac{1}{18}\begin{bmatrix} 3 & 4 \\ 3 & 1 \end{bmatrix}\begin{Bmatrix} 6.325 \\ 1.166 \end{Bmatrix} + \frac{1}{18}\begin{Bmatrix} 30 \\ 30 \end{Bmatrix} = \begin{Bmatrix} 0.243 \\ 0.5211 \end{Bmatrix}$$

Hence

$$\{\delta\} = \begin{Bmatrix} z_1 \\ z_2 \end{Bmatrix} = \begin{Bmatrix} 0.243 \\ 0.521 \\ 6.325 \\ 1666 \end{Bmatrix}$$

Computation of $[P_1]$, $\{q_1\}$ and $\{q_2\}$ can be done by avoiding inversion. Using Gaussian elimination for computing $[P_1]$ and $\{q_1\}$ treating coefficients of $[B_1]$ as elements of L.H.S. and $\{g_1\}$ and $[A_1]$ as elements of R.H.S., we have

$$\left[\begin{array}{c|c} 4 & 2 \\ \hline 2 & 10 \end{array}\right] \quad \left\{\begin{array}{c} 10 \\ 20 \end{array}\right\} \quad \left[\begin{array}{cc} 1 & 1 \\ 2 & 1 \end{array}\right] \quad \begin{array}{c} (1) \\ (2) \end{array}$$

Elements of

	$\downarrow$			$\downarrow$			$\downarrow$	
	B_1			g_1			A_1	

4	2		10		1	1	(1)
0	9		15		1.5	0.5	$(2) - \dfrac{(1)}{2}$

If $\qquad q_1 = \left\{\begin{array}{c} q_{11} \\ q_{21} \end{array}\right\} \qquad P_1 = \left[\begin{array}{cc} p_{11} & p_{12} \\ p_{21} & p_{22} \end{array}\right]$

we have

$$9q_{21} = 15 \qquad\qquad 9p_{21} = 1.5 \qquad\qquad 9p_{22} = 0.5$$

$$\therefore \qquad q_{21} = 1.66 \qquad\qquad p_{21} = 0.166 \qquad\qquad p_{22} = 0.055$$

$$4q_{11} + 2q_{21} = 10 \qquad 4p_{11} + 2p_{21} = 1 \qquad 4p_{12} + 2p_{22} = 1$$

$$\therefore \quad q_{11} = \frac{10 - 2 \times 1.66}{4} \quad P_{11} = \frac{1 - 2 \times 0.166}{4} \quad P_{12} = \frac{1 - 2 \times 0.055}{4}$$

$$q_{11} = 1.666 \qquad P_{11} = 0.1666 \qquad p_{12} = 0.222$$

$$\therefore \quad \{q_1\} = \left\{\begin{array}{c} 1666 \\ 1666 \end{array}\right\} \qquad [P_1] = \left[\begin{array}{cc} 0.1666 & 0.222 \\ 0.1666 & 0.055 \end{array}\right]$$

They are the same as the ones obtained earlier.

To exploit symmetry, Gaussian elimination scheme can be replaced by Cholesky scheme.

➤ 2.8   SUBMATRICES ELIMINATION SCHEME

In Section 2.7, Potters' scheme involved inversion of square matrices at the level of every substructure. It is known for all problems using the finite element method $C_{i+1}^T = A_i$ and this property was not made use of in the solution scheme. Let us consider again the formulation exploiting this property. The original problem can be expressed in the following form.

$$\begin{bmatrix} B_1 & A_1 & & & \\ A_1^T & B_2 & A_2 & & \\ & A_2^T & B_3 & A_3 & \\ \cdots & \cdots & \cdots & \cdots & \\ & & & A_{n-1}^T & B_n \end{bmatrix} \begin{Bmatrix} z_1 \\ z_2 \\ \cdots \\ \cdots \\ z_n \end{Bmatrix} = \begin{Bmatrix} g_1 \\ g_2 \\ \cdots \\ \cdots \\ g_n \end{Bmatrix} \qquad (2.57)$$

Recasting Eq. (2.57.), one can write

$$\begin{bmatrix} [L_1 L_1^T] & [L_1 P_1] & & \\ & [L_2 L_2^T] & [L_2 P_2] & \\ & & & [L_n L_n^T] \end{bmatrix} \begin{Bmatrix} z_1 \\ z_2 \\ z_n \end{Bmatrix} = \begin{Bmatrix} L_1 q_1 \\ L_2 q_2 \\ L_n q_n \end{Bmatrix} (2.58)$$

Expanding the first equations of Eqs (2.57) and (2.58), we have

$$[B_1]\{z_1\} + [A_1]\{z_2\} = \{g_1\}$$

$$[L_1][L_1^T]\{z_1\} + [L_1][P_1]\{z_2\} = [L_1]\{q_1\}$$

Comparing

$$[L_1][L_1^T] = [B_1]; \quad [L_1][P_1] = [A_1]; \quad [L_1]\{q_1\} = \{g_1\} \qquad (2.59)$$

Also $$[L_1]^T\{z_1\} + [P_1]\{z_2\} = \{q_1\} \qquad (2.60)$$

Hence $$\{z_1\} = [L_1]^{-T}\{q_1\} - [L_1]^{-T}[P_1]\{z_2\} \qquad (2.61)$$

Expanding the second equations of Eqs (2.57) and (2.58), we have

$$[A_1]^T\{z_1\} + [B_2]\{z_2\} + [A_2]\{z_3\} = \{g_2\} \qquad (2.62)$$

$$[L_2][L_2]^T\{z_2\} + [L_2][P_2]\{z_3\} = [L_2]\{q_2\} \qquad (2.63)$$

Substituting for $\{z_1\}$ from Eq. (2.61) and for $[A_1]$ from Eq. (2.59) in Eq. (2.62), we have

$$-[P_1]^T[L_1]^T[L_1]^{-T}[P_1]\{z_2\} + [B_2]\{z_2\} + [A_2]\{z_3\}$$

$$= \{g_2\} - [P_1]^T[L_1]^T[L_1]^{-T}\{q_1\}$$

or

$$([B_2] - [P_1]^T[P_1])\{z_2\} + [A_2]\{z_3\} = \{g_2\} - [P_1]^T\{q_1\}$$

Comparing this with Eq. (2.63), we have

$$[L_2][L_2]^T = [B_2] - [P_1]^T[P_1]$$

$$[L_2][P_2] = [A_2] \qquad (2.64)$$

$$[L_2]\{q_2\} = \{g_2\} - [P_1]^T\{q_1\}$$

Extending this argument to any other substructure, one can write the following recurrence relationships

$$[B_i - P_{i-1}^T P_{i-1}] = [L_i][L_i]^T$$

$$[A_i] = [L_i][P_i] \qquad (2.65)$$

$$\{g_i\} - [P_{i-1}^T]\{q_{i-1}\} = [L_i]\{q_i\}$$

and

$$[L_i]^T\{z_i\} + [P_i]^T\{z_{i+1}\} = \{q_i\} \qquad (i = 2, 3, ..., n - 1) \quad (2.66)$$

For the nth substructure

$$[B_n - P_{n-1}^T P_{n-1}] = [L_n][L_n]^T$$

$$\{g_n - P_{n-1}^T q_{n-1}\} = [L_n]\{q_n\} \qquad (2.67)$$

and

$$[L_n]^T\{z_n\} = \{q_n\} \qquad (2.68)$$

For calculations in the forward direction, $[L_i]$, $[P_i]$ and $\{q_i\}$ are to be computed and stored for each substructure and in the reverse direction, $\{z_i\}$ are to be calculated. For problems involving iteration (Chapters 3 and 4), $[P_i]$ and $[L_i]$ will be computed only for the first iteration whereas $\{q_i\}$ and $\{z_i\}$ for every iteration, reducing the computer time enormously. The number of substructures n and the size of each substructure can be conveniently chosen depending on the computer capability. A slightly modified form of this scheme is presented by Melosh and Banaford [2.8].

Example 2.10
Calculate the bending moment and deflection of the beam shown in Fig. 2.4 by the following methods using ten intervals.

   (i) Direct inversion
   (ii) Gaussian elimination
  (iii) Potters' scheme

P = intensity of loading = 100 N/cm.
Young's modulus $E = 2.1 \times 10^7$ N/cm^2.
   With reference to Eqs (1.8), (1.9) and (1.10) choosing a as 100,

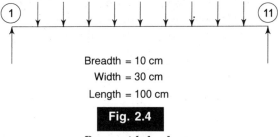

Breadth = 10 cm
Width = 30 cm
Length = 100 cm

Fig. 2.4

Beam with loads

$h = 0.1$ since $n = 10$. At stations (1) and (11) the boundary conditions are $w = 0$ and $(m_\xi) = 0$. Expressing $\{z_i\}$ as $\{w/m_\xi\}_i$ we have the following equations at $i = 1$ and $i = 11$,

$$\{z_1\} = \{0\}$$

$$\{z_{11}\} = \{0\}$$

For any $i(2 \leq i \leq 10)$, we have from Eq. {1.116}.

$$\frac{w_{i+1} - 2w_i + w_{i-1}}{\Delta^2} + (m_\xi)_i = 0$$

$$\frac{(m_\xi)_{i+1} - 2(m_\xi)_i + (m_\xi)_{i-1}}{\Delta^2} = \frac{P_i a^3}{EI}$$

or rearranging (for $2 \leq i \leq 10$)

$$\begin{bmatrix} \dfrac{1}{\Delta^2} & 0 \\ 0 & \dfrac{1}{\Delta^2} \end{bmatrix} \{z_{i+1}\} + \begin{bmatrix} -\dfrac{2}{\Delta^2} & 1 \\ 0 & -\dfrac{2}{\Delta^2} \end{bmatrix} \{z_i\}$$

$$+ \begin{bmatrix} \dfrac{1}{\Delta^2} & 0 \\ 0 & \dfrac{1}{\Delta^2} \end{bmatrix} z_{i-1} = \left\{ \begin{array}{c} 0 \\ \dfrac{Pa^3}{EI} \end{array} \right\}$$

$$\Delta = 0.1$$

These constitute 22 equations in 22 unknowns. They can be solved by the three methods listed above. In the case of direct inversion, a matrix of size 22 × 22 was directly inverted. In the Gaussian elimination procedure semi-bandwidth was assumed to be 4. In Potter's scheme, 11 sets of matrices [A], [B] and [C] of 2 × 2 were identified. The following are the results.

Maximum deflection at the centre of the beam = 27.557 mm

Maximum bending moment at half span = -1.25×10^5 N m

To summarise the following observations can be made on the methods discussed. 'Inversion is not desirable unless inverse is really needed. Gaussian elimination with multiple constant vectors and LU decomposition are recommended in core, non-banded procedures. In all cases, double precision should be employed if the storage permits it' [2.9].

Example 2.11
Compute the deflected shape of the boiler frame shown in Fig. 2.5. The properties of the frame are given in Table 2.1.

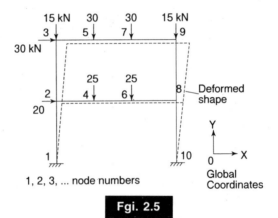

Fgi. 2.5

Boiler frame with loads specified

Table 2.1

Sectional Properties

Member	Area in sq. cm	Length in cm	Flexural rigidity N cm^2
1–2	45	200	1×10^{11}
2–3	45	200	1×10^{11}
2–4	60	100	2×10^{11}
4–6	60	100	2×10^{11}
6–8	60	100	2×10^{11}
3–5	60	100	2×10^{11}
5–7	60	100	2×10^{11}
7–9	60	100	2×10^{11}
9–8	45	200	1×10^{11}
8–10	45	200	1×10^{11}

Solve the problem by the following methods:

(i) By inverting the stiffness matrix
(ii) By Potters' scheme
(iii) By Cholesky's factorisation

From Fig. 2.5 it is quite clear that the lengthwise directions of the ten members in the frame are not always parallel to the global X-axis. The element stiffness matrices as shown in Sec. 1.2.1 are expressed in terms of local co-ordinate axes ox and oy along and at right angles to the beam length. In general the global co-ordinate axes can be inclined to the local coordinates by an angle θ as shown in Fig. 2.6. We can express the global co-ordinates (X, Y) of any P in the following way.

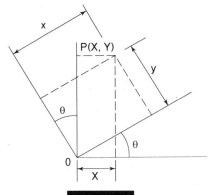

Fig. 2.6

Local and global systems of co-ordinates

$$X = (x \cos \theta - y \sin \theta)$$
$$Y = (x \sin \theta - y \cos \theta)$$

This can be rewritten as

$$\begin{Bmatrix} X \\ Y \end{Bmatrix} = \begin{bmatrix} \cos \theta & -\sin \theta \\ \sin \theta & \cos \theta \end{bmatrix} \begin{Bmatrix} x \\ y \end{Bmatrix}$$

or

$$\begin{Bmatrix} x \\ y \end{Bmatrix} = \begin{bmatrix} \cos \theta & \sin \theta \\ -\sin \theta & \cos \theta \end{bmatrix} \begin{Bmatrix} X \\ Y \end{Bmatrix}$$

Extending this argument to the deformations and force in the element one can write

$$\left\{ \begin{array}{c} u_1 \\ v_1 \\ \theta_1 \\ u_2 \\ v_2 \\ \theta_2 \end{array} \right\}_{\text{local}}$$

$$= \begin{bmatrix} \cos\theta & \sin\theta & & & & \\ -\sin\theta & \cos\theta & & & & \\ & & 1 & & & \\ & & & \cos\theta & \sin\theta & \\ & & & -\sin\theta & \cos\theta & \\ & & & & & 1 \end{bmatrix} \left\{ \begin{array}{c} u_1 \\ v_1 \\ \theta_1 \\ u_2 \\ v_2 \\ \theta_2 \end{array} \right\}_{\text{global}}.$$

or
$$\{\delta\}_{\text{local}} = [T]\{\delta\}_{\text{global}}$$

Similarly, for the forces
$$\{f\}_{\text{local}} = [T]\{f\}_{\text{global}}$$

Since
$$[K_e]\{\delta\}_{\text{local}} = \{f\}_{\text{local}} \tag{2.3}$$

We can write
$$[T]^T[K_e][T]\{\delta\}_{\text{global}} = \{f\}_{\text{global}}$$

or
$$[K]_{\text{global}}\{\delta\}_{\text{global}} = \{f\}_{\text{global}}$$

where
$$[K]_{\text{global}} = [T]^T[K_e][T]$$

For the problem given above, let us assume that all the degrees of freedom are zero at nodes (1) and (10). The stiffness matrix will be of size (24 × 24), when solution by inversion is resorted to. When solved by Cholesky's factorisation, the size of the problem will be 24 and semibandwidth 9. When solved by Potters' method one can choose four partitions, each of size 6, Program 2.8 explains this problem. The deformed shape of the frame is also shown in Fig. 2.5.

For problems connected with local co-ordinates in three orthogonal directions, the transformation matrix is given in Appendix II.

A summary of core and labour is given in Table 2.2.

Table 2.2

Summary of Core and Labour Involved (Static Problems)

Method	Core	Labour	Remarks
Cholesky factorisation uniform bandwidth	$nb + 2n$	$\dfrac{nb^2}{2}$ for factorisation $2nb$ substitution	$[K]$ and $[L]$ are stored as rectangular arrays
Cholesky factorisation, non-uniform bandwidth	Elements below sky line + $2n + n$ for principal diagonal	$\sum\limits_{i=1}^{n} \dfrac{b_i}{2}(b_i + 1)$ $+\,2\sum\limits_{i=1}^{n} b_i$	Very efficient for in core solution
Potters' scheme Gaussian elimination for solution within a substructure	$2(s_i \times s_{i+1}) + s_i^2$ $+2(s_i \times s_{i-1}) +2\,s_i$	$\sum\limits^{m} \left[s_{i-1}s_i^2 + \dfrac{s_i^3}{3}\,s_i^2 s_{i+1} \right.$ $\left. + s_i^2 + s_i s_{i-1} + s_i s_i + 1 \right]$	Best out of core solution for unsymmetric $[K]$
Potters' scheme Cholesky factorisation for solution within	$s_i^2 + 2(s_i s_{i+1}) + 2s_i$	$\sum\limits_{2}^{m} \left[s_{i-1}s_i^2 + \dfrac{s_i^3}{3}\,s_i^2 s_{i+1} \right.$ $\left. + s_i^2 + 2\,s_i s_{i-1} + 2\,s_i s_{i+1} \right]$	Best out of core solution for symmetric $[K]$

n—size of the problem, b—semibandwidth, s_i—size of ith substructure, m—number of sub-structures, b_i—bandwidth of ith row.

➤ 2.9 SPARSE MATRIX

Most of the engineering problems encountered lead to matrices which are sparse (not densely populated). One way of effectively reducing the labour involved in computation is to limit the bandwidth (semi) to the minimum. This can be explained through the following example.

$$a_{11}x_1 + a_{12}x_2 + a_{14}x_4 = f_1$$

$$a_{21}x_1 + a_{22}x_2 + a_{23}x_3 + a_{24}x_4 + a_{26}x_6 = f_2$$

$$a_{32}x_2 + a_{33}x_3 + a_{36}x_6 + a_{3,10}x_{10} = f_3$$

$$a_{41}x_1 + a_{42}x_2 + a_{44}x_4 + a_{46}x_6 + a_{45}x_5 = f_4$$

$$a_{54}x_4 + a_{55}x_5 + a_{56}x_6 + a_{57}x_7 = f_5$$

$$a_{62}x_2 + a_{63}x_3 + a_{64}x_4 + a_{65}x_5 + a_{66}x_6$$

$$+ a_{67}x_7 + a_{69}x_9 + a_{6,10}x_{10} = f_6 \qquad (2.65)$$

$$a_{75}x_5 + a_{76}x_6 + a_{77}x_7 + a_{78}x_8 + a_{79}x_9 = f_7$$

$$a_{87}x_7 + a_{88}x_8 + a_{89}x_9 = f_8$$

$$a_{97}x_7 + a_{98}x_8 + a_{99}x_9 + a_{9,10}x_{10} = f_9$$

$$a_{10,3}x_3 + a_{10,6}x_6 + a_{10,9}x_9 + a_{10,10}x_{10} = f_{10}$$

(10×10) square matrix of coefficients can be graphically represented by Fig. 2.7 if one uses the following convention.

(a) All the principal diagonals are present.
(b) Firm line connecting any two points i and j signifies non-zero coefficient a_{ij}. The number of unknowns n, for this problem is $n = 10$.

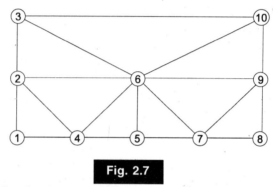

Fig. 2.7

Graphical representation of the given problem

The maximum semi-bandwidth, $b = 8$. (Maximum difference in numbers between the extremities of any line +1.)

The number of multiplications involved if Cholesky's factorisation with uniform bandwidth is used

$$= (nb^2/2) + 2nb = 10(8)^2/2 + 2 \cdot 10 \cdot 8 = 320 + 160 = 480.$$

Figure 2.7 can easily be rearranged with a view to reduce the bandwidth. In our actual problem it amounts to rearranging the order in which the unknowns x_1 to x_{10} are evaluated. This is shown in Fig. 2.8. For the Fig. 2.8, $n = 10$, $b = 5$.
Hence the number of multiplications

$$= 10(5)^2/2 + 2 \cdot 10 \cdot 5 = 125 + 100 = 225$$

There is a substantial reduction in the multiplications involved.

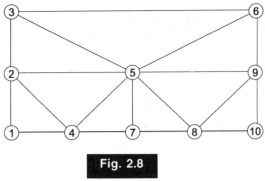

Fig. 2.8

Rearranged nodes

➤ 2.10 VARIABLE BANDWIDTH

Quite often matrices which are symmetric, but which have variable bandwidth are encountered in practice [2.10]. For the typical problem shown in Fig. 2.9, the bandwidth varies from row to row if lower triangular matrix is used or from column to column if upper triangular matrix is used. It has widths of 1, 1, 2, 1, 5 and 3 respectively as one progresses from the first row to the sixth row. Bearing in mind the fact that the matrix is symmetric, there are only 13 non-zero elements to be stored. They are a_{11}, a_{22}, a_{32}, a_{33}, a_{44}, a_{51}, a_{52}, a_{53}, a_{54}, a_{55}, a_{64}, a_{65}, a_{66} if the lower triangular matrix is used. Of these 13, the diagonal elements have to be identified. They can be identified by the integer array ID(1) to ID(6) in the following way: ID(1) = 1; ID(2) = 2; ID(3)

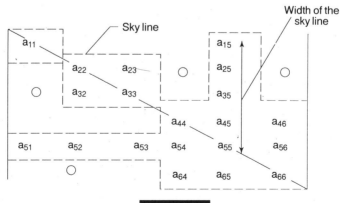

Fig. 2.9

Matrix with variable bandwidth

= 4; ID(5) = 10; ID(6) = 13. When the non-zero elements a_{11} to a_{66} are stored in the order a_{11}, a_{22}, a_{23}, etc. up to a_{66}; the principal diagonals are the 1st, 2nd, 4th, 5th, 10th and 13th elements. Off-diagonal elements are given by

$$a_{ji} = A[ID(i) + j - i]$$

This procedure results in considerable saving in core and time. The Cholesky factorisation of stiffness matrix also results in a lower triangular matrix of variable bandwidth. The method of solving problems with variable bandwidth is given in Prog. 2.7. Every a_{ji} gives rise to a corresponding l_{ij}.

The problem given below will explain the basic difference in ordering.

Example 2.12

The static problem as given by Fig. 2.10 is as follows:

$$
\begin{bmatrix}
1 & 0.1 & 0.1 & & & & & \\
0.1 & 2 & 0 & 0.2 & & & & \\
0.1 & 0 & 3 & 0 & 0.3 & & & \\
& 0.2 & 0 & 4 & 0 & 0.4 & & \\
& & 0.3 & 0 & 3 & 0 & 0.3 & \\
& & & 0.4 & 0 & 2 & 0 & 0.2 \\
& & & & 0.3 & 0 & 2 & 0.2 \\
& & & & & 0.2 & 0.2 & 1
\end{bmatrix}
\begin{Bmatrix}
\delta_1 \\
\delta_2 \\
\cdot \\
\cdot \\
\cdot \\
\cdot \\
\cdot \\
\delta_8
\end{Bmatrix}
=
\begin{Bmatrix}
1.4 \\
4.7 \\
7.0 \\
14.0 \\
10.8 \\
10.2 \\
9.9 \\
6.6
\end{Bmatrix}
$$

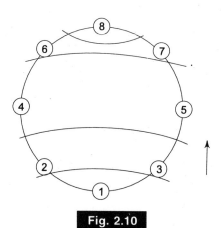

Fig. 2.10

Frontal ordering for a ring network

Using submatrices elimination scheme with $s = 2$, we have

$$[A_1] = \begin{bmatrix} 0.1 & 0 \\ 0 & 0.2 \end{bmatrix} \quad [B_1] = \begin{bmatrix} 1 & 0.1 \\ 0.1 & 2 \end{bmatrix}$$

$$[C_2] = \begin{bmatrix} 0.1 & 0 \\ 0 & 0.2 \end{bmatrix} \quad [B_2] = \begin{bmatrix} 3 & 0 \\ 0 & 4 \end{bmatrix} \quad [A_2] = \begin{bmatrix} 0.3 & 0 \\ 0 & 0.4 \end{bmatrix}$$

$$[C_3] = \begin{bmatrix} 0.3 & 0 \\ 0 & 0.4 \end{bmatrix} \quad [B_3] = \begin{bmatrix} 3 & 0 \\ 0 & 2 \end{bmatrix} \quad [A_3] = \begin{bmatrix} 0.3 & 0 \\ 0 & 0.2 \end{bmatrix}$$

$$[C_4] = \begin{bmatrix} 0.3 & 0 \\ 0 & 0.2 \end{bmatrix} \quad [B_4] = \begin{bmatrix} 2 & 0.2 \\ 0.2 & 1 \end{bmatrix}$$

$$[P_1] = \begin{bmatrix} 0.1 & 0.01 \\ -0.007 & 0.141 \end{bmatrix} \quad \{q_1\} = \begin{Bmatrix} 1.4 \\ 3.22 \end{Bmatrix} \quad [L_1] = \begin{bmatrix} 1 & 0 \\ 0.1 & 1.41 \end{bmatrix}$$

$$[P_2] = \begin{bmatrix} 0.173 & 0 \\ 0.0005 & 0.2 \end{bmatrix} \quad \{q_2\} = \begin{Bmatrix} 3.96 \\ 6.82 \end{Bmatrix} \quad [L_2] = \begin{bmatrix} 1.73 & 0 \\ 0.0006 & 1.99 \end{bmatrix}$$

$$[P_3] = \begin{bmatrix} 0.17 & 0 \\ 0 & 0.14 \end{bmatrix} \quad \{q_3\} = \begin{Bmatrix} 5.88 \\ 6.31 \end{Bmatrix} \quad [L_3] = \begin{bmatrix} 1.72 & 0 \\ 0 & 1.4 \end{bmatrix}$$

$$\{q_4\} = \begin{Bmatrix} 6.36 \\ 4.90 \end{Bmatrix} \quad [L_4] = \begin{bmatrix} 14 & 0 \\ 0.14 & 0.28 \end{bmatrix}; \text{ hence } \{z_4\} = \begin{Bmatrix} 4 \\ 5 \end{Bmatrix}$$

$$\{z_3\} = \begin{Bmatrix} 3.00 \\ 4.04 \end{Bmatrix}$$

$$\{z_2\} = \begin{Bmatrix} 2.02 \\ 3.01 \end{Bmatrix}$$

$$\{z_1\} = \begin{Bmatrix} 1.00 \\ 2.02 \end{Bmatrix}$$

The same problem renumbered (Fig. 2.11) is as follows:
Here $n = 8$, b for the first row = 1, for the next six rows = 2 and b for the last row = 8.

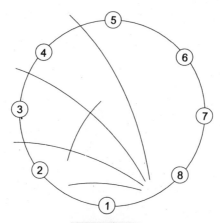

Fig. 2.11

Progressive rotations use of variable bandwidth

$$
\begin{bmatrix}
1 & 0.1 & & & & & & 0.1 \\
0.1 & 2 & 0.2 & & & & & 0 \\
 & 0.2 & 4 & 0.4 & & & & 0 \\
 & & 0.4 & 2 & 0.2 & & & 0 \\
 & & & 0.2 & 1 & 0.2 & 0 & 0 \\
 & & & & 0.2 & 2 & 0.3 & 0 \\
 & & & & & 0.3 & 3 & 0.3 \\
0.1 & 0 & 0 & 0 & 0 & 0 & 0.3 & 3
\end{bmatrix}
\begin{Bmatrix}
\delta_1 \\ \cdot \\ \cdot \\ \cdot \\ \cdot \\ \cdot \\ \cdot \\ \delta_8
\end{Bmatrix}
=
\begin{Bmatrix}
1.4 \\ 4.7 \\ 14.0 \\ 10.2 \\ 6.6 \\ 9.9 \\ 10.8 \\ 7.0
\end{Bmatrix}
$$

The solution to the problem by the variable bandwidth approach is
as follows:
Expressing the stiffness matrix as

$$[K] = [L][L]^T$$

[L] is given by

$$
\begin{bmatrix}
1 & & & & & & & \\
0.1 & 1.41 & & & & & & \\
 & 0.14 & 1.99 & & & & & \\
 & & 0.20 & 1.40 & & & & \\
 & & & 0.14 & 0.99 & & & \\
 & & & & 0.20 & 1.40 & & \\
 & & & & & 0.20 & 1.72 & \\
0.1 & 0.0007 & 0 & 0 & 0 & 0.175 & 0.175 & 1.73
\end{bmatrix}
$$

It may be observed that wherever the elements of K are zero, the elements of L are also zero.

Let $$\{y\} = [L]^{-1}\{f\}$$

$\{f\}$ is the given force vector.

$$\{y\} = \begin{Bmatrix} 1.4 \\ 3.23 \\ 6.84 \\ 6.30 \\ 5.77 \\ 6.25 \\ 5.55 \\ 3.42 \end{Bmatrix}$$

$\{\delta\}$, the resultant displacement vector

$$= [L]^{-T}\{y\} = \begin{Bmatrix} 1 \\ 2 \\ 3 \\ 4 \\ 5 \\ 4 \\ 3 \\ 2 \end{Bmatrix}$$

The time taken for the two schemes in IPL 4440 is as given below:

Frontal technique = 0.53 s
Variable bandwidth = 0.48 s

Using Gaussian elimination in order to avoid inversion in the first scheme will further reduce the time.

➤ 2.11 BANDWIDTH REDUCTION

The discussion on numbering as shown in Figs 2.7 and 2.8 indicates that computational labour can be considerably reduced by properly numbering the nodes. Let us explain it through a simple illustration. Shown in Fig. 2.12 are two schemes of numbering pipelines. Problems of this type are encountered in flexibility analysis of pipelines connected with petrochemical plants. For simplicity let us assume one degree of freedom per node. With Cholesky factorisation method having uniform width, the first scheme has a maximum bandwidth of 102 whereas the second has a bandwidth of 3. The labour involved in Cholesky factorisation for the two schemes are respectively

$$1.5 \times 10^6 = \left(\frac{300 \times 102^2}{2} \right) \text{ and } 1350 = \left(\frac{300 \times 3^2}{2} \right)$$

Let us now consider the variable bandwidth algorithm.

Scheme-I—Number of multiplications

$$\text{Cholesky factorisation} = 1 + \frac{298}{2} (2 \times 3) + \frac{(102 \times 103)}{2}$$
$$= 1 + 894 + 5253 = 6148 \text{ multiplications.}$$

Forward and backward substitution = 2 [(298 × 2) + 102 + 1]
$$= 1198 \text{ multiplications}$$

Hence,   total labour = 7346 multiplications

Scheme-II—Number of multiplications

$$\text{Cholesky factorisation} = 1 + \frac{100}{2} (2 \times 3) + \frac{199}{2} \times (3 \times 4)$$
$$= 1 + 300 + 1194 = 1495 \text{ multiplications}$$

Forward and backward substitution = 2 [1 + (100 × 2) + (199 × 3)]
$$= 1596 \text{ multiplications}$$

Hence,   total labour = 3091 multiplications

The second scheme takes less than half the time taken by the first scheme. This clearly indicates the necessity for an efficient numbering scheme. Saving becomes enormous for practical problems since the degrees of freedom per node will be six.

There are many schemes of numbering available for optimising the bandwidth. Let us consider one of the very effective and simple methods, known as Cuthill—McKee algorithm. (R.C.M algorithm) [2.11].

The method can be explained thus. Designate any node as 1. Name the nodes connected to 1 as 2, 3, 4 and so on in the increasing order. After exhausting node 1, proceed to node 2 and number the nodes connected to node 2, not so far numbered, in the increasing order from where we have stopped. This procedure is to be continued till the entire configuration is covered. When all the n nodes have been

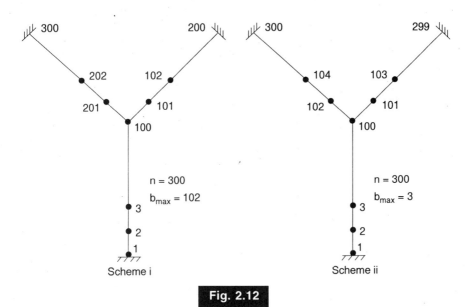

Fig. 2.12

Numbering pipeline network

numbered, the numbering is reversed replacing number i by $(n - i + 1)$ (where n is the total number of nodes). Reversal does not reduce the bandwidth, but reduces the total number of elements below the skyline.

This can be explained with reference to the matrix of variable bandwidth as given by Fig. 2.10. This is graphically represented by Fig. 2.13. If numbering is commenced from node 6 and Cuthill—McKee method is adopted, the numbers shown within dotted circles will appear. If they are reversed, the nodes as shown within the full circles appear. From the matrix shown in Fig. 2.10, it is clear that this corresponds to minimum number of elements below the skyline.

Example 2.13
Rearrange the nodal numbering given in Fig. 2.14 by the Cuthill—McKee approach. Compare the bandwidth given by Fig. 2.14 with the modified one. (Assume one degree of freedom at each node.)

For Figs 2.14, 2.15 and 2.16 the maximum bandwidths are 7, 6 and 6 respectively. The matrix corresponding to Fig. 2.14 is shown in Fig. 2.17 and the one corresponding to 2.16 in Fig. 2.18. It can be seen that the number of elements below the skyline (called profile) for Fig. 2.14 is 116, for Fig. 2.15 is 94 and for Fig. 2.16 is 85. The number of multiplications in Cholesky factorisation corresponding to Fig. 2.17 is 391 and for Fig. 2.18 to 215.

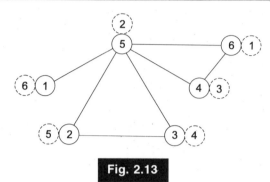

Fig. 2.13

Graphical representation of Fig. 2.10

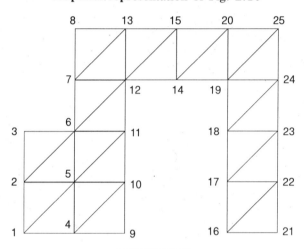

Fig. 2.14

Nodal numbering

➤ 2.12 FIGURES OF SIGNIFICANCE

The overall problem of statics reported in this chapter reduces to solving linear simultaneous equations in many unknowns to evaluate deformations and slopes at specified nodes. These are used to subsequently calculate the strains and the stresses experienced by the machine elements. One should not forget the fact that problems of practical significance are solved and hence, the deformations, the strains and the stresses must be within elastic limit. The dimensions of members analysed may vary from a few millimetres to several metres. The deformations experienced will vary from about 10^{-3} to 10^{-1} mm. The slopes at any of the nodes may not exceed about 1°, which is around .02 radians. The materials under question will be

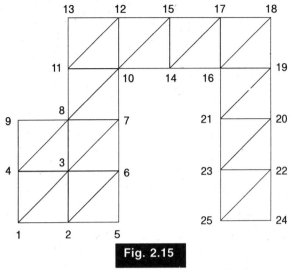

Fig. 2.15

Rearrangement of nodal numbering

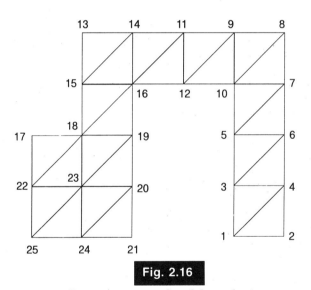

Fig. 2.16

Rearrangement of nodal numbering

mostly mild steel, alloy steel or cast iron. The maximum direct stress (tension or compression) is unlikely to exceed 150 MPa (N/mm²), but usually be less than 100 MPa and the shear stress less than 50 MPa. The direct strains are likely to be less than 500 microns. In case the values computed exceed the ones specified above by several times, it

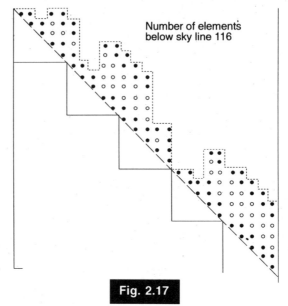

Number of elements below sky line 116

Fig. 2.17

Stiffness matrix corresponding to Fig. 2.14

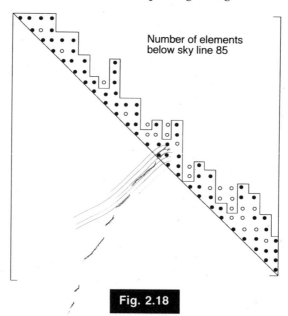

Number of elements below sky line 85

Fig. 2.18

Stiffness matrix corresponding to Fig. 2.16

is advisable to recheck the data thoroughly. Otherwise, one is likely to become a slave to the numbers printed out by the computer which may turn out to be illogical. We should not forget as engineers, that we handle practical problems which always have meaningful results.

➤ 2.13   INDUSTRIAL APPLICATIONS

2.13.1   Multilayer Pressure Vessel Stress Analysis

The deformation and the stress experienced by a multilayer pressure vessel as shown in Fig. 2.19 have been obtained by Potters' scheme [2.12]. The pressure vessel is 23.4 m tall and 3 m in diameter. The thickness of the cylindrical portion is 183 mm. Its normal operating pressure is 220 atm and operating temperature 457°C. The finite difference analysis uses the governing differential equations as given in Chapter 1 for the analysis. The three components of deflection and the meridional bending moment are computed from the basic simultaneous equations and the stress resultants (other than the meridional bending moment) are obtained from them. The number of discrete points chosen for the analysis is about 300. The sub-matrices are of size 4. The time taken for the entire analysis is just 20 s in IBM 370. The analysis was subsequently extended to asymmetric loading due to wind. The details of the finite difference scheme are shown in Fig. 2.19. These results are compared with the results obtained by the finite element scheme treating the pressure vessel as a rotationally symmetric object. Finite element discretisation is shown in Fig. 2.20 and the comparison of results in Figs 2.21 and 2.22.

2.13.2   Stress Analysis of the Main Housing of Hydraulic Press

The investigation is connected with the stress analysis of the C frame of open front hydraulic press of 40 tonne capacity (Fig. 2.23). This problem is solved by using Potters' scheme (out-of-core solution). Figure 2.24 gives an idea of the number of substructures used. The element used is the triangular plate element with six degrees of freedom at each node. The number of unknowns handled in each substructure did not exceed 120 (20 nodes). The program was run on a PC. For the size of the problem handled, the program needed 13 mega bytes of memory for storing the relevant data out of the main core. Typical stress distribution on one of the faces of the press is shown in Fig. 2.25.

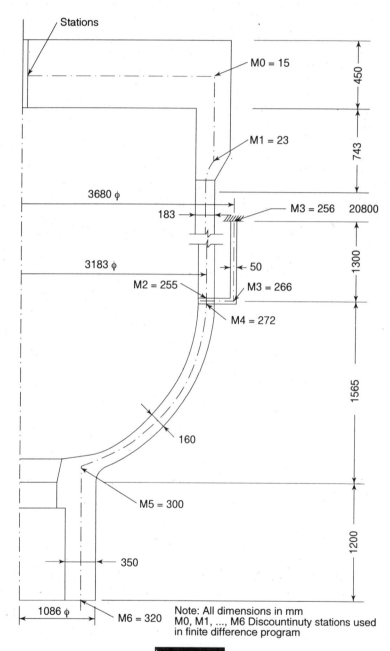

Note: All dimensions in mm
M0, M1, ..., M6 Discountinuty stations used
in finite difference program

Fig. 2.19

Dimensions of ammonia convertor

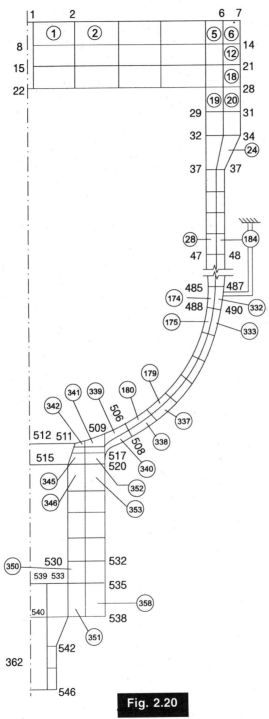

Fig. 2.20

Finite element discretization

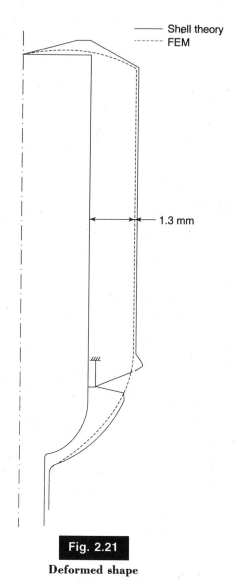

Fig. 2.21

Deformed shape

2.13.3 Rigidity and Stress Analysis of Two-stage Reduction Gearbox

Figure 2.26 gives the elevation and plan view of the fabricated gearbox. It consists of six plates welded at the edges to form a box. The strength of the box is reinforced by having another parallel plate bent at the edges and welded to faces $s1$ and $s2$ of the basic box. The

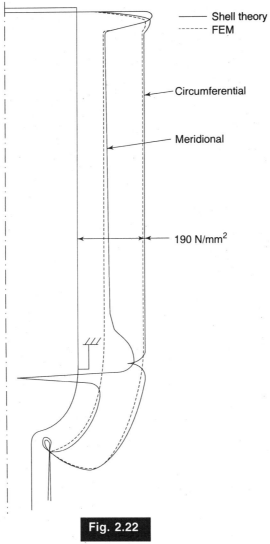

Fig. 2.22

Stress distribution

bearing block is supported at its two ends by the sides $s1$ and $s7$ on one side and $s2$ and $s8$ on the other [2.13]. The edges of the gearbox are strengthened by having a box-type construction at the four corners bent in a press brake.

Figure 2.27 gives the elevation and plan view of the cast gearbox. It is a cuboid with bearing blocks projecting on either sides of faces $s1$ and $s2$. The manufacture of this box requires a pattern. Cores are to be placed in moulds to fill in voids in the casting.

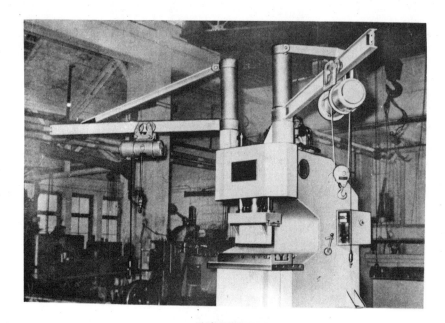

Fig. 2.23

C Frame hydraulic press

For the purpose of analysis, a two-stage reduction helical gearbox transmitting 1940 hp input speed of 1500 rpm was considered. The size of the gearbox is shown in Table 2.4. The specifications for the gearbox model were obtained from a standard gear manufacturer's catalogue. The antifriction bearings of input, intermediate and output shafts were respectively 30224, 29436 and 29352. Both the fabricated and cast constructions are analysed.

Geometric Model

Both the gearboxes are made in two halves and bolted at the centre. For the purpose of analysis, the two halves are considered as being integral, as shown. The geometric model of the casing was discretised into a number of triangular plate element (Type 1.2.3), with six degrees of freedom at each node. The use of 3D brick element is not attempted since this would involve enormous labour.

1. The bearing holes are modelled as octagons for both constructions.
2. For the fabricated construction, bearing blocks are modelled as triangular plate elements perpendicular to the face of the

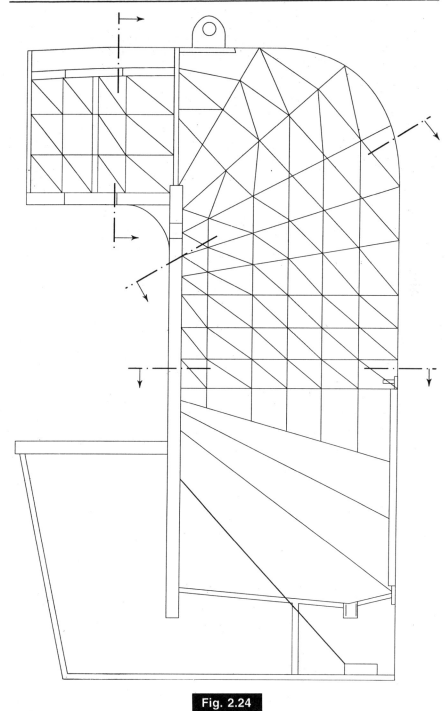

Fig. 2.24

40 tonne *C* frame hydraulic press

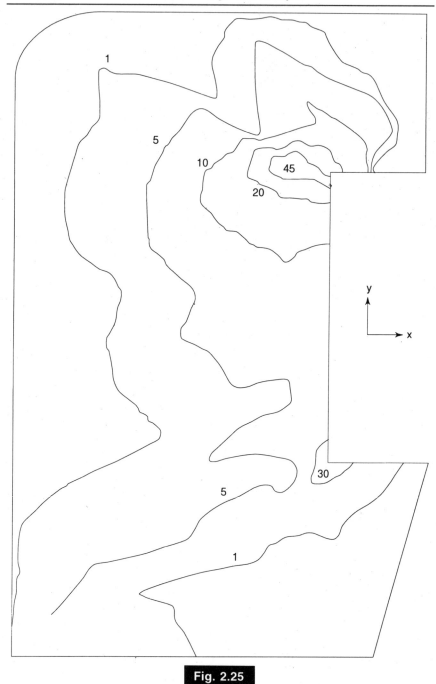

Fig. 2.25

Isostress σ_y in N/mm^2— front face of the press

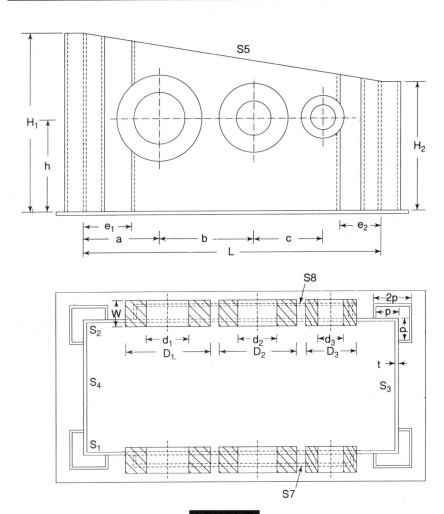

Fig. 2.26

Fabricated gearbox

Table 2.3

Gearbox Dimensions (All Dimensions in mm)

Case	HP	i	L	H1	H2	B	h	a
I	1940	8	25301500	·1000	1000	65	65

HP—Horse power transmitted in HP
i—Speed ratio

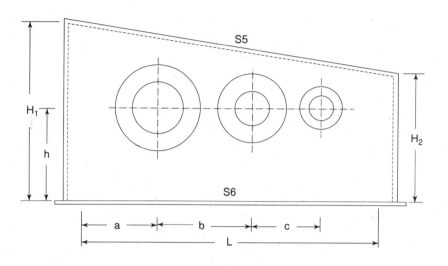

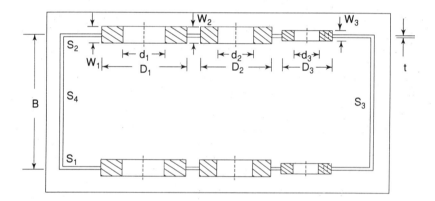

Cast gearbox

Table 2.4

Gear Dimensions

Gear	z	m	d (mm)	b (mm)	a	h
1	45	7	315	240	20°	11°
2	120	7	840	240	20°	11°
3	40	10	1400	320	20°	18°
4	120	10	1200	320	20°	18°

z—number of gear tooth; m—module; d—pitch diameter; b—width of gears; a—pressure angle; h—helix angle

gearbox with thickness about 0.8 times the outer diameter of the bearing. For the cast construction, since the bearing projections are less, the thickness of the elements are considered perpendicular to the sides.

3. The bottom face of the gearbox s_6 is fixed. This is taken care of in modelling by suppressing all the degrees of freedom in those nodes.

The variable bandwidth algorithm was used to solve the above problem. Once the displacements are found, stress values in each element are determined. The average stress value at each of the nodes is determined as the weighted average of contribution from all the elements surrounding the node. The iso-stress plot in cast gear casing is shown in Fig. 2.28.

The material saving obtained by adopting the fabricated construction has also been found and the results are given in Table 2.5.

2.13.4   Design of Heavyduty Moulding Boxes

In this section the design aspects of moulding boxes used in heavy engineering foundry of sizes ranging form 2 m × 2 m to 4 m × 4 m is reported [2.14]. The primary motivating factor for this study is the buckling of many of these boxes while in operation. The details of these boxes are given in Fig. 2.29. They are made of steel plates suitably welded to form hollow squares in plan and channels in section. The depth of these boxes varies from 0.5 m to 1.0 m. The loads these carry including the moulding sand and casting, may vary from 2 tonnes to 16 tonnes. They are lifted by two cylindrical lugs as shown in Fig. 2.29. While the moulding boxes swing about a horizontal axis (when lifted), the sides of the boxes are subjected to loads of varying intensities. These boxes are analysed to determine the elastic deformation and the bending and torsional stresses experienced. Finite element method is employed to estimate these values. Castigliano's theorem is used to verify these results before a parametric study is attempted.

2.13.4.1   Statement of the Problem

The moulding box along with its load, is to function in any plane rotating about the axis AD (Figs 2.29 and 2.30). The load experienced by the box is the dead load due to the moulding sand and the casting inside (this may be hot or cold). The channel section of the moulding box is to withstand twisting and bending due to this load. Both the deformation and stress experienced are of concern. Before the finite

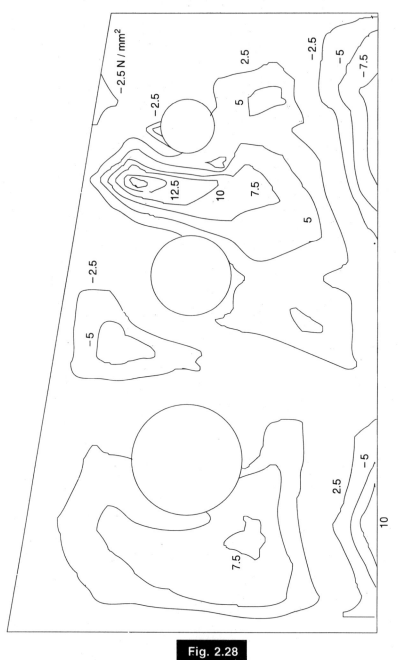

Fig. 2.28

Iso-stress *sx* for face *l* of cast gearbox

Table 2.5

Comparison of Gearbox Casings

	Parameters (in mm)	Maximum deflection (in mm)	Maximum slope at bearing	Maximum stress (in N/mm^2)	Percentage saving in material
Fabricated					
1.1	$t = 8$			27	
	$p = 200$	1.16	0.27°		33%
Cast					
1.2	$t = 22$	0.645	0.106°	38	

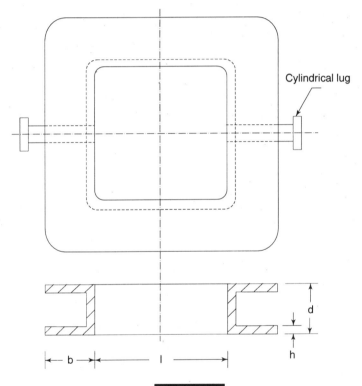

Fig. 2.29

Heavyduty moulding box

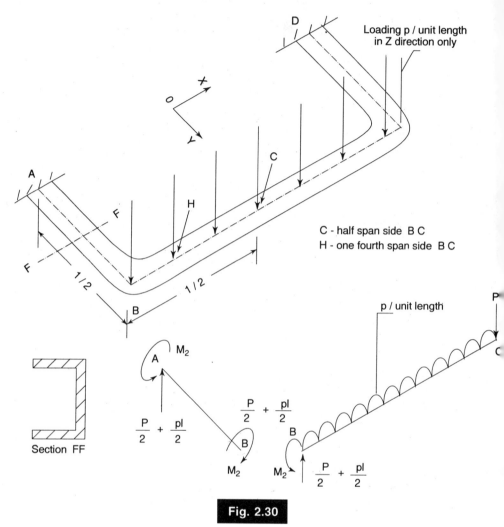

Fig. 2.30

Free body diagram-load Z direction only

element analysis is attempted, Castigliano's theorem is used to propose a simplified approach.

2.13.4.2   Use of Castigliano's Theorem

For case (i), the load is assumed to act with uniform intensity in the direction of minimum moment of inertia (Z direction) along the line BC (Fig. 2.30). For case (ii), the load is assumed to act with uniform intensity in the Y direction only. However, in practice, both the components of the load will be present. Let E, g and μ be the modulus of elasticity, rigidity and Poisson's ratio of the material.

Case (i) Load in the Z direction only

Referring to Fig. 2.29, the bending moment M_{BC} and the twisting moment T in any cross-section along the length BC is given by

$$M_{BC} = M_2 - plx/2 + px^2/2, \quad T_{BC} = 0 \qquad (2.66)$$

Bending moment M_{AB} and twisting moment T_{AB} in any cross-section along the length AB is given by [2.15, 2.16]

$$M_{AB} = ply/2, \quad T_{AB} = M_2 \qquad (2.67)$$

If we assume that the moulding box is clamped at A, then by Castigliano's theorem,

$$\frac{\partial U}{\partial M_2} = 0 \qquad (2.68)$$

where U is the total strain energy due to stretching, bending and twisting [2.14, 2.15]. From Eq. (2.66)

$$\frac{\partial M_{BC}}{\partial M_2} = 1, \quad \frac{\partial T_{BC}}{\partial M_2} = 0 \qquad (2.69)$$

From Eq. (2.67)

$$\frac{\partial M_{AB}}{\partial M_2} = 0, \quad \frac{\partial T_{AB}}{\partial M_2} = 1 \qquad (2.70)$$

Ignoring the contribution from stretching,

$$\frac{\partial U}{\partial M_2} = \frac{1}{EI_1} \int_0^{1/2} M_{BC} \frac{\partial M_{BC} X}{\partial M_2} \, dx + \frac{1}{GJ} \int_0^{1/2} T_{AB} \frac{\partial T_{AB}}{\partial M_2} \, dy$$

$$= \frac{1}{EI_1} \int_0^{1/2} (M_2 - plx/2 + px^2/2) \, dx + \frac{1}{EJ} \int_0^{1/2} M_2 \, dy = 0 \quad (2.71)$$

where I, and J are the major principal moment of inertia and polar moment of inertia of the channel.
Solving for M_2, one gets,

$$M_2 = \frac{Wl}{12 [1 + 2 (1 + \mu) I_1/J]} = \frac{Wlc_1}{12} \qquad (2.72)$$

where
$$c_1 = \frac{1}{[1 + 2 (1 + \mu) I_1/J]} \qquad (2.73)$$

Bending moment M at any section along BC is given by,

$$M = -M_2 + plx/2 - \frac{px^2}{2} \qquad (2.74)$$

In order to compute the deflection at C, let us introduce a fictitious load P at C. Proceeding as before, the introduction of this load gives rise to the following,

$$\frac{\partial M_{AB}}{\partial P} = \frac{y}{2}, \; \frac{\partial M_{BC}}{\partial P} = \frac{x}{2}, \; \frac{\partial T_{AB}}{\partial P} = 0, \; \frac{\partial T_{BC}}{\partial P} = 0 \qquad (2.75)$$

Hence

$$\partial_C = \frac{\partial U}{\partial P} \, (P = 0) = 2 \left[\frac{1}{EI_1} \int_0^{1/2} (P/2 + pl/2) \, y^2/2 \, dy \right.$$

$$\left. + \frac{1}{EI_1} \int_0^{1/2} [-M_2 + (P/2 + pl/2) \, x - px^2/2] \, x/2 \, dx \right],$$

$$= 2W \, [1 - 1c + 1\_1] = Wl \, \{6.5 - 2c_1\}/(192EI_1) \qquad (2.76)$$

Case (ii) Load in the Y direction only

Referring to Fig. 2.31,

$$M_{BC} = M_3 + plx/2 - px^2/2, \; M_{AB} = M_3 \qquad (2.77)$$

and the twisting moments are zero.

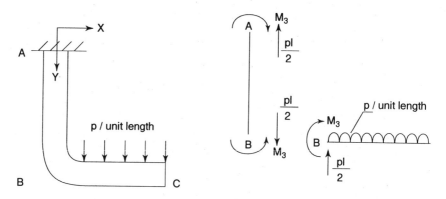

Fig. 2.31

Free body diagram—load Y direction only

Hence
$$\frac{\partial U}{\partial M_3} = M_3 l/2 + pl^3/16 - pl^3/48 + M_3 l/2 = 0 \qquad (2.78)$$

Hence
$$M_3 = -pl^2/24 \qquad (2.79)$$

The bending moment M in any section is given by,

$$M_{BC} = M_3 + plx/2 - px^2/2 \qquad (2.80)$$

Proceeding on lines similar to case (i) deflection at C is given by,

$$\delta_C = Wl^3/384\, EI_2\, I_2 \qquad (2.81)$$

where I_2 is the minor principal moment of inertia of the channel.

2.13.4.3 Finite Element Approach

To do a detailed analysis and to determine the deformed shape of the moulding box, finite element method was chosen. In all, 208 triangular plate elements were used to cover the portion AC of the box. The associated nodes were 129. Six degrees of freedom were allowed at each node. At the location A all the degrees of freedom were arrested, and at location C symmetric conditions were assumed. ($u = 0$, $\theta_y = 0$, $\theta_z = 0$). The overall size of the problem was (693, 108). Incore solution using Cholesky factorisation was employed. The response of the moulding box when the box was tilted from the horizontal to the vetical plane was determined Before taking up detailed investigation, the results obtained, for one sample case, by the finite element method were compared with the results by the use of Catigliano's theorem. The example considered had the following data: $l = 3$ m, $d = 1$ m, $b = 250$ mm, $h = 16$ mm, total load in the Z direction 4 tonnes and in the Y direction 4 tonnes. $I_1 = 32.2 \times 10^8$ mm^4, $I_2 = 0.559 \times 10^8$ mm^4. Thickness of the web and flanges was deliberately chosen to be comparable to the height of the web to make sure that the simple bending theory is, by and large, valid to justify the use of Castigliano's theorem for verification of results. The deflection at C and the stresses at the outer fibre at

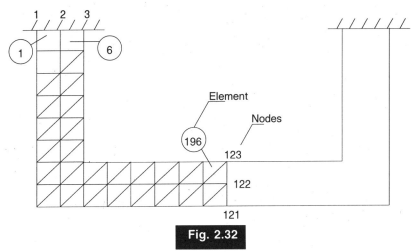

Fig. 2.32

Finite element discretisation

Table 2.6

Comparison of Results

Method employed	Load Y direction			Load Z direction		
	δ_C mm	σ_C N/mm^2	σ_H N/mm^2	δ_C mm	σ_C N/mm^2	σ_H N/mm^2
F.E.M.	.036	1.00	.47	.017	.252	.176
Castigliano	.043	.73	.47	.017	.27	.20

C (mid span) and H (quarter span), obtained by both the methods, are compared in Table 2.6. The agreement between the two approaches is good.

2.13.4.4 Parametric Study

Three typical moulding boxes of sizes 3 m, 2.25 m and 1.5 m had been taken up for the analysis. Their leading dimensions, the input loads considered, along with the results are shown in Table 2.7. The stress and deformation experienced by the moulding box as it is

Table 2.7

Performance of Moulding Boxes

l (m)	d (m)	b (m)	h (mm)	load (tons)	δ_C max (mm)	σ_C stress (N/mm^2)
3	1	.25	16	8	2.8	48
2.25	0.75	.188	12	6	2.8	64
1.50	0.50	.125	8	4	2.8	96

turned about the axis of the lugs (Fig. 2.33) are determined. The deformed shape of midsection CC as the moulding box is tilted as shown in Fig. 2.34 for the largest size moulding box. It may be observed that when $\theta = 22.5°$, there is considerable twisting besides bending, but when $\theta = 90°$, there is pure bending. The bending stress experienced by the side BC for the three sizes of the box is shown in Fig. 2.35. It may be observed from Figs 2.34 and 2.35 that the bending stress values are as high as 90 N/mm^2 (for the small size box) and the deflections are as high as 28 mm in the web portion and the two flanges open out by 4.5 mm when $\theta = 90°$. Isostress lines for the inner face of the moulding box for the web portion is shown in

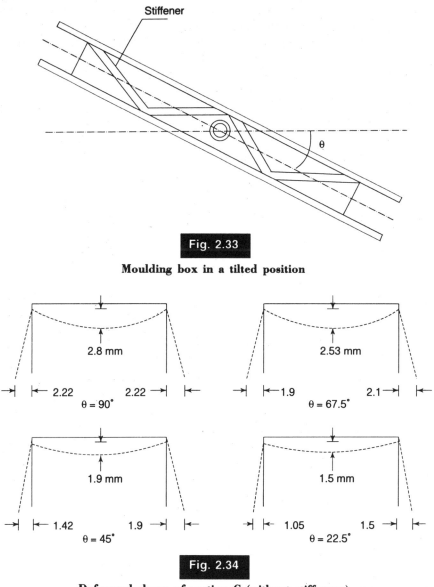

Fig. 2.33

Moulding box in a tilted position

Fig. 2.34

Deformed shape of section C (without stiffeners)

Fig. 2.36 when the moulding box is tilted through an angle of 45°. The critical zone of stress is around midspan, close to section CC. In order to reduce distortion of the moulding box, the channel sections are stiffened, as shown in Fig. 2.33. The influence of these stiffeners has also been studied. The big size moulding box is stiffened by a

16 mm thick plate diagonally as shown. One could observe that this has considerably improved both the strength and the rigidity of the moulding box. The maximum stress has come down to 60% of the original value (Fig. 2.35), and the deflection throughout has dropped sharply, (Fig. 2.37). Opening out of the flanges observed in the case of a box with no stiffeners is not at all present.

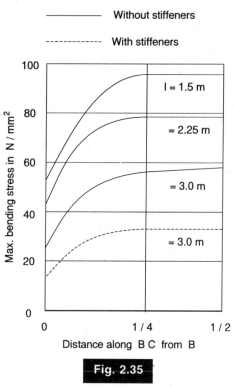

Fig. 2.35

Stress distribution along BC

2.13.4.5 Conclusions

1. Heavyduty moulding boxes of sizes reported here do really perform arduous duties (stress values as high as 90 N/mm^2) warranting detailed analysis to optimise the steel used in manufacture. If may be pointed out that the weight of the biggest moulding box reported here is around 2 tonnes. Since the top and bottom surfaces have to be machined for proper sitting the weight before machining would be much more.
2. The procedure outlined gives a detailed picture of the deformation and stress experienced by the moulding box.

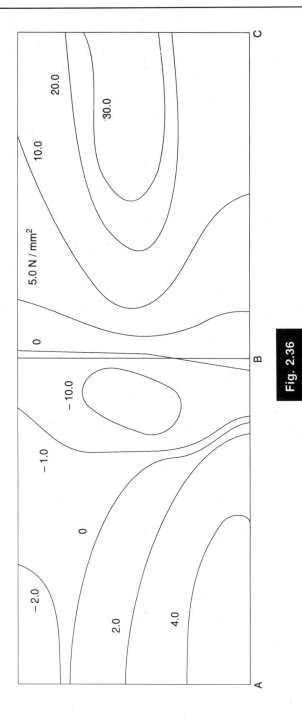

Fig. 2.36

Iso-stress lines for σ_Y (web ABC), $\theta = 45°$

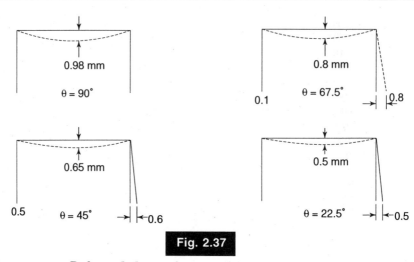

Fig. 2.37

Deformed shape of section CC (with stiffeners)

3. Introduction of the stiffeners increases the weight of the moulding box by about 20%, but it has increased the strength to twice its original value. The distortion of the box (opening out of the flanges) has completely disappeared. Hence one can make a judicious combination of channel dimensions and the number and size of ribs to optimise the weight of the box.

REFERENCES

2.1 Kadar, I., Three-dimensional structural program in every day usage, *Comp. Struct.*, vol. 6, no. 6, 1976, pp. 481–487.

2.2 Salvadori, M.G. and M.L. Baron, *Numerical Methods in Engineering*, Prentice-Hall Englewood Cliffs, N.J., 1961.

2.3 Hovanassian, S.A. and L.A. Pipes, *Digital Computer Methods in Engineering*, McGraw-Hill, New York, 1969.

2.4 James, M.L., G.M. Smith and J.C. Wolford, *Applied Numerical Methods for Digital Computation,* 2nd, edn, Harper and Row, New York, 1977.

2.5 McCormick, J.M. and M.G. Salvadori, *Numerical Methods in ForTran*, Prentice-Hall of India, New Delhi, 1985.

2.6 Bhirud, L.L., *Matrix Operations on the Computer*, Oxford and IBH, New Delhi, 1975.

2.7 Potters, M.L., A matrix method for the solution of a second order difference equation in two variables, Mathematische Centrum, Amsterdam, Holland, *Report No. MR 19*, 1955.

2.8 Melosh, R.J. and R.M. Banaford, Efficient solution of load deflection equations, *Proc. A.S.C.E.*, ST4, pp. 661–676, 1969.

2.9 Segui, W.T., Computer programs for the solution of systems of linear algebraic equations, *Inst. J. Num. Methods in Engg.*, vol. 7, 1973, pp. 479–490.

2.10 Jennings, A., *Matrix Computation for Engineers and Scientists*, John Wiley and Sons, London, 1977.

2.11 Cook, R.D., *Concepts and Applications of Finite Element Analysis*, John Wiley and Sons, New York, 1981.

2.12 Ramamurti, V. and P. Seshu, Dynamic analysis of a pressure vessel, *Computers and Structures*, vol. 29, no. 1, pp. 161–170, 1988.

2.13 Ramamurti, V., P.S. Arul Kumar and K. Jayaraman, Performance comparison of cast and fabricated gear boxes, *Computers and Structures*, vol. 37, no. 3, pp. 353–359, 1990.

2.14 Ramamurti, V., Design Aspects of heavyduty moulding boxes, *Jou. Inst. Engr. India* (ME) vol. 76, pp. 225-228 (1996).

2.15 Shigley J.E. and C. Mischke,*Mechanincal Engg. Design*, McGraw-Hill, New York, 1989.

2.16 Singer F.L. and A. Pytel, *Strength of Materials*, Harper and Row, New York, 1987.

EXERCISES

2.1 Find out the inverse of the square matrix given below by (i) method of partitioning, and (ii) Gauss Jordan method.

$$[K] = \begin{bmatrix} 12 & -6 & -6 & -1.5 \\ -6 & 4 & 3 & 0.5 \\ -6 & 3 & 6 & 1.5 \\ -1.5 & 0.5 & 1.5 & 1 \end{bmatrix}$$

Ans.

$$\begin{bmatrix} \frac{4}{9} & \frac{5}{9} & \frac{1}{9} & \frac{2}{9} \\ \frac{5}{9} & \frac{10}{9} & -\frac{1}{9} & \frac{4}{9} \\ \frac{1}{9} & -\frac{1}{9} & \frac{4}{9} & -\frac{4}{9} \\ \frac{2}{9} & \frac{4}{9} & -\frac{4}{9} & \frac{16}{9} \end{bmatrix}$$

2.2 Solve the problem $[K]\{\delta\} = \{f\}$ by Crout's procedure. $[K]$ is as given in Exercise 2.1.

$$\{f\} \text{ is given by} \begin{Bmatrix} 1 \\ 2 \\ 3 \\ 4 \end{Bmatrix}$$

Ans.
$$\{\delta\} = \begin{Bmatrix} 2.775 \\ 4.22 \\ -0.555 \\ 6.86 \end{Bmatrix}$$

2.3 Solve the problem given in Exercise 2.2 by Cholesky's factorisation, followed by forward and backward substitutions.

Ans.
$$\begin{bmatrix} 3.4641 & 0 & 0 & 0 \\ -1.732 & 1.00 & 0 & 0 \\ -1.732 & 0 & 1.732 & 0 \\ -0.433 & -0.2499 & 0.433 & 0.75 \end{bmatrix} = [L]$$

$$\{y\} = \begin{Bmatrix} 0.2886 \\ 2.499 \\ 2.0272 \\ 5.1664 \end{Bmatrix} \qquad \{\delta\} = \begin{Bmatrix} 2.776 \\ 4.219 \\ -0.554 \\ 6.885 \end{Bmatrix}$$

2.4 Solve the problem given in Exercise 2.2 by expressing $[K]$ as $[L][D][L]^T$.

2.5 Solve the problem given below by Cholesky's factorisation and variable bandwidth.

$$\begin{bmatrix} 5 & -4 & 1 & 0 & 0 \\ -4 & 6 & -4 & 1 & 0 \\ 1 & -4 & 6 & -4 & 1 \\ 0 & 1 & -4 & 6 & -4 \\ 0 & 0 & 1 & -4 & 5 \end{bmatrix} \begin{Bmatrix} \delta_1 \\ \delta_2 \\ \delta_3 \\ \delta_4 \\ \delta_5 \end{Bmatrix} = \begin{Bmatrix} 1 \\ 2 \\ 3 \\ 2 \\ 1 \end{Bmatrix}$$

2.6 Compute the deflection $\{\delta\}$ of the problem given below by Potters' scheme $[K]\{\delta\} = \{f\}$. Use three partitions, each of size 2.

$$[K] = \begin{bmatrix} 2 & 1 & 0 & 0 & 0 & 0 \\ & 4 & 3 & 2 & 0 & 0 \\ & & 4 & 3 & 1 & 0 \\ & \text{Sym} & & 6 & 2 & 1 \\ & & & & 4 & 2 \\ & & & & & 2 \end{bmatrix}; \qquad \{f\} = \begin{Bmatrix} 4 \\ 24 \\ 29 \\ 36 \\ 19 \\ 9 \end{Bmatrix}$$

2.7 Assume one degree freedom per node for the problem given in Fig. 2.38. The method chosen to solve this static problem is Cholesky's approach with variable bandwidth.

2.7.1 Compute the profile for this problem.

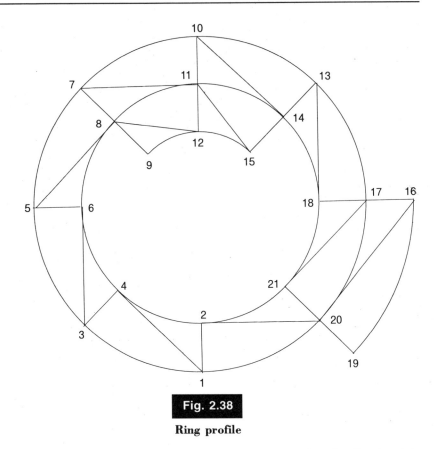

Fig. 2.38

Ring profile

    2.7.2 Estimate the labour involved (number of multiplications) in solving this problem.

2.8 Using reverse Cuthill McKee (R.C.M.) algorithm, rearrange the numbering.

    2.8.1 Compute the profile for this problem.

    2.8.2 Estimate the labour involved (number of multiplications) in solving this problem.

2.9 For the rearranged scheme of numbering, estimate the maximum number of substrcutures and associated degrees of freedom for each substructure if Potters' scheme is used for the solution.

    Note on computer programs connected with Chapter 2.

    The program lists attached attempt to bring out the utility of different methods discussed. Each program has a main program and the associated subroutines. To save space, same subroutines being called in different programs are printed out only once during their first occurrence.

    The problems are chosen to illustrate the different types of modelling and also the type of discretisation as the case may be.

2.1 Inversion by Partitioning

2.2 Gauss Jordan Method

2.3 Gaussian Elimination

2.4 Crout's Procedure

2.5 Potters' Scheme

Programs 2.6 and 2.7 solve the ring problem of Example 2.12 (finite element modelling) by employing different node numbering schemes.

2.10 Employs frontal scheme with discretisation as illustrated in Fig. 2.10.

2.11 Employs variable bandwidth algorithm with discretisation as shown in Fig. 2.11.

The application of Potters' algorithm of finite element modelling and employing partitioning concept is brought out by solving the boiler frame problem of Example 2.11 in Program 2.8.

Program 2.9 combines out of core solution and Cholesky scheme and solves Example 2.12.

Program 2.10 uses RCM algorithm to rearrange the given numbering for optimum bandwidth. The problem solved is the one given in Fig. 3.20.

3

Eigenvalue Problems

Classical books dealing with vibration studies present methods proposed by Holzer, Mykelstad or Prohl for solving multidegree of freedom systems problems. But these methods are extremely difficult to implement when modelling a continuous system with finite degrees of freedom. Even Transfer Matrix approach [3.1] is not very elegant when a finite number of frequencies and associated mode shapes are desired. The complexities are really challenging when the geometry combines plates, beams and brick elements. Today it is possible to predict the first few natural frequencies of units like bladed discs with lacing rods, gear boxes, rotating shafts, with non-uniform diameter, shafts with discs very accurately. General methods of approach to such large size problems will be discussed in this chapter.

General formulation of such problems will be of the type:

$$[M]\{\ddot{u}\} + [K]\{u\} = 0.$$

Eigenvalues of the above problem are the natural frequencies of the multidegree freedom system represented by $[K]$ and $[M]$. These are the frequencies to be avoided in actual operation. Besides these eigenvalues, associated eigenvectors are needed to estimate the dynamic response of a multidegree freedom system by modal superposition technique to be discussed in Chapter 5. The emphasis in this chapter is mainly on large-size problems of degrees of freedom n more than 100, but having bandwidths much smaller than n. Excellent presentation of this subject has been reported in [3.2], [3.3] and [3.4].

It is essential to stress the concepts of eigenvalue problems before discussing the solutions. With this end in view, let us discuss the properties of matrices, eigenvalues and eigenvectors of interest.

> ### 3.1   DEFINITION

3.1.1   Trace of a Matrix

The trace of a square matrix is defined as the sum of its principal diagonal elements.

Example 3.1

$$[A] = \begin{bmatrix} 1 & 2 & 3 \\ 2 & 3 & 4 \\ 3 & 4 & 5 \end{bmatrix}$$

Its diagonal elements are 1, 3 and 5. Its trace is 9.

3.1.2   Minor

The minor of a square matrix [A] is the determinant obtained from the elements of the matrix left after striking out certain rows and columns.

3.1.3   Principal Minor

The principal minor of a square matrix [A] is the determinant obtained by striking out certain i rows and the very same i columns.

3.1.4   Positive Definite Matrix

A matrix is said to be positive definite if all its eigenvalues (Sec. 3.2.1) are positive. A positive semidefinite (non-negative definite) matrix is a matrix which admits zero and positive eigenvalues.

Symmetric positive definite matrices admit only positive eigenvalues and positive determinants for themselves and for their principal minors. Similarly semidefinite matrices admit zero and positive eigenvalues and zero and positive determinants for themselves and for their principal minors.

Example 3.2

$$\text{Matrix } [A] = \begin{bmatrix} 2 & -1 & 0 \\ -1 & 3.5 & -1.5 \\ 0 & -1.5 & 2.5 \end{bmatrix}$$

Its eigenvalues are 1, 2.633, 4.366.

The value of det [A] is 11.5. Its 2 × 2 principal minors are

$$
\begin{bmatrix} 2 & -1 \\ -1 & 3.5 \end{bmatrix} \quad
\begin{bmatrix} 2 & 0 \\ 0 & 2.5 \end{bmatrix} \quad
\begin{bmatrix} 3.5 & -1.5 \\ -1.5 & 2.5 \end{bmatrix}
$$

Their eigenvalues are

$$
4, \ 1.5 \qquad\qquad 2, \ 2.5 \qquad\qquad 3 \pm \sqrt{10}
$$

3.1.5 Complex Matrix

If the elements of a matrix consist of complex numbers, the matrix is said to be complex.

Conjugate of a Matrix

If the elements of a matrix which are complex are replaced by their conjugates, the resulting matrix is the conjugate of the original matrix.

3.1.6 Hermitian Matrix

If the transpose of a complex matrix is its conjugate, then the matrix is said to be Hermitian.
Example 3.3

$$
[A] = \begin{bmatrix} 2 & 3+i \\ 3-i & 4 \end{bmatrix} \quad
[A]^* = \begin{bmatrix} 2 & 3-i \\ 3+i & 4 \end{bmatrix}
$$

$$
[A]^T = \begin{bmatrix} 2 & 3-i \\ 3+i & 4 \end{bmatrix}
$$

Hence $[A]^* = [A]^T$
$[A]$ is the complex conjugate of $[A]^*$.

3.1.7 Orthogonal Matrix

A square matrix $[A]$ is orthogonal if

$$
[A][A]^T = [A]^T[A] = [I] \quad \text{or} \quad [A]^T = [A]^{-1}
$$

Example 3.4

$$
[A] = \begin{bmatrix} \cos\theta & \sin\theta \\ -\sin\theta & \cos\theta \end{bmatrix}
$$

Here $[A][A]^T = [I]$

3.1.8   Unitary Matrix

A square complex matrix is unitary if $[A][A]*^T = [I]$.

3.1.9   Rotation Matrix

Consider the matrix $[P]$ given by

$$\begin{array}{cc} & i\text{th col.}\quad j\text{th col.} \end{array}$$

$$\begin{array}{c} i\text{th row} \\ j\text{th row} \end{array} \begin{bmatrix} \cos\theta & -\sin\theta \\ \sin\theta & \cos\theta \end{bmatrix}$$

All rows other than ith and jth have principal diagonal equal to unity (others are zero). ith row, jth column is $-\sin\theta$, jth row ith column is $\sin\theta$, (i, i) and (j, j) are $\cos\theta$. This is an orthogonal matrix since $[P]^{-1} = [P]^T$.

3.1.10   Reflection Matrix

Consider a vector $\{v\}$ of size $(n \times 1)$.

Consider a matrix $[P] = [I] - \alpha\{v\}\{v\}^T$

$$\alpha = 2/\{v\}^T\{v\}$$

Here $[P]$ is of size $(n \times n)$, unit matrix $[I]$ of size $(n \times n)$, and α is a number Matrix $[P]$ is an orthogonal matrix.
Example 3.5

Use $[v] = \begin{Bmatrix} 3 \\ 2 \end{Bmatrix}$. Form an orthogonal matrix of size (2×2)

$$\{v\}^T\{v\} = [3\ 2]\begin{Bmatrix} 3 \\ 2 \end{Bmatrix} = 13$$

∴ $\alpha = 2/13$

$$\{v\}\{v\}^T = \begin{Bmatrix} 3 \\ 2 \end{Bmatrix}[3\ 2] = \begin{bmatrix} 9 & 6 \\ 6 & 4 \end{bmatrix}$$

∴ $[P] = \begin{bmatrix} 1 & 0 \\ 0 & 1 \end{bmatrix} - \dfrac{2}{13}\begin{bmatrix} 9 & 6 \\ 6 & 4 \end{bmatrix} = \begin{bmatrix} -\dfrac{5}{13} & -\dfrac{12}{13} \\ -\dfrac{12}{13} & -\dfrac{5}{13} \end{bmatrix}$

$$[P^T] = \begin{bmatrix} -5/13 & -12/13 \\ -12/13 & 5/13 \end{bmatrix}$$

$$[P]^T = [P]^{-1}$$

Hence $[P]$ is orthogonal.

➤ 3.2   PROPERTIES OF EIGENVALUES AND EIGENVECTORS

3.2.1   Standard Eigenvalue Problem

Let $\qquad\qquad [A]\{u\} = \lambda\{u\}$ $\qquad\qquad\qquad$ (3.1)

where $[A]$ is a symmetric square matrix of size $(n \times n)$. There exist n non-trivial solutions (not null vectors) for $\{u\}$ of size $(n \times 1)$ each corresponding to n distinct λ values. Values of λ are known as eigenvalues and values of $\{u\}$ eigenvectors. In other words, there are n eigen pairs as solutions (λ_1, u_1), (λ_2, u_2) and so on.

Equation (3.1) is known as the standard eigenvalue problem since the square matrix is only one side of the equation (Ref. Eq. (3.29)).

3.2.2   Characteristic Polynomial

Equation (3.1) can be rewritten as

$$[A - \lambda I]\{u\} = 0$$

If λ is a known value and if the right hand side is replaced by a column vector $\{f\}$ then this gets modified as follows

$$[A - \lambda I]\{u\} = \{f\} \qquad\qquad (3.2)$$

The replacement vector $\{u\}$ can be evaluated by Cramer's rule as

$$u_i = \frac{\Delta_i}{\Delta} \qquad\qquad (3.3)$$

u_i is the ith element of the vector. Δ is the determinant of coefficients of the L.H.S. and Δ_i is the determinant of coefficients replacing the ith column of the L.H.S. by $\{f\}$. For all the terms of the displacement vector, the common denominator Δ is the determinant formed by the square matrix $[A - \lambda I]$. If this determinant is zero, all the displacements will become infinity.

If $\qquad\qquad p(\lambda) = \det (A - \lambda I)$ $\qquad\qquad\qquad$ (3.4)

for specific values of λ, $\det (A - \lambda I) = 0$. These values are the eigenvalues $p(\lambda)$ is known as the characteristic polynomial of A.

Det $[A - \lambda I]$ is the same as det $[A^T - \lambda I]$. Hence the eigenvalues of $[A]$ and $[A]^T$ are the same.

Example 3.6

Find out the eigenvalues and associated vectors of the matrix $[A]$ given by

$$[A] = \begin{bmatrix} 2 & -1 & -1 \\ -1 & 2 & -1 \\ -1 & -1 & 2 \end{bmatrix}$$

$$[A - \lambda I] = \begin{bmatrix} (2-\lambda) & -1 & -1 \\ -1 & (2-\lambda) & -1 \\ -1 & -1 & (2-\lambda) \end{bmatrix}$$

$$p(\lambda) = (2-\lambda)[(2-\lambda)^2 - 1] - 1[1 + (2-\lambda)] - 1[(2-\lambda) + 1]$$

$$= (8 - 12 + 6\lambda^2 - \lambda^3 - 2 + \lambda - 2 - 4 + 2\lambda)$$

$$= (-9\lambda + 6\lambda^2 - \lambda^3) = 0$$

The eigenvalues are 0, 3, 3.
The eigenvectors are

$$\begin{Bmatrix} 1/\sqrt{3} \\ 1/\sqrt{3} \\ 1/\sqrt{3} \end{Bmatrix} \quad \begin{Bmatrix} 1/\sqrt{6} \\ 1/\sqrt{6} \\ -2/\sqrt{6} \end{Bmatrix} \quad \begin{Bmatrix} 1/\sqrt{2} \\ 1/\sqrt{2} \\ 0 \end{Bmatrix}$$

The eigenvectors have been normalised such that the sum of the square of elements is equal to 1.

3.2.3   Orthogonal Transformation

Let
$$[A]\{u\} = \lambda\{u\} \qquad \text{[From Eq. (3.1)]}$$
be a standard eigenvalue problem.

Let
$$\{u\} = [P]\{\bar{u}\} \qquad (3.5)$$
where $[P]$ is a orthogonal matrix
  Premultiplying (3.1) by $[P]^T$ we have

$$[P]^T[A][P]\{\bar{u}\} = \lambda[P]^T[P]\{\bar{u}\}$$

$$[\bar{A}]\{\bar{u}\} = \lambda\{\bar{u}\} \qquad (3.6)$$

where
$$[\bar{A}] = [P]^T[A][P]$$

For the standard eigenvalue problem given by Eq. (3.1), λ is the eigenvalue. Transforming $[A]$ to $[\bar{A}]$ is known as the orthogonal similarity transformation. λ is also the eigenvalue for the transformed matrix $[\bar{A}]$ as seen from Eq. (3.6).

Example 3.7

$$[A] = \begin{bmatrix} 1 & 2 \\ 2 & 4 \end{bmatrix}$$

$p(\lambda)$ is given by

$$(1 - \lambda)(4 - \lambda) - 4 = (4 - 5\lambda + \lambda^2 - 4)$$

Values of λ are 0 and 5.
The eigenvectors are

$$\begin{Bmatrix} 2/\sqrt{5} \\ -1/\sqrt{5} \end{Bmatrix} \quad \text{and} \quad \begin{Bmatrix} 1/\sqrt{5} \\ 2/\sqrt{5} \end{Bmatrix}$$

Consider an orthogonal matrix

$$[P] = \begin{bmatrix} 0 & -1 \\ -1 & 0 \end{bmatrix}$$

$$[\bar{A}] = [P^T][A][P] = \begin{bmatrix} 0 & -1 \\ -1 & 0 \end{bmatrix} \begin{bmatrix} 1 & 2 \\ 2 & 4 \end{bmatrix} \begin{bmatrix} 0 & -1 \\ -1 & 0 \end{bmatrix}$$

$$= \begin{bmatrix} 0 & -1 \\ -1 & 0 \end{bmatrix} \begin{bmatrix} -2 & -1 \\ -4 & -2 \end{bmatrix} = \begin{bmatrix} 4 & 2 \\ 2 & 1 \end{bmatrix}$$

The eigenvalues of $[A]$ are 0 and 5.
Eigenvectors $\{\bar{u}\}$ are

$$\begin{Bmatrix} 1/\sqrt{5} \\ -2/\sqrt{5} \end{Bmatrix} \qquad \begin{Bmatrix} -2/\sqrt{5} \\ -1/\sqrt{5} \end{Bmatrix}$$

$$\{\bar{u}\} = \begin{Bmatrix} 1/\sqrt{5} \\ -2/\sqrt{5} \end{Bmatrix}; \quad \text{corresponding} \quad \{u\} = \begin{Bmatrix} 2/\sqrt{5} \\ -1/\sqrt{5} \end{Bmatrix}$$

$$[P]\{\bar{u}\} = \begin{bmatrix} 0 & -1 \\ -1 & 0 \end{bmatrix} 1/\sqrt{5} \begin{Bmatrix} 1 \\ -2 \end{Bmatrix} = \begin{Bmatrix} 2/\sqrt{5} \\ -1/\sqrt{5} \end{Bmatrix} = \{u\}$$

When

$$\{\bar{u}\} = 1/\sqrt{5} \begin{Bmatrix} -2 \\ -1 \end{Bmatrix}, [P]\{\bar{u}\} = \begin{bmatrix} 0 & -1 \\ -1 & 0 \end{bmatrix} 1/\sqrt{5} \begin{Bmatrix} -2 \\ -1 \end{Bmatrix}$$

$$= 1/\sqrt{5} \begin{Bmatrix} 1 \\ 2 \end{Bmatrix} = \{u\}$$

3.2.4   Eigenvalues with a Shift

Let us subtract $\mu(u)$ (μ is a constant) from the standard eigenvalue problem given by Eq. (3.1)

$$[A]\{u\} = \lambda\{u\}$$

Then $$[A - \mu I]\{u\} = (\lambda - \mu)\{u\} \tag{3.7}$$

Comparing Eq. (3.4) with Eq. (3.1) we find that the eigenvalues of the matrix $[A - \mu I]$ are $(\lambda - \mu)$, where λ are the eigenvalues of $[A]$. But the eigenvectors of Eqs (3.1) and (3.7) are the same. This procedure can be used to accelerate convergence in computing eigenvalues.

Example 3.8
Let us consider Example 3.6. Let us use a value of $\mu = 1$. Then

$$[A - \lambda I] = \begin{bmatrix} 1 & -1 & -1 \\ -1 & 1 & -1 \\ -1 & -1 & 1 \end{bmatrix}$$

$$p(\lambda) = (1 - \lambda)[(1 - \lambda)^2 - 1] - 2[1 + (1 - \lambda)]$$

$$= 1 - 3\lambda + 3\lambda^2 - \lambda^3 - 1 + \lambda - 4 + 2\lambda$$

$$= (-4 + 3\lambda^2 - \lambda^3)$$

The eigenvalues are $-1, 2, 2$.
The eigenvectors are

$$\begin{Bmatrix} 1/\sqrt{3} \\ 1/\sqrt{3} \\ 1/\sqrt{3} \end{Bmatrix} \begin{Bmatrix} 1/\sqrt{6} \\ 1/\sqrt{6} \\ -2/\sqrt{6} \end{Bmatrix} \begin{Bmatrix} 1/\sqrt{2} \\ -1/\sqrt{2} \\ 0 \end{Bmatrix}$$

It may also be observed from $[A - \mu I]$ and $[A]$ that their traces are equal to sum of the eigenvalues.

Trace of [A] is equal to 2 + 2 + 2 = 6
Sum of the λ's of [A] is equal to 0 + 3 + 3 = 6
Similarly, trace of $[A - \mu I]$ equals 3
Sum of λ's of $[A - \mu I]$ equals $-1 + 2 + 2 = 3$
This is an important property.

3.2.5 Eigenvalues of a Real Symmetric Matrix

The eigenvalues of a real symmetric matrix are real. Let us consider a symmetric matrix of the order n. Let us assume it admits a complex eigenvalue, $\lambda = v + i\delta$. If this is possible its complex conjugate $\bar{\lambda} = v - i\delta$ should also be an eigenvalue since the trace of the matrix, which is real, must be equal to the sum of the eigenvalues. Then

$$[A]\{u\} = \lambda\{u\} \tag{3.8}$$

$$[A]\{\bar{u}\} = \bar{\lambda}\{\bar{u}\} \tag{3.9}$$

$\{u\}$ and $\{\bar{u}\}$ are complex conjugates associated with λ and $\bar{\lambda}$. Premultiplying Eq. (3.8) by $\{\bar{u}\}^T$ and post multiplying the transpose of Eq. (3.9) by $\{u\}$ and subtracting, we have

$$\{\bar{u}\}^T[A]\{u\} - [A\bar{u}]^T\{u\} = (\lambda - \bar{\lambda})\{\bar{u}\}^T\{u\}$$

or

$$\{\bar{u}\}^T[A]\{u\} - \{\bar{u}\}^T[A]\{u\} = 0 = (\lambda - \bar{\lambda})\{\bar{u}\}^T\{u\} \text{ since } [A] \text{ is symmetric.}$$

or

$$2i\delta\{\bar{u}\}^T\{u\} = 0$$

Since $\{\bar{u}\}^T\{u\}$ is not zero, δ has to be zero. Hence eigenvalues will also be real.

3.2.6 Eigenvectors of a Real Symmetric Matrix

Let λ_i and λ_j be two distinct eigenvalues and $\{u_i\}$ and $\{u_j\}$ be two distinct eigenvectors of a real symmetric matrix [A].
Then

$$[A]\{u_i\} = \lambda_i\{u_i\} \tag{3.10}$$

$$[A]\{u_j\} = \lambda_j\{u_j\} \tag{3.11}$$

Premultiplying the first by $\{u_j\}^T$ and post multiplying the transpose of the second by $\{u_i\}$ and subtracting, we have

$$\{u_j\}^T[A]\{u_i\} - [Au_j]^T\{u_i\} = (\lambda_i - \lambda_j)\{u_j\}^T\{u_i\}$$

Since $[A]$ is symmetric, we have

$$(\lambda_i - \lambda_j)\{u_j\}^T\{u_i\} = 0$$

Since $\qquad\qquad \lambda_i \neq \lambda_j, \{u_j\}^T\{u_i\} = 0$

But when $i = j$, $\{u_i\}^T\{u_i\} \neq 0$. But $\{u_i\}$ can be normalised such that

$$\{u_i\}^T\{u_i\} = 1$$

Let $[U] = [u_1, u_2, u_3 \ldots u_n]$ where each of these columns is normalised. Let $[\Lambda]$ be a diagonal matrix whose diagonal elements are $\lambda_1, \lambda_2, \lambda_3 \ldots$ Then it follows

$$\underset{\text{size } (n \times n)}{[A][U]} = \underset{\text{size } (n \times n)}{[A][\Lambda]} ; [U]^T[U] = [I] \qquad (3.12)$$

If the first of these equations is premultiplied by $\{U\}^T$, using the second, we can write

$$\{U\}^T[A]\{U\} = [\Lambda]$$

Example 3.9
For Prob. 3.6, $\lambda_1 = 0$, $\lambda_2 = 3$ and $\lambda_3 = 3$
The eigenvectors are

$$\left\{\begin{array}{c} 1/\sqrt{3} \\ 1/\sqrt{3} \\ 1/\sqrt{3} \end{array}\right\} \quad \left\{\begin{array}{c} 1/\sqrt{6} \\ 1/\sqrt{6} \\ -2/\sqrt{6} \end{array}\right\} \quad \left\{\begin{array}{c} 1/\sqrt{2} \\ -1/\sqrt{2} \\ 0 \end{array}\right\}$$

$\therefore [U]^T[A]\{U\} =$

$$\begin{bmatrix} 1/\sqrt{3} & 1/\sqrt{3} & 1/\sqrt{3} \\ 1/\sqrt{6} & 1/\sqrt{6} & -2/\sqrt{6} \\ 1/\sqrt{2} & -1/\sqrt{2} & 0 \end{bmatrix} \begin{bmatrix} 2 & -1 & -1 \\ -1 & 2 & -1 \\ -1 & -1 & 2 \end{bmatrix} \begin{bmatrix} 1/\sqrt{3} & 1/\sqrt{6} & 1/\sqrt{2} \\ 1/\sqrt{3} & 1/\sqrt{6} & -1/\sqrt{2} \\ 1/\sqrt{3} & -2/\sqrt{6} & 0 \end{bmatrix}$$

$$= \begin{bmatrix} 1/\sqrt{3} & 1/\sqrt{3} & 1/\sqrt{3} \\ 1/\sqrt{6} & 1/\sqrt{6} & -2/\sqrt{6} \\ 1/\sqrt{2} & -1/\sqrt{2} & 0 \end{bmatrix} \begin{bmatrix} 0 & 3/\sqrt{6} & 3/\sqrt{2} \\ 0 & 3/\sqrt{6} & -3/\sqrt{2} \\ 0 & -6/\sqrt{6} & 0 \end{bmatrix}$$

$$= \begin{bmatrix} 0 & 0 & 0 \\ 0 & 3 & 0 \\ 0 & 0 & 3 \end{bmatrix} = [\Lambda]$$

3.2.7 Eigenvalues of a Hermitian Matrix

From Section 3.1.6 we know if $[\bar{A}]$ is the complex conjugate of $[A]$ (if $[A]$ is Hermitian) $[A] = [\bar{A}]^T = [A]^H$.

Let us presuppose that the eigenvalue of the hermitian matrix $[A]$ be complex given by $\lambda = \gamma + i\delta$.

Let $[A]\{u\} = \lambda\{u\}$ [from Eq. (3.1)]

Let $\bar{\lambda} = \gamma - i\delta$ such that

$$[\bar{A}]\{\bar{u}\} = \bar{\lambda}\{\bar{u}\} \tag{3.13}$$

Premultiplying Eq. (3.1) by $\{\bar{u}\}^T$ and post multiplying the transpose of Eq. (3.13) by $\{u\}$ and subtracting,

$$\{\bar{u}\}^T[A]\{u\} - [\bar{A}\bar{u}]^T\{u\} = (\lambda - \bar{\lambda})\{\bar{u}\}^T\{u\} = 2i\delta\{u\}^T\{u\}$$

or

$$\{\bar{u}\}^T[A]\{u\} - \{\bar{u}\}^T[A]\{u\} = 0 = 2i\delta\{u\}^T\{u\} \quad \text{[since $[A]$ is Hermitian]}$$

Since $\{\bar{u}\}^T\{u\}$ is positive, $\delta = 0$.

Hence the eigenvalues of a Hermitian matrix are real.

Example 3.10

Let $[A] = \begin{bmatrix} 2 & (3+i) \\ (3-i) & 4 \end{bmatrix}$ $[\bar{A}] = \begin{bmatrix} 2 & (3-i) \\ (3+i) & 4 \end{bmatrix} = [A]^T$

Hence $[A]$ is Hermitian

$$p(\lambda) = (2 - \lambda)(4 - \lambda) - (3 + i)(3 - i)$$

$$= 8 - 6\lambda + \lambda^2 - 10$$

$$= (-2 - 6\lambda + \lambda^2)$$

$\therefore$ $\lambda_1 = 3 - \sqrt{11}, \quad \lambda_2 = 3 + \sqrt{11}$

For $[\bar{A}]$ also the eigenvalues are

$$\lambda_1 = 3 - \sqrt{11}, \quad \lambda_2 = 3 + \sqrt{11}$$

3.2.8 Eigenvectors of a Real Nonsymmetric Matrix

Consider an eigenvalue problem

$$[A]\{u\} = \lambda\{u\} \quad \text{[from Eq. (3.1)]}$$

where $[A]$ is real and nonsymmetric of size $(n \times n)$.

Let $[A]^T$ be the transpose of $[A]$. Its eigenvalues will also be λ (Sec. 3.2.2). Let its eigenvalue problem be

$$[A]^T \{v\} = \lambda\{v\} \qquad (3.14)$$

For a distinct λ, say λ_i and λ_j, we have

$$[A]\{u_i\} = \lambda_i\{u_i\} \qquad (3.15)$$

$$[A]^T\{v_j\} = \lambda_j\{v_j\} \qquad (3.16)$$

Taking the transpose of either side in Eq. (3.16), we have

$$\{v_j\}^T[A] = \lambda_j\{v_j\}^T \qquad (3.17)$$

As before premultiplying Eq. (3.15) by $\{v_j\}^T$ and post multiplying Eq. (3.17) by $\{u_i\}$ and subtracting

$$\{u_j\}^T[A]\{u_i\} - \{u_j\}^T[A]\{u_i\} = 0 = (\lambda_i - \lambda_j)\{v_j\}^T\{u_i\}$$

When $\lambda_i \neq \lambda_j$, we have $\qquad \{v_j\}^T\{u_i\} = 0$

$$(i, j = 1, 2, \ldots n)$$

Proceeding along lines similar to 3.2.6 one can write

$$[V]^T[U] = [I] \qquad (3.18)$$

where $\qquad [V] = [v_1, v_2, \ldots v_n], \; [U] = [u_1, u_2 \ldots u_n] \qquad (3.19)$

If $\{v_j\}^T\{u_j\} = 1 \qquad (j = 1, 2 \ldots n)$

From Eq. (3.18), we have $\quad [V]^T = [U]^{-1} \qquad (3.20)$

Also $\qquad\qquad\qquad [U]^T = [V]^{-1} \qquad (3.21)$

From Eq. (3.12) we have

$$[A][U] = [U][\Lambda] \qquad (3.22)$$

Premultiplying Eq. (3.22) by $[V]^T$ we have

$$[V]^T[A][U] = [U]^{-1}\Lambda[U] = [\Lambda] \qquad (3.23)$$

Example 3.11
Let us consider the following problem.

$$[A]\{u\} = \lambda\{u\}$$

where $\qquad\qquad [A] = \begin{bmatrix} 4 & 1 \\ 5 & 8 \end{bmatrix}$

$[A]$ is a real unsymmetric matrix. The values of λ are given by

$$p(\lambda) = (4 - \lambda)(8 - \lambda) - 5 = 0$$

$$\therefore \qquad\qquad\qquad \lambda = 3 \quad \text{or} \quad 9$$

When
$$\lambda_1 = 3, \{u_1\} = \left\{ \begin{array}{c} 1/\sqrt{6} \\ -1/\sqrt{6} \end{array} \right\}$$

$$\lambda_2 = 9, \{u_2\} = \left\{ \begin{array}{c} 1/\sqrt{6} \\ 5/\sqrt{6} \end{array} \right\}$$

$$[A]^T = \begin{bmatrix} 5 & 5 \\ 1 & 8 \end{bmatrix} \qquad \text{Let } [A]^T\{v\} = \lambda\{v\}$$

Since λ's are the same,

When
$$\lambda_1 = 3, \{v_1\} = \left\{ \begin{array}{c} 5/\sqrt{6} \\ -1/\sqrt{6} \end{array} \right\}$$

$$\lambda_2 = 9, \{v_2\} = \left\{ \begin{array}{c} 1/\sqrt{6} \\ 1/\sqrt{6} \end{array} \right\}$$

Only these normalised values will satisfy Eq. (3.18)

$$[V]^T[U] = 1/\sqrt{6} \begin{bmatrix} 5 & -1 \\ 1 & 1 \end{bmatrix} \begin{bmatrix} 1 & 1 \\ -1 & 5 \end{bmatrix} 1/\sqrt{6} = \begin{bmatrix} 1 & 0 \\ 0 & 1 \end{bmatrix}$$

$$[V]^T[A][U] = 1/\sqrt{6} \begin{bmatrix} 5 & -1 \\ 1 & 1 \end{bmatrix} \begin{bmatrix} 4 & 1 \\ 5 & 8 \end{bmatrix} \begin{bmatrix} 1 & 1 \\ -1 & 5 \end{bmatrix} 1/\sqrt{6}$$

$$= \begin{bmatrix} 3 & 0 \\ 0 & 9 \end{bmatrix} = [\Lambda] \qquad \text{[from Eq. (3.23)]}$$

It may be observed that when $[A]$ is not symmetric, $[V]^T[V] \neq [I]$,

$$[U]^T[U] \neq [I], \text{ only } [V]^T[U] = [U]^T[V] = [I],$$

and also
$$[V]^T = [U]^{-1}[U]^T = [V]^{-1}$$

Example 3.12
Consider

$$[K] = \begin{bmatrix} (25 - 7\cos\psi) & (9 - 9\cos\psi - i\sin\psi) \\ (9 - 9\cos\psi + i\sin\psi) & (17 - 9\cos\psi) \end{bmatrix};$$

$$\{f\} = \left\{ \begin{array}{c} -1 - 1.736i \\ 0 \end{array} \right\}$$

Let $\psi = 72°$

Solve the problem $[K]\{\delta\} = \{f\}$

$$[K] = \begin{bmatrix} 22.837 & (6.219 - i(0.959)) \\ (6.219 + i(.959)) & 14.219 \end{bmatrix}$$

$[K]$ is Hermitian. It can be Cholesky factorised and decomposed as $[L][L]^H$ or $[U]^H[U]$.

$$[L] = \begin{bmatrix} 4.778 & 0 \\ 1.301 + .199i & 3.558 \end{bmatrix}$$

$$[L]^H = \begin{bmatrix} 4.778 & 1.301 - i.199 \\ 0 & 3.558 \end{bmatrix}$$

Forward and backward substitution using $\{f\}$ yields

$$[\delta] = \begin{Bmatrix} -.0498 - .068i \\ 0.0257 + i.034 \end{Bmatrix}$$

It will be shown in Chapter 4 that the stiffness matrix for a repeated structure will be Hermitian. For complex loads, deflections will be complex. However, loads on repeated structures will be groups of complex conjugates and hence, the net displacement will be real.

Let us consider the corresponding eigenvalue problem

$$[K]\{u\} = \omega^2[M]\{u\}$$

Let $[K]$ be as before and $[M] = \begin{bmatrix} .01 & 0 \\ 0 & 0.01 \end{bmatrix}$

$[K]$ in N/mm and $[M]$ in $N \cdot Sec^2/mm$.

The determinant formed by the governing equation is given by

$$\begin{bmatrix} (22.837 - .01\omega^2) & (6.219 - .951i) \\ (6.219 + .951i) & (14.219 - 0.01\omega^2) \end{bmatrix} = 0$$

$\therefore$ $\omega^2 = 2616$ or 1090

$\omega = 33.01$ or 51.14 rad/sec.

Here, the two natural frequencies are real.

Example 3.13
Consider the problem given in Fig. 3.1.

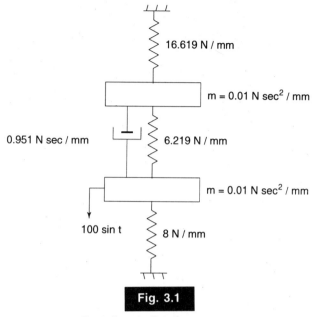

Fig. 3.1

Two degree freedom system

The governing determinant for natural frequencies takes the following form

$$\begin{bmatrix} (22.837 - .01\omega^2 + i.951\omega) & (-6.219 - .951\omega i) \\ (-6.219 - .951\omega i) & (14.218 - .01\omega^2 + .951\omega i) \end{bmatrix}$$

The square matrix is complex but symmetric; unlike in the earlier case. If damping is ignored when the determinant is expanded, one gets the equation

$$\omega^4 \times 10^{-4} - 10^{-2}\omega^2[37.056] + (324.72 - 38.76) = 0$$

The two roots of ω^2 are 1090 and 2616. For the forced vibration problem, we have

$$\begin{bmatrix} (22.837 - .011 + .951i) & (-6.219 - .951i) \\ (-6.219 - .951i) & (14.219 - .01 + .951i) \end{bmatrix} \begin{Bmatrix} X_1 \\ X_2 \end{Bmatrix} = \begin{Bmatrix} 0 \\ 100 \end{Bmatrix}$$

The square matrix on the L.H.S. is symmetric, but complex. This can be Cholesky factorised and X_1 and X_2 can be evaluated.

➤ 3.3 TYPICAL PROBLEM FORMULATIONS

3.3.1 Torsional or Longitudinal Vibration Problems

Let us consider a four-mass system shown in Fig. 3.2 subjected to forcing functions $F_1 \sin \omega t$, $F_2 \sin \omega t$, etc. acting on the different masses. The governing differential equations for the problem can be written as

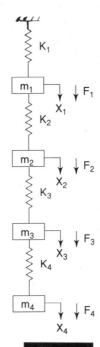

Fig. 3.2

Four mass longitudinal system

$$+ \, m_1 \ddot{x}_1 + k_1 x_1 + k_2(x_1 - x_2) = F_1 \sin \omega t$$

$$+ \, m_2 \ddot{x}_2 + k_2(x_2 - x_1) + k_3(x_2 - x_3) = F_2 \sin \omega t$$

$$+ \, m_3 \ddot{x}_3 + k_3(x_3 - x_2) + k_4(x_3 - x_4) = F_3 \sin \omega t$$

$$+ \, m_4 \ddot{x}_4 + k_4(x_4 - x_3) = F_4 \sin \omega t \qquad (3.24)$$

Seeking the solution as

$$x_i = X_i \sin \omega t \qquad (i = 1, 2, 3, 4) \qquad (3.25)$$

we have

$$
\begin{bmatrix}
(-m_1\omega^2+k_1+k_2) & -k_2 & & \\
(-k_2) & (-m_2\omega^2+k_2+k_3) & (-k_3) & \\
& (-k_3) & (-m_3\omega^2+k_3+k_4) & -k_4 \\
& & -k_4 & (-m_4\omega^2+k_4)
\end{bmatrix}
$$

$$
\begin{Bmatrix} X_1 \\ X_2 \\ X_3 \\ X_4 \end{Bmatrix} = \begin{Bmatrix} F_1 \\ F_2 \\ F_3 \\ F_4 \end{Bmatrix}
\qquad (3.26)
$$

or it is of the form

$$[A]\{u\} = \{f\}$$

The square matrix $[A]$ is tridiagonal (semi-bandwidth is 2). It is symmetric and real. If the forces and the frequency of excitation are known, (steady state vibration problem) computing the response is just solving Eq. (2.26). The methods discussed in Chapter 2 can be used for solving lage-size problems. If the natural frequencies of the system are to be evaluated, then Eq. (3.26) gets modified to

$$
\begin{bmatrix}
(k_1 + k_2) & -k_2 & & \\
-k_2 & (k_2 + k_3) & -k_3 & \\
& -k_3 & (k_3 + k_4) & -k_4 \\
& & -k_4 & k_4
\end{bmatrix}
\begin{Bmatrix} X_1 \\ X_2 \\ X_3 \\ X_4 \end{Bmatrix}
$$

$$
= \omega^2
\begin{bmatrix}
m_1 & & & \\
& m_2 & & \\
& & m_3 & \\
& & & m_4
\end{bmatrix}
\begin{Bmatrix} X_1 \\ X_2 \\ X_3 \\ X_4 \end{Bmatrix}
$$

or it is of the form

$$[K]\{u\} = \omega^2[M]\{u\} \qquad (3.27)$$

This is a non-standard eigenvalue problem since square matrices $[K]$ and $[M]$ are appearing on L.H.S. and R.H.S.

The assembled mass matrix on the R.H.S. of Eq. (3.27) is of different sizes for different elements used. If it is a torsional vibration problem, the element stiffness and mass matrices will be of size 2×2 having one degree of freedom (rotation along the axis) at each node of an element. This corresponds to rows/columns 4 and 10 of Eq. (1.22) for the stiffness matrix and mass moments of inertia about the axis along the principal diagonal of the mass matrix. This is used in the case study 6.2.

In the case of lateral vibration problems in one plane, the stiffness and mass matrices will be of size 4×4, having two degrees of freedom at each node. This corresponds to rows/columns 2, 3, 5 and 6 of Eq. (1.19) for stiffness matrix, and the corresponding mass matrix from Eq. (1.20), (Application 3.13.4). In the case of lateral vibration problems with gyroscopic effect included, the mass matrix includes the inertia effect of the disc and is expressed by Eq. (3.144) (Application 3.13.1). This approach has a very wide application as compared to the influence coefficient approach discussed in Sec. 3.3.2. For all other cases, appropriate element mass matrices have to be derived using Eq. (1.12). These are of size 12×12 for space frames (Application 6.5), 18×18 for triangular shell elements (Application 6.4) and 24×24 for brick elements. It may be observed that the torsional vibration problems, lateral vibration problems in one plane and lateral vibration problems in 2 planes at right angles to each other in rotor dynamics can be derived from this single 3D beam element (Eq. 123).

The system shown in Fig. 3.3 is a four-rotor system having four inertias connected by three torsional stiffnesses. The behaviour of this system is in no way different from the system shown in Fig. 3.2. For a steady-state torsional vibration problem this gives rise to a square matrix which is real, symmetric and tridiagonal. For the eigenvalue problem this gives rise to an equation similar to Eq. (3.27). This system has practical application in multicylinder engines. But multicylinder engines have damping (source of dissipation of energy) in the crankshafts (hysterisis damping) and in inertias (engine damping, damping in torsional vibration dampers and in propellers). Since damping present is normally small in magnitude, for determining the natural frequencies it is ignored. Then th᠄ eigenvalue problem to be solved is similar to Eq. (3.29). But for

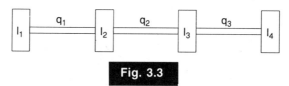

Fig. 3.3

Four rotor torsional systems

steady-state stress and response determination damping terms are introduced at the appropriate locations. Then the resultant Eq. (3.27) have coefficients of the square matrix [A] which are complex. This can be solved by declaring the response {u} as complex. This implies that the different inertias have response with varying phase angles.

3.3.2 Lateral Vibration Problems

Before the advent of finite element method influence coefficient approach was used for solving lateral vibration problems. This method can be used only for because of uniform cross-section. Hence it does not have wide application. Consider a uniform beam shown in Fig. 3.4 subjected to transverse oscillations in the plane of the paper. Let

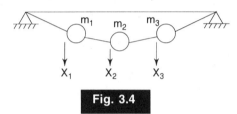

Fig. 3.4

Lateral oscillations with three masses

x_1, x_2 and x_3 be the dynamic displacements of the beam reckoned positive downwards. The inertia forces at the location of the masses amount to $m_1\ddot{x}_1$, $m_2\ddot{x}_2$ and $m_3\ddot{x}_3$ acting upwards. Let α_{ij} be the influence coefficient of the beam (deflection at the ith location due to unit load at the jth location). Then the governing differential equation is given by

$$- \alpha_{11}m_1\ddot{x}_1 - \alpha_{12}m_2\ddot{x}_2 - \alpha_{13}m_3\ddot{x}_3 = x_1$$
$$- \alpha_{21}m_1\ddot{x}_1 - \alpha_{22}m_2\ddot{x}_2 - \alpha_{32}m_3\ddot{x}_3 = x_2 \qquad (3.28)$$
$$- \alpha_{31}m_1\ddot{x}_1 - \alpha_{32}m_2\ddot{x}_2 - \alpha_{33}m_3\ddot{x}_3 = x_2$$

These equations are based on the assumption that the dynamic deflection curve is similar to the static deflection curve and only one degree of freedom, viz. vertical deflection is permitted at every location.

Assuming $x_i = X_i \sin \omega t$, we have

$$(\alpha_{11}m_1 - 1/\omega^2)X_1 + \alpha_{12}m_2X_2 + \alpha_{13}m_3X_2 = 0$$
$$\alpha_{21}m_1X_1 + (\alpha_{22}m_1 - 1/\omega^2)X_2 + \alpha_{23}m_3X_3 = 0$$
$$\alpha_{31}m_1X_1 + \alpha_{32}m_2X_2 + (\alpha_{33}m_3 - 1/\omega^2)X_3 = 0 \qquad (3.29)$$

Rewriting, we have

$$[A]\{u\} = \lambda\{u\} \qquad (3.30)$$

where

$$[A] = \begin{bmatrix} \alpha_{11}m_1 & \alpha_{12}m_2 & \alpha_{13}m_3 \\ \alpha_{21}m_1 & \alpha_{22}m_2 & \alpha_{23}m_3 \\ \alpha_{31}m_1 & \alpha_{32}m_2 & \alpha_{33}m_3 \end{bmatrix} \qquad (3.31)$$

and

$$\{u\} = \begin{Bmatrix} X_1 \\ X_2 \\ X_3 \end{Bmatrix}, \qquad \lambda = \frac{1}{\omega^2}$$

By Maxwell-Betti's theorem $\alpha_{ij} = \alpha_{ji}$. Only $m_1 = m_2 = m_3$, the square matric [A] becomes real and symmetric. For a general case it is non-symmetric. The matrix is not tridiagonal as in the case of torsional vibration problems. Besides vertical deflections, rotations (slopes) at these locations can also be included. The resulting problem will also be far coupled and the resulting eigenvalue problem will have the matrix [A] non-symmetric and real.

Example 3.13
Formulate the eigenvalue problem for the cantilever beam shown in Fig. 3.5. The influence coefficients are

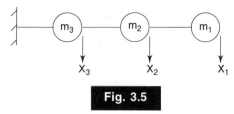

Fig. 3.5

Cantilever with three masses

$$\alpha_{11} = \frac{27l^3}{3EI}, \qquad \alpha_{21} = \alpha_{12} = \frac{14}{27}\alpha_{11}$$

$$\alpha_{22} = \frac{8}{27}\alpha_{11}, \; \alpha_{23} = \alpha_{32} = \frac{2.5}{27}\alpha_{11}$$

$$\alpha_{33} = \frac{\alpha_{11}}{27}, \qquad \alpha_{13} = \alpha_{31} = \frac{4}{27}\alpha_{11}$$

where $3l$ is the total length of the beam.

The matrix equation can be written as

$$\begin{Bmatrix} X_1 \\ X_2 \\ X_3 \end{Bmatrix} = \frac{\omega^2 \alpha_{11}}{27} \begin{bmatrix} 27 & 14 & 4 \\ 14 & 8 & 2.5 \\ 4 & 2.5 & 1 \end{bmatrix} \begin{bmatrix} m_1 & 0 & 0 \\ 0 & m_2 & 0 \\ 0 & 0 & m_3 \end{bmatrix} \begin{Bmatrix} X_1 \\ X_2 \\ X_3 \end{Bmatrix}$$

For a typical set of values of $m_1 = 1$, $m_2 = 2$ and $m_3 = 3$ and $\alpha_{11} = 27$, we have

$$\begin{bmatrix} 27 & 28 & 12 \\ 14 & 16 & 7.5 \\ 4 & 5 & 3 \end{bmatrix} \{u\} = \lambda \{u\}$$

where $\lambda = 1/\omega^2$.

This is a nonsymmetric real eigenvalue problem with the characteristic polynomial

$$p(\lambda) = 19.5 - 83.2\lambda + 46\lambda^2 - \lambda^3 = 0$$

Hence λ values are 0.277, 1.63, 44.10.
The eigenvectors are given by

$$\{u_1\} = \begin{Bmatrix} 0.608 \\ -0.948 \\ 0.823 \end{Bmatrix} \quad \{u_2\} = \begin{Bmatrix} 0.550 \\ -0.325 \\ -0.400 \end{Bmatrix} \quad \{u_3\} = \begin{Bmatrix} 0.951 \\ 0.515 \\ 0.155 \end{Bmatrix}$$

The transpose of the given matrix is

$$\begin{bmatrix} 27 & 14 & 4 \\ 28 & 16 & 5.0 \\ 12 & 7.5 & 3.0 \end{bmatrix}$$

Its eigenvalues are

$$\lambda_1 = 0.277 \qquad \lambda_2 = 1.63 \qquad \lambda_3 = 44.10$$

Its eigenvectors are

$$\{v_1\} = \begin{Bmatrix} 0.151 \\ -0.456 \\ 0.608 \end{Bmatrix} \quad \{v_2\} = \begin{Bmatrix} 0.550 \\ -0.648 \\ -1.21 \end{Bmatrix} \quad \{v_3\} = \begin{Bmatrix} 0.628 \\ 0.685 \\ 0.314 \end{Bmatrix}$$

The following may easily be verified:

$$[V]^T[A][U] = [\Lambda] \quad \text{as per (Eq. 3.23)}$$
$$[V]^T[U] = [U]^T[V] = [I] \quad \text{as per Eq. (3.18)}$$

➤ 3.4   STURM SEQUENCE

Let us consider the generalised eigenvalue problem $[K][\phi] = \omega^2[M][\phi]$. Let us assume that for a shift of μ, Cholesky's factorisation of $[K - \mu M]$ into $[L][D][L]^T$ can be obtained. One of the important characteristics of $[D]$ is that the number of negative in the principal diagonal of $[D]$ in $[D]$ is equal to the number of eigenvalues smaller than μ. This method is extremely useful in very quickly determining the approximate locations of eigenvalues.

For an assumed μ_1, find out the number of negatives, say n_1 in $[D]$ after factorisation. For an assumed $\mu_2 (> \mu_1)$, find out the number of negatives say n_2. Then as a first approximation for every increase in interval from μ_1 to μ_2, by an amount $(\mu_2 - \mu_1)/(n_2 - n_1)$ there will be an additional negatives. A detailed discussion on Sturm sequence and an exhaustive computer program has been given in [3.5].

Referring to Sec. 2.4 on Gaussian elimination, let us assume that $a_{11}, a'_{22}, a''_{33}$, etc. appearing as the first non-zero terms in the respective rows, after the L.H.S. is fully triangularised be equal to d_{11}, d_{22}, d_{33}, etc.

Then
$$d_{11} = a_{11} = k_{11}$$
$$d_{22} = a'_{22} = k_{22} - k_{12}^2/d_{11}$$
$$d_{33} = a''_{33} = k_{33} - k_{13}^2/d_{11} - k_{23}^2/d_{22}$$

if the coefficients of the L.H.S. are the elements of the symmetric real stiffness matrix $[K]$.

Now referring to Sec. 2.6 on Cholesky's scheme, if K is expressed as $[K] = [L][D][L]^T$, the elements of $[D]$ from Eq. (2.40) are

$$d_{11} = k_{11}$$

$$d_{22} = k_{22} - l_{21}g_{21} = k_{22} - \frac{k_{12}^2}{d_{11}}$$

$$d_{33} = k_{33} - l_{31}g_{31} - l_{32}g_{32} = k_{33} - \frac{k_{13}^2}{d_{11}} - \frac{k_{23}^2}{d_{22}}$$

These are the same as the ones obtained when the Gaussian elimination procedure was adopted. Hence to obtain the elements of $[D]$, triangularisation in the Gaussian elimination process has to be achieved.

Example 3.14
Compute the eigenvalues of the four mass system shown in Fig. 3.6 by Sturm sequence (all K's in N/mm and M's in N sec^2/mm). The eigenvalue problem of this example is given by

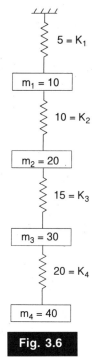

Fig. 3.6

Four mass system

$$\begin{bmatrix} 15 & -10 & & \\ -10 & 25 & -15 & \\ & -15 & 35 & -20 \\ & & -20 & 20 \end{bmatrix} \{u\} = \omega^2 \begin{bmatrix} 10 & & & 0 \\ & 20 & & \\ & & 30 & \\ 0 & & & 40 \end{bmatrix} \{u\}$$

Let us try μ_1 of 0.025 and μ_2 of 2.50.

When $\qquad\qquad \mu_1 = 0.025,$

$$[K - \mu_1 M] = \begin{bmatrix} 14.75 & -10 & & \\ -10 & 24.50 & -15 & \\ & -15 & 34.25 & -20 \\ & & -20 & 19 \end{bmatrix}$$

Following the Gaussian elimination procedure, this can be written as

$$
\begin{bmatrix}
14.75 & -10 & & \\
0 & 17.72 & -15 & \\
& & 21.55 & -20 \\
& & & 0.438
\end{bmatrix}
$$

All the elements of [D] are positive.
When $\mu_2 = 2.50$,

$$
[K - \mu_2 M] =
\begin{bmatrix}
-10 & -10 & & \\
-10 & -25 & -15 & \\
& -15 & -40 & -20 \\
& & -20 & -80
\end{bmatrix}
$$

By Gaussian elimination this can be rewritten as

$$
\begin{bmatrix}
-10 & -10 & & \\
0 & -15 & -15 & \\
& 0 & -25 & -20 \\
& & 0 & -64
\end{bmatrix}
$$

All the elements of [D] are negative. Hence all the four natural frequencies ω^2 are less than 2.5.
Value of det $[K - \mu M]$ (when $\mu = 0$) is got from [K]. By Gaussian elimination [K] is modified as

$$
\begin{bmatrix}
15 & -10 & & \\
0 & 18.33 & -15 & \\
& 0 & 22.72 & -20 \\
& & 0 & 2.39
\end{bmatrix}
$$

$\therefore$ det $[K - \mu M]$ when $\mu = 0 = (15 \times 18.33 \times 22.72 \times 2.39) = 14930$.

When $\mu_1 = 0.025$,

$$\text{det } [K - \mu_1 M] = 14.75 \times 17.72 \times 21.55 \times 0.438$$

$$= 2467.0$$

Referring to Fig. 3.7 if the two values of determinants at $\mu = 0$ and $\mu = 0.025$ are extrapolated the first natural frequency ω^2 is approximately equal to

$$0.025 + \frac{0.025 \times 2467}{(14930 - 2467)} = 0.029$$

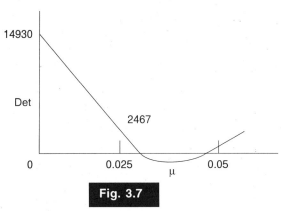

Fig. 3.7

Method of bisection

By bisection the other natural frequencies ω^2 are

$$0.029 + \frac{(2.5 - 0.029)}{3} \qquad 0.029 + \frac{2}{3}(2.5 - 0.029) \qquad (2.5)$$

They are 0.852, 1.675 and 2.5 rad/sec.

Exact values by other methods are 0.03, 0.599, 1.51 and 2.26. It may be observed that the intermediate eigenvalues are erroneous since μ_{max} of 2.5 is far off the mark. The values obtained are very low since the masses assumed are very heavy.

Example 3.15

Calculate the fundamental frequency of torsional oscillation of the system given below by Sturm sequence (Fig. 3.8). $I_1 = I_2 = I_3 = I_4 = 1$, $I_5 = 2$, $I_6 = I_7 = 5$, $q = 10$. (All I's in Nmm sec^2 and q's in Nmm/rad).

Let us first approximately compute the natural frequency by using the energy principle. Let us assume that the deflected mode shape is given by $\theta_1 = 1.0, \theta_2 = 0.5$, $\theta_3 = 1.0$, $\theta_4 = 0.5$, $\theta_5 = 0.25$, $\theta_6 = -0.25$ and $\theta_7 = -0.50$. This gives rise to one node between inertias 5 and 6.

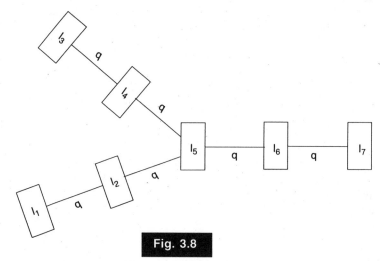

Fig. 3.8

Branched systems

Total kinetic energy at the vibrating frequency

$$= \Sigma \frac{1}{2} I \omega^2 \theta^2$$

$$= \frac{\omega^2}{2} [1(1)^2 + 1(0.5)^2 + 1(1)^2 + 1(0.5)^2 + 2(0.25)^2$$

$$+ 5(0.25)^2 + 5(0.5)^2]$$

$$= \frac{\omega^2}{2}(4.2)$$

Strain energy $= \frac{1}{2} \Sigma \, q_i (\theta_i - \theta_{i-1})^2$

$$= \frac{10}{2} [(0.5)^2 + (0.25)^2 + (0.5)^2(0.25)^2 + (0.5)^2 + (0.25)^2]$$

$$= 9.375/2$$

Equating the two,

$$\omega^2 = \frac{9.375}{4.2} \approx 2.1$$

Let us try a value of $\mu = 2.0$.

$[K - \mu M]$ is given by

$$
\begin{bmatrix}
8 & -10 & 0 & 0 & 0 & 0 & 0 \\
-10 & 18 & 0 & 0 & -10 & 0 & 0 \\
0 & 0 & 8 & -10 & 0 & 0 & 0 \\
0 & 0 & 10 & 18 & -10 & 0 & 0 \\
0 & -10 & 0 & -10 & 26 & -10 & 0 \\
0 & 0 & 0 & 0 & -10 & 10 & -10 \\
0 & 0 & 0 & 0 & 0 & -10 & 0
\end{bmatrix}
$$

Eliminating the bottom half by Gaussian elimination scheme, we have

$$
\begin{bmatrix}
8 & -10 & 0 & 0 & 0 & 0 & 0 \\
0 & 5.5 & 0 & 0 & -10 & 0 & 0 \\
0 & 0 & 8 & -10 & 0 & 0 & 0 \\
0 & 0 & 0 & 5.5 & -10 & 0 & 0 \\
0 & 0 & 0 & 0 & -3.10 & -10 & 0 \\
0 & 0 & 0 & 0 & 0 & 20 & -10 \\
0 & 0 & 0 & 0 & 0 & 0 & -5
\end{bmatrix}
$$

There are two negative signs in $[D]$. Since it is a free system there is one non-trivial frequency below 2.0. ($\omega = 0$ is the other solution.) When $\mu = 1.0$, $[K - \mu M]$ is given by

$$
\begin{bmatrix}
9 & -10 & 0 & 0 & 0 & 0 & 0 \\
-10 & 19 & 0 & 0 & -10 & 0 & 0 \\
0 & 0 & 9 & -10 & 0 & 0 & 0 \\
0 & 0 & -10 & 19 & -10 & 0 & 0 \\
0 & -10 & 0 & -10 & 28 & -10 & 0 \\
0 & 0 & 0 & 0 & -10 & 15 & -10 \\
0 & 0 & 0 & 0 & 0 & -10 & 5
\end{bmatrix}
$$

This when tridiagonalised gives

$$
\begin{bmatrix}
.9 & -10 \\
 & 7.89 & 0 & 0 & -10 \\
 & & 9 & -10 \\
 & & & 7.89 & -10 \\
 & & & & 2.4 & -10 \\
 & & & & & -26.6 & -10 \\
 & & & & & 0 & 1.2
\end{bmatrix}
$$

There is only one negative in $[D]$. Hence the fundamental frequency is higher than 1.0. The value of the determinant $\approx -4 \times 10^{+5}$. Choosing μ of 1.6, $[K - \mu M]$ is given by

$$
\begin{bmatrix}
8.4 & -10 & 0 & 0 & 0 & 0 & 0 \\
-10 & 8.4 & -10 & 0 & 0 & 0 & 0 \\
0 & 0 & 8.4 & -10 & 0 & 0 & 0 \\
0 & 0 & -10 & 18.4 & -10 & 0 & 0 \\
0 & -10 & 0 & -10 & 26.80 & -10 & 0 \\
0 & 0 & 0 & 0 & -10 & 12 & -10 \\
0 & 0 & 0 & 0 & 0 & -10 & 2
\end{bmatrix}
$$

Triangularising

$$
\begin{bmatrix}
8.4 & -10 & 0 & 0 & 0 & 0 & 0 \\
0 & 6.5 & 0 & -10 & -10 & 0 & 0 \\
0 & 0 & 8.4 & -10 & 0 & 0 & 0 \\
0 & 0 & 0 & 6.5 & -10 & 0 & 0 \\
0 & 0 & 0 & 0 & -3.96 & -10 & 0 \\
0 & 0 & 0 & 0 & 0 & 37.2 & -10 \\
0 & 0 & 0 & 0 & 0 & 0 & -0.68
\end{bmatrix}
$$

The value of the determinant $\approx 4 \times 10^4$.

This value of ω^2 could be further improved. Values of θ for this ω are $\theta_1 = -1.51$, $\theta_2 = -1.27$, $\theta_3 = -1.51$, $\theta_4 = 1.27$, $\theta_5 = -0.826$, $\theta_6 = 0.268$ and $\theta_7 = 1.000$.

➤ 3.5 JACOBI, GIVENS AND HOUSEHOLDER'S TRANSFORMATIONS

Torsional vibration and longitudinal vibration problems (close coupled systems) discussed in Sec. 3.3.1 have stiffness matrices which are tridiagonal. Lateral oscillation problems formulated using influence coefficients have stiffness matrices fully populated. These are known as far-coupled systems as distinct from the close-coupled systems connected with torsional or longitudinal oscillations. The finite element approach yields stiffness matrices which are banded but not tridiagonal. The labour involved in solving tridiagonal matrices is much less than the one connected with banded matrices. One of the methods of solving eigenvalue problems is to first reduce the semibandwidth (without changing the eigenvalue) to one or two. When the semiband-width is one, the matrix is diagonal, when it is two, it is tridiagonal. These are accomplished by a series of orthogonal transformations using reflection or rotation matrices. We will now discuss three different transformations; Jacobi's which reduces the matrix to diagonal form, Given's and Householder's which reduce the matrix to the tridiagonal form.

3.5.1 Jacobi Transformation

This method uses the rotation matrix defined in Sec. 3.1.9 to transform the given square matrix [A] which is real and symmetric to diagonal form. The final diagonal matrix gives the eigenvalues of the given problem. The disadvantage of this approach is that this procedure is iterative and hence time consuming for large size problems. The orthogonal matrix [P] used has the following form:

$$
[P] = \begin{bmatrix} 1 & & 0 & & 0 \\ & 1 & & & \\ & & \cos\theta_k & -\sin\theta_k & \\ & & +\sin\theta_k & \cos\theta_k & \\ & & 0 & 0 & 1 \end{bmatrix} \qquad \overset{\displaystyle i\text{th} \qquad\quad j\text{th}}{} \tag{3.32}
$$

Angle θ_k chosen for the kth iteration is given by

$$
\tan 2\theta_k = \frac{2a_{ij}^{(k-1)}}{a_{ii}^{(k-1)} - a_{jj}^{(k-1)}} \tag{3.33}
$$

The original matrix $[A]$ get modified during the kth iteration when premultiplied by $[P]^T$ and post multiplied by $[P]$ as follows:

$$a_{ii}^k = a_{ii}^{k-1} \cos^2 \theta_k + 2a_{ij}^{k-1} \sin \theta_k \cos \theta_k + a_{jj}^{k-1} \sin^2 \theta_k$$

$$a_{jj}^k = a_{ii}^{k-1} \sin^2 \theta_k - 2a_{ij}^{k-1} \sin \theta_k \cos \theta_k + a_{jj}^{k-1} \cos^2 \theta_k$$

$$a_{ij}^k = - (a_{ii}^{k-1} - a_{jj}^{k-1}) \sin \theta_k \cos \theta_k + a_{ij}^{k-1} (\cos^2 \theta_k - \sin^2 \theta_k) \quad (3.34)$$

$$a_{mj}^k = - a_{mi}^{k-1} \sin \theta_k + a_{mj}^{k-1} \cos \theta_k \qquad m \neq i, j \qquad (3.35)$$

$$a_{mi}^k = a_{mi}^{k-1} \cos \theta_k + a_{mj}^{k-1} \sin \theta_k$$

$$a_{in}^k = a_{in}^{k-1} \cos \theta_k + a_{jn}^{k-1} \sin \theta_k$$

$$n \neq i, \ j$$

$$a_{jn}^k = - a_{in}^{k-1} \sin \theta_k + a_{jn}^{k-1} \cos \theta_k$$
$$a_{mn}^k = a_{mn}^{k-1} \qquad m, n \neq i, \ j \qquad (3.36)$$

The straightforward procedure for sequence of rotation is to choose (1, 2), then (1, 3), (1, 4) ... (1, n), then (2, 3), (2, 4) ... This procedure needs $2n^3$ multiplications for one complete cycle [3, 3]. A figure of $10n^3$ for all the off-diagonal elements to become negligibly small will be in order. Naturally, this is unsuitable for large-size problems.

Example 3.16

Eliminate (1, 2) and (2, 1) of matrix A given by

$$\begin{bmatrix} 4 & 3 & 2 & 1 \\ 3 & 8 & 6 & 2 \\ 2 & 6 & 8 & 3 \\ 1 & 2 & 3 & 4 \end{bmatrix}$$

without altering its eigenvalues.

$$a_{11} = 4 \qquad a_{12} = 3 \qquad a_{22} = 8$$

$$\tan 2\theta = \frac{(2 \times 3)}{(4 - 8)} = - \frac{6}{4} \qquad 2\theta = - 56.31° \qquad \theta = - 28.15°$$

$$\sin \theta = - 0.4717 \qquad \cos \theta = 0.8817$$

$$\therefore \ a_{11}^2 = a_{11} \cos^2 \theta + 2a_{12} \sin \theta \cos \theta + a_{22} \sin^2 \theta$$

$$= 3.11 - 2.5 + 1.78 = 2.39$$

$$a_{12}^2 = - (4.8) \sin \theta \cos \theta + 3 (\cos^2 \theta - \sin^2 \theta)$$

$$= - 1.66 + 1.66 = 0$$

$a_{22}^2 = (4 \times 0.2222) - 6 \times 0.777$

$\quad = 0.8888 + 2.50416 \times 6.216 = 9.600$

$a_{13}^2 = a_{13} \cos \theta + a_{23} \sin \theta = (2 \times 0.881) + (- 0.472) = - 1.072$

$a_{14}^2 = a_{14} \cos \theta + a_{24} \sin \theta = 0.882 - 2 \times 0.472 = 0.062$

$a_{23}^2 = - a_{13} \sin \theta + a_{23} \cos \theta = (2 \times 0.472) + 6 \times 0.882 = 6.266$

$a_{24}^2 = - a_{14} \sin \theta + a_{24} \cos \theta = + 0.472 - 2 \times 0.882 = - 1.288$

$$\therefore \quad \text{Modified } [A] = \begin{bmatrix} 2.39 & 0 & -1.072 & -0.062 \\ 0 & 9.600 & 6.266 & -1.288 \\ -1.072 & 6.266 & 8 & 3 \\ -0.062 & -1.288 & 3 & 4 \end{bmatrix}$$

3.5.2 Givens' Transformation

Jacobi's method is iterative, but finally results in a diagonal matrix. Givens' method produces a tridiagonal matrix. This procedure is not iterative, but produces it in a finite number of steps. The procedure again uses rotation matrix for orthogonal transformation summarised as below

To annihilate, $\qquad a_{iq}^k, \qquad i \neq p, q$ $\qquad\qquad$ (3.37)

$$a_{iq}^k = - a_{ip}^{k-1} \sin \theta_k + a_{iq}^{k-1} \cos \theta_k \qquad i \neq p, q$$

from Eq. (3.35)

$$\sin \theta_k = \frac{a_{iq}^{k-1}}{r_k} \qquad \cos \theta_k = \frac{a_{ip}^{k-1}}{r_k} \qquad\qquad (3.38)$$

where $\qquad\qquad r_k = [(a_{ip}^{k-1})^2 + (a_{iq}^{k-1})^2]^{1/2} \qquad\qquad (3.39)$

Example 3.17

Eliminate the first row last two columns and first column last two rows of the given matrix [A]

$$[A] = \begin{bmatrix} 4 & 3 & 2 & 1 \\ 3 & 8 & 6 & 2 \\ 2 & 6 & 8 & 3 \\ 1 & 2 & 3 & 4 \end{bmatrix}$$

I Stage

Let $\qquad i = 1, \quad p = 2, \quad q = 4$

Rotation is in 2, 4 plane.

$$\sin \theta = \frac{a_{14}}{r} \qquad \cos \theta = \frac{a_{12}}{r}$$

$$r^2 = a_{12}^2 + a_{14}^2 = 3^2 + 1^2 = 10$$

$\therefore \qquad \sin \theta = 1/\sqrt{10}, \qquad \cos \theta = 3/\sqrt{10}$

$\therefore$

$$[P_1] = \begin{bmatrix} 1 & 0 & 0 & 0 \\ 0 & 3/\sqrt{10} & 0 & -1/\sqrt{10} \\ 0 & 0 & 1 & 0 \\ 0 & 1/\sqrt{10} & 0 & 3/\sqrt{10} \end{bmatrix}$$

$$[A_2] = [P_1]^T [A] [P_1] = \begin{bmatrix} 4 & 3.16 & 2 & 0 \\ 3.16 & 8.8 & 6.64 & 0.4 \\ 2 & 6.64 & 8 & 0.945 \\ 0 & 0.4 & 0.945 & 3.2 \end{bmatrix}$$

II Stage

Let us eliminate (1, 3) and (3, 1)

$$i = 1, \quad p = 2, \quad q = 3$$

We have

$$r^2 = a_{12}^2 + a_{13}^2 = 10 + 4 = 14$$

$$\sin \theta = \frac{2}{\sqrt{14}} \qquad \cos \theta = \frac{\sqrt{10}}{\sqrt{14}}$$

Hence

$$[P_2] = \begin{bmatrix} 1 & 0 & 0 & 0 \\ 0 & 0.845 & -0.5345 & 0 \\ 0 & +0.5345 & 0.845 & 0 \\ 0 & 0 & 0 & 1 \end{bmatrix}$$

Hence

$$[A_3] = [P_2]^T [A_2] [P_2] = \begin{bmatrix} 4.00 & 3.74 & 0 & 0 \\ 3.74 & 14.56 & -4.45 & 0.843 \\ 0 & -4.45 & 3.30 & 0.586 \\ 0 & 0.843 & 0.586 & 3.20 \end{bmatrix}$$

The total number of multiplications required in Givens' method to reduce to a tridiagonal form is $4n^3/3$. Eigenvalues of the tridiagonal matrix can be found by Sturm sequence.

The original matrix $[A_1]$ and final matrix $[A_{n+1}]$ have the same eigenvalues since the original matrix is subjected to a series of orthogonal transformations. $[A_{n+1}]$ can be expressed as

$$[A_{n+1}] = [P_n^T P_{n-1}^T \dots P_1^T][A_1][P_1 \dots P_n] \tag{3.40}$$

$$[A_{n+1}] = [P]^T[A_1][P] \tag{3.41}$$

where
$$[P] = [P_1 P_2 \dots P_n] \tag{3.42}$$

Let the original eigenvalue problem be

$$[A_1]\{u\} = \lambda\{u\} \tag{3.43}$$

if
$$[P]\{v\} = \{u\}$$

then
$$\{v\} = [P]^T\{u\}$$

since $[P]$ is orthogonal.

$\therefore$
$$[A_1]\{u\} = [A_1][P]\{v\} = \lambda\{u\}$$

Premultiplying by $[P]^T$, we have

$$[P]^T[A_1][P]\{v\} = \lambda[P]^T\{u\}$$

or

$$[A_n]\{v\} = \lambda\{v\} \tag{3.44}$$

Eigenvector $\{v\}$ and $\{u\}$ are related by the equation

$$[P]\{v\} = \{u\} \tag{3.45}$$

Having got $\{v\}$ for the tridiagonal matrix, one can obtain $\{u\}$ for the original matrix by using Eq. (3.45).

3.5.3 Householder's Transformation

Householder's method reduces the real symmetric matrix to tridiagonal form by using reflection matrices (Sec. 3.1.10). It is much more efficient than Given's method since it needs only one-half [3.6] of the multiplications as of Givens. There are totally $(n - 2)$ transformations needed for tridiagonalisation. The reflection matrix (orthogonal) is of the form

$$[P] = [I] - 2\{w\}\{w\}^T \tag{3.46}$$

where $\{w\}$ is column vector $(n \times 1)$ whose Euclidean norm is unity.

That is,

$$\{w\}^T\{w\} = 1 \qquad (3.47)$$

It is easy to prove that

$$[P]^T[P] = [I]$$

Let the given problem be

$$[A]\{u\} = \lambda\{u\} \qquad (3.48)$$

After the first stage of operation

$$[A_2] = [P_1]^T[A][P_1] \qquad (3.49)$$

If the vector

$$\{w\} = \begin{Bmatrix} 0 \\ w_2 \\ w_3 \\ \dots \\ w_n \end{Bmatrix} \qquad (3.50)$$

Then

$$[P_1] = \begin{bmatrix} 1 & & & \\ & 1 - 2w_2^2 & -2w_2w_3 & \dots \\ & -2w_2w_3 & & \\ & \vdots & & \\ & -2w_2w_n & & 1 - 2w_n^2 \end{bmatrix} \qquad (3.51)$$

The off-diagonal elements of the first row of $[A_2]$ except for the superdiagonal element a_{12}^2 must be zero. Comparing the coefficients

$$a_{12}^2 = a_{12} - 2w_2c = r$$
$$a_{13}^2 = a_{13} - 2w_3c = 0 \qquad (3.52)$$
$$\dots \quad \dots \quad \dots$$
$$a_{1n}^2 = a_{1n} - 2w_nc = 0$$

where

$$c = a_{12}w_2 + a_{13}w_3 + \dots + a_{1n}w_n \qquad (3.53)$$

From Eq. (3.47), we have

$$w_2^2 + w_3^2 + \dots + w_n^2 = 1 \qquad (3.54)$$

By squaring Eq. (3.52) and adding

$$r^2 = a_{12}^2 + a_{13}^2 + \ldots a_{1n}^2 \qquad (3.55)$$

By multiplying Eq. (3.43) by a_{12}, a_{13} respectively and adding

$$2c^2 = r^2 - a_{12}r \qquad (3.56)$$

For any row k,

$$r = -(\text{sign } a_{k,k+1})\sqrt{a_{k,k+1}^2 + a_{k,k+2}^2 + \ldots + a_{k,n}^2}$$

$$2c^2 = r^2 - ra_{k,k+1}$$

$$\text{vector } (v) = (0, \ldots 0, a_{k,k+1} - r, a_{k,k+2}, \ldots a_{k,k+n})^T \qquad (3.57)$$

Then

$$[P] = [I] - \frac{1}{2c^2}\{v\}\{v\}^T \qquad (3.58)$$

Addition of sign of $a_{k,k+1}$ in r enables us to get larger of the two values for $(a_{k,k+1} - r)$. This in turn gives rise to $\{v\}^T\{v\}$ which will avoid division by a small quantity in Eq. (3.58).

Equation (3.58) can be verified easily. Since

$$\{v\}^T\{v\} = (a_{12} - r)^2 + a_{13}^2 + \ldots a_{1n}^2 \qquad \text{when } k = 1$$

$$= a_{12}^2 + \ldots a_{1n}^2 - 2a_{12}r + r^2$$

$$= 2(r^2 - a_{12}r) \text{ from Eq. (3.55)}$$

$$= 4c^2 \text{ from Eq. (3.56)}$$

Therefore, Eq. (3.58) can be written as

$$[P] = [I] - \frac{2}{\{v\}^T\{v\}} \cdot \{v\}\{v\}^T \qquad (3.59)$$

This is the reflection matrix given in Sec. 3.1.10.

As in Givens' method there is no loss of accuracy in this method. The method is not iterative. Wilkinson [3.6] has shown that the tridiagonal matrices obtained by both the methods are the same but for change of signs in some rows and columns.

Total number of multiplications involved is $2/3n^3 + 3/2n^2$. Unfortunately even for a banded matrix, zero elements will no more be zero after the first stage of modification and hence sparseness of the matrix offers no special advantage.

Example 3.17
Tridiagonalise the given matrix [A] by Householder's method.

$$[A] = \begin{bmatrix} 4 & 3 & 2 & 1 \\ 3 & 8 & 6 & 2 \\ 2 & 6 & 8 & 3 \\ 1 & 2 & 3 & 4 \end{bmatrix}$$

I Stage

Eliminating a_{13}, a_{14}, a_{31}, a_{41}

$$\{\bar{v}_1\} = \begin{bmatrix} 3 \\ 2 \\ 1 \end{bmatrix} + (\text{sign } a_{12})\sqrt{a_{12}^2 + a_{13}^2 + a_{14}^2}\begin{bmatrix} 1 \\ 0 \\ 0 \end{bmatrix}$$

$$= \begin{bmatrix} 3 \\ 2 \\ 1 \end{bmatrix} + \sqrt{3^2 + 2^2 + 1^2}\begin{bmatrix} 1 \\ 0 \\ 0 \end{bmatrix} = \begin{bmatrix} 6.7416 \\ 2 \\ 1 \end{bmatrix}$$

$$\therefore \quad \{v_1\} = \begin{bmatrix} 0 \\ 6.7416 \\ 2 \\ 1 \end{bmatrix} \qquad \theta_1 = \frac{2}{\{v_1\}^T \{v_1\}} = 0.03976$$

$$\therefore \quad [P_1] = \begin{bmatrix} 1 & 0 & 0 & 0 \\ 0 & 1 & 0 & 0 \\ 0 & 0 & 1 & 0 \\ 0 & 0 & 0 & 0 \end{bmatrix} - 0.03976$$

$$\begin{bmatrix} 0 \\ 6.7416 \\ 2 \\ 1 \end{bmatrix} \begin{bmatrix} 0 & 6.7416 & 2 & 1 \end{bmatrix}$$

$$= \begin{bmatrix} 1 & 0 & 0 & 0 \\ 0 & -0.808 & -0.579 & -0.270 \\ 0 & -0.579 & 0.840 & -0.086 \\ 0 & -0.270 & -0.086 & 0.960 \end{bmatrix}$$

and hence

$$[A_2] = [P_1]^T [A] [P_1] = \begin{bmatrix} 4 & 0.374 & 0.00 & 0.00 \\ -3.74 & 14.57 & -4.43 & -0.843 \\ 0 & -4.43 & 3.30 & 0.590 \\ 0 & -0.843 & 0.590 & 3.18 \end{bmatrix}$$

But for the change of sign in the same rows it is essentially the same as the Givens' transformation worked out earlier.

II Stage

Elimination of a_{24}, a_{42},

$$(\bar{v}_2) = \begin{bmatrix} -4.43 \\ 0.843 \end{bmatrix} - \sqrt{4.43^2 + 0.843^2} \begin{bmatrix} 1 \\ 0 \end{bmatrix} = \begin{bmatrix} -8.93 \\ -0.843 \end{bmatrix}$$

$$\therefore \quad (v_2) = \begin{bmatrix} 0 \\ 0 \\ -8.93 \\ -0.843 \end{bmatrix} \quad \frac{2}{\{v_2\}^T \{v_2\}} = \theta_2 = 2.5 \times 10^{-2}$$

$$\therefore \quad [P_2] = [I] - \theta_2 \{v_2\}\{v_2\}^T = \begin{bmatrix} 1 & 0 & 0 & 0 \\ 0 & 1 & 0 & 0 \\ 0 & 0 & -0.999 & -0.018 \\ 0 & 0 & -0.015 & +0.999 \end{bmatrix}$$

$$\therefore \quad [A_3] = [P_2]^T [A_2] [P_2] \begin{bmatrix} 4.000 & -3.740 & 0 & 0 \\ -3.740 & 14.57 & -4.42 & 0 \\ 0 & -4.42 & 3.25 & 0.590 \\ 0 & 0 & 0.590 & 3.14 \end{bmatrix}$$

➤ 3.6 FORWARD AND INVERSE ITERATION SCHEMES

3.6.1 Iteration Concept

Let $$[A]\{u\} = \lambda\{u\} \tag{3.60}$$

be the given eigenvalue problem of the symmetric matrix $[A]$. Let its eigenvalues and associated eigenvectors be given by $\lambda_1, \lambda_2 ...$, etc. and $\{u_1\}, \{u_2\}, ...$, etc. respectively.

Let $$\lambda_1 > \lambda_2 > \lambda_3 > \lambda_4 > \ldots > \lambda_n$$

Let us choose a trial vector $\{u^0\}$.

Let $$\{u^0\} = c_1\{u_1\} + c_2(u_2) + \ldots c_n(u_n) \tag{3.61}$$

(c_1, c_2, ... are constants.)

Premultiplying Eq. (3.61) by $[A]$ and designating $[A]\{u^0\}$ as $\{u^1\}$, we have

$$\{u^1\} = [A]\{u^0\} = c_1[A]\{u_1\} + c_2[A]\{u_2\} + \ldots c_n[A]\{u_n\}$$

$$= c_1\lambda_1\{u_1\} + c_2\lambda_2\{u_2\} + \ldots + c_n\lambda_n\{u_n\} \tag{3.62}$$

If this premultiplication is done m times, we have

$$\{u^m\} = c_1\lambda_1^{\,m}\{u_1\} + c_2\lambda_2^{\,m}\{u_2\} + \ldots + c_n\lambda_n^{\,m}\{u_n\}$$

Since $$\lambda_1 > \lambda_2 > \lambda_3 \ldots, \quad \text{we have}$$

$$\lambda_1^{\,m} > \lambda_2^{\,m} > \lambda_3^{\,m} \ldots$$

Hence $$\{u^m\} \approx c_1\lambda_1^{\,m}\{u_1\}$$

Since the eigenvectors can be arbitrarily scaled, we can divide the premultiplied vector every time by $[\{u\}^T\{u\}]^{1/2}$ (normalising). When $\{u_1\}$ ultimately becomes the eigenvector, $\{u_1\}^T\{u_1\}$ will become unity.

Now the steps can be modified and written as

$$\{v\}^k = [A]\{u^k\}$$

$$\{u^{k+1}\} = \frac{\{v^k\}}{\alpha} \tag{3.63}$$

where $$\alpha^2 = \{v^k\}^T\{v^k\}$$

(Normalising $\{v^k\}$.)

At every stage the following quotient can be computed.

$$\frac{\{u^k\}^T[A]\{u^k\}}{\{u^k\}^T\{u^k\}} \tag{3.64}$$

This is known as Rayleigh's quotient. This ends to λ_1 when $\{u^k\}$ has fully converged to $\{u_1\}$. The numerator will then tend to λ_1 and the denominator to 1.

In case λ_1 and λ_2 are very close to each other ($\lambda_1 > \lambda_2$), then

$$\{u^k\} = c_1\{u_1\} + c_2\{u_2\} \tag{3.65}$$

Then $$\{u^k\}^T\{u^k\} = c_1^2 + c_2^2 \tag{3.66}$$

since $\{u_1\}^T = \{u_1\} = 1$, $\{u_2\}^T\{u_2\} = 1$, $\{u_2\}^T\{u_1\} = \{u_1\}^T\{u\}_2 = 0$

$$\{u^k\}^T[A]\{u^k\} = c_1^2\lambda_1 + c_2^2\lambda_2 \tag{3.67}$$

Hence Rayleigh quotient $= \dfrac{\{u^k\}^T [A]\{u^k\}}{\{u^k\}^T \ \{u^k\}} = \dfrac{c_1^2 \lambda_1 + c_2^2 \lambda_2}{c_1^2 + c_2^2}$

$$= \lambda_1 \dfrac{c_1^2 (1 + (c_2^2/c_1^2)(\lambda_2/\lambda_1))}{c_1^2 (1 + (c_2^2/c_1^2))}$$

$$= \lambda_1 \left[\left(1 + \dfrac{c_2^2}{c_1^2} \dfrac{\lambda_2}{\lambda_1}\right)\left(1 + \dfrac{c_2^2}{c_1^2}\right)^{-1} \right]$$

If $\qquad\qquad\qquad \dfrac{c_2}{c_1} \ll 1$, then

$$\text{Rayleigh quotient} = \lambda_1 \left[1 - \dfrac{c_2^2}{c_1^2} \dfrac{(\lambda_1 - \lambda_2)}{\lambda_1} \right] \qquad (3.68)$$

The error in Rayleigh's quotient is of the order of c_2^2/c_1^2. It may also be observed that Rayleigh's quotient underestimates the eigenvalue [3.1].

Example 3.18
Compute the largest natural frequency of the system in Example 3.14.

$$\begin{bmatrix} 15 & -10 & & \\ -10 & 25 & -15 & \\ & -15 & 35 & -20 \\ & & -20 & 20 \end{bmatrix}\{u\} = \omega^2 \begin{bmatrix} 10 & & & \\ & 20 & & \\ & & 30 & \\ & & & 40 \end{bmatrix}\{u\}$$

Approach I

Rewriting the problem as

$$[A]\{u\} = \omega^2 \{u\}$$

we have $\qquad\qquad\qquad [A] = [M]^{-1}[K]$

$$\begin{bmatrix} 1.5 & 1.0 & & \\ -0.5 & 1.25 & -0.75 & \\ & -0.5 & 1.166 & -0.660 \\ & & -0.5 & 0.5 \end{bmatrix}\{u\} = \omega^2 \{u\}$$

or $\qquad\qquad\qquad [A]\{u\} = \omega^2 \{u\}$

Iteration 1

Let
$$\{u^1\} = \begin{Bmatrix} 1 \\ 0 \\ 0 \\ 0 \end{Bmatrix}$$

$$\{v^1\} = [A]\{u^1\} = \begin{Bmatrix} 1.5 \\ -0.5 \\ 0 \\ 0 \end{Bmatrix}$$

Normalising dividing $\{v^1\}$ by $\sqrt{(1.5)^2 + (0.5)^2}$

$$\{u^2\} = \begin{Bmatrix} 0.9045 \\ -0.3015 \\ 0 \\ 0 \end{Bmatrix}$$

Iteration 2

$$\{v^2\} = [A]\{u^2\}$$

$$\therefore \quad \{v^2\} = \begin{Bmatrix} 1.65825 \\ -0.82600 \\ 0.4500 \\ 0 \end{Bmatrix}$$

Normalising (dividing $\{v^2\}$ by $\sqrt{(1.65)^2 + (0.83)^2 + (0.45)^2}$)

$$\{u^3\} = \begin{Bmatrix} 0.8910 \\ -0.4460 \\ 0.0810 \\ 0 \end{Bmatrix}$$

Iteration 3

$$\{v^3\} = [A]\{u^3\} = \begin{Bmatrix} 1.7700 \\ -1.00000 \\ 0.31700 \\ -0.0400 \end{Bmatrix}$$

Normalising

$$\{u^4\} = \begin{Bmatrix} 0.7435 \\ -0.4436 \\ 0.1324 \\ -0.0169 \end{Bmatrix}$$

Iteration 4

$$\{v^4\} = [A]\{u^4\} = \begin{Bmatrix} 1.558 \\ -1.025 \\ 0.3875 \\ -0.0765 \end{Bmatrix}$$

Normalising

$$\{u^5\} = \begin{Bmatrix} 0.6967 \\ -0.4583 \\ 0.1732 \\ -0.0333 \end{Bmatrix}$$

Rayleigh's quotient $= \dfrac{\{u^4\}^T [A]\{u^4\}}{[u^4]^T \{u^4\}} = 2.237$

The exact value is 2.28 which can be obtained after some more iterations.

General Remarks

The matrix $[A]$ involved in the foregoing calculation is real, but unsymmetric.

$$[A] = \begin{bmatrix} 1.5 & -1.0 & 0 & 0 \\ -0.5 & 1.25 & -0.75 & 0 \\ 0 & -0.5 & 1.166 & -0.666 \\ 0 & 0 & -0.5 & +0.5 \end{bmatrix}$$

$\{u\}^T\{u\} = 1$ is valid only for real symmetric matrices. Hence the eigenvector obtained is not the correct solution. Let us now redo the same problem by modifying $[A]$ to the symmetric form.

Approach II

The given problem is

$$[K]\{u\} = \omega^2[M]\{u\} \tag{3.69}$$

where

$$[K] = \begin{bmatrix} 15 & -10 & & \\ -10 & 25 & -15 & \\ & -15 & 35 & -20 \\ & & -20 & 20 \end{bmatrix}$$

$$[M] = \begin{bmatrix} 10 & 0 & 0 & 0 \\ 0 & 20 & 0 & 0 \\ 0 & 0 & 30 & 0 \\ 0 & 0 & 0 & 40 \end{bmatrix}$$

Premultiplying both sides of Eq. (3.69) by $[M]^{-1/2}$ we have

$$[M]^{-1/2}[K]\{u\} = \omega^2[M]^{+1/2}\{u\}$$

Let

$$\{u\} = M^{-1/2}(q) \tag{3.70}$$

then

$$[M]^{-1/2}[K][M]^{-1/2}\{q\} = \omega^2\{q\} \tag{3.71}$$

If

$$[A] = [M]^{-1./2}[K][M]^{-1/2} \tag{3.72}$$

then the resulting standard eigenvalue problem

$$[A]\{q\} = \omega^2\{q\}$$

involves a symmetric matrix. When $[M]$ is diagonal this approach can be used elegantly to modify a non-standard eigenvalue problem to a standard one.

For the given problem.

$$[M]^{-1/2} = \begin{bmatrix} 0.316 & & & \\ & 0.224 & & \\ & & 0.182 & \\ & & & 0.158 \end{bmatrix}$$

Hence using Eq. (3.74)

$$[A] = \begin{bmatrix} 1.50 & -0.707 & & \\ -0.707 & 1.25 & -1.612 & \\ & -1.612 & 1.166 & -0.575 \\ & & -0.575 & 0.500 \end{bmatrix}$$

Let
$$\{q^1\} = \begin{Bmatrix} 1 \\ 0 \\ 0 \\ 0 \end{Bmatrix}$$

$$\{q^2\} = [A]\{q^1\} = \begin{Bmatrix} 1.5 \\ -0.707 \\ 0 \\ 0 \end{Bmatrix}$$

Normalising

$$\{q^2\} = \begin{Bmatrix} 0.907 \\ -0.422 \\ 0 \\ 0 \end{Bmatrix}$$

$$\{q^3\} = [A]\{q^2\} = \begin{Bmatrix} 1.66 \\ -1.178 \\ 0.258 \\ 0 \end{Bmatrix}$$

Normalising

$$\{q^3\} = \begin{Bmatrix} 0.798 \\ -0.566 \\ 0.125 \\ 0.0 \end{Bmatrix}$$

$$\{q^4\} = [A]\{q^3\} = \begin{Bmatrix} 1.592 \\ -1.348 \\ 0.492 \\ 0.072 \end{Bmatrix}$$

Rayleigh's quotient $(\{q^3\}^T[A]\{q^3\})/(\{q^3\}^T\{q^3\}) = \{q^3\}^T\{q^4\} = 2.20$. The exact value is 2.28.

Normalising

$$\{q^4\} = \begin{Bmatrix} 0.742 \\ -0.628 \\ 0.229 \\ -0.034 \end{Bmatrix}$$

Hence

$$\{u\} = [M]^{-1/2}\{q\} = \begin{Bmatrix} 0.234 \\ -0.141 \\ 0.042 \\ -0.0054 \end{Bmatrix}$$

This could be further improved by continuing with a few more iterations. It could be seen from $\{u\}$ that the eigenvector corresponds to the fourth mode.

Approach III

Let us work out the same problem retaining its non-standard form. Let us assume

$$\{u^1\} \text{ as } \begin{Bmatrix} 1 \\ 0 \\ 0 \\ 0 \end{Bmatrix}$$

I cycle
Orthonormalising it with respect to $[M]$

$$[u^1]^T[M]\{u^1\} = [I]$$

Hence

$$\{u^1\} = \begin{Bmatrix} 1/\sqrt{10} \\ 0 \\ 0 \\ 0 \end{Bmatrix}$$

$$[K]\{u^1\} = \begin{Bmatrix} 15/\sqrt{10} \\ 10/\sqrt{10} \\ 0 \\ 0 \end{Bmatrix}$$

Writing $$[K]\{u^1\} = [M]\{v^1\}$$

$\{v^1\}$ can be obtained as

$$\{v^1\} = \begin{Bmatrix} 0.474 \\ -0.158 \\ 0 \\ 0 \end{Bmatrix}$$

II cycle
Orthonormalising $\{v^1\}$ with respect to $[M]$, we have

$$\{u^2\} = \begin{Bmatrix} 0.286 \\ -0.095 \\ 0 \\ 0 \end{Bmatrix}$$

Computing $\{v^2\}$ as before, which is

$$\{v^2\} = [M]^{-1}[K]\{u^2\} = \begin{Bmatrix} 0.524 \\ -0.262 \\ 0.048 \\ 0 \end{Bmatrix}$$

III cycle
Orthonormalising with respect to $[M]$

$$\{u^3\} = \begin{Bmatrix} 0.256 \\ -0.128 \\ 0.023 \\ 0 \end{Bmatrix}$$

Hence $$\{v^3\} = \begin{Bmatrix} 0.512 \\ 0.309 \\ 0.093 \\ -0.012 \end{Bmatrix}$$

IV cycle
Orthonormalising

$$\{u^4\} = \left\{ \begin{array}{c} 0.240 \\ -0.135 \\ 0.047 \\ -0.006 \end{array} \right\}$$

Rayleigh's quotient = $(\{u^4\}^T[K]\{u^4\})/(\{u^4\}^T[M]\{u^4\})$ = 2.23

3.6.2 Forward and Inverse Iteration

Let us consider the eigenvalue problem

$$[K]\{u\} = \omega^2[M]\{u\} \tag{3.73}$$

This can be written in the following two ways:

$$[M]^{-1}[K]\{u\} = \omega^2\{u\} \tag{3.74}$$

or

$$(1/\omega^2)\{u\} = [K]^{-1}[M]\{u\} \tag{3.75}$$

Equation (3.74) can be written as

$$[A]\{u\} = \omega^2\{u\} \tag{3.76}$$

where

$$[A] = [M]^{-1}[K] \tag{3.77}$$

Equation (3.75) can be written as

$$[A]\{u\} = \lambda\{u\} \tag{3.78}$$

where

$$[A] = [K]^{-1}[M] \qquad \lambda = \frac{1}{\omega^2} \tag{3.79}$$

If the iteration concepts are used in Eq. (3.76), ω^2 converges to its largest value. If it is applied in Eq. (3.78), λ converges to its largest value and hence ω^2 to its smallest value. The first one is known as *forward iteration* and the second *inverse iteration*.

The following procedure is used to solve problems by forward iteration.

$$[K]\{u\} = \omega^2[M]\{u\}$$

Assuming $\{u^1\}$ by straightforward multiplication $[K]\{u^1\}$ is obtained. This vector is premultiplied by $[M]^{-1}$ to get $\{v^1\}$. Since $[M]$ is usually a diagonal matrix, $[M]^{-1}$ can be easily computed as explained in Approach III of Example 3.18.

To get $\{u^2\}$, $\{v^1\}$ is normalised by dividing by $\{v^1\}^T\{v^1\}$. This procedure is continued for the next cycle.

Problems to be solved by inverse iteration, follow a different approach. It is this approach which is very widely used. For large size problems it is only the lower frequencies which are of concern. In Eq. (3.81) $[K]^{-1}$ is needed. This is not easy to obtain for large-size problems. This alternate approach is used. Assuming $\{u^1\}$, $[M]\{u^1\}$ is computed. Let it be $\{y^1\}$. Next we have to compute $[K]^{-1}\{y^1\}$. Let it be $\{v^1\}$, or

$$[K]\{v^1\} = \{y^1\} \tag{3.80}$$

Since $\{y^1\}$ is known computing $\{v^1\}$ is to solve simultaneous equations in n unknowns if $[K]$ is of size $(n \times n)$. Since $[K]$ is banded the procedures outlined in Chapter 2 can be used. Either Potters' method or Cholesky's scheme can be used. Having found $\{v^1\}$, after ortho-normalising it, with respect to $[M]$, one can proceed to the second cycle of iteration.

➤ 3.7 GRAM SCHMIDT DEFLATION

So far we have discussed the cases where irrespective of the trial vector chosen, one is in a position to obtain the lowest or the highest natural frequency. In practice we will be interested in the first few frequencies. Logical extension of this iteration technique to other modes is desirable. The only way to achieve it is to eliminate the influence of the first mode from the iteration when one computes the second, to remove the influence of the first and the second, when one computes the third, and so on.

This may be explained by the following analogy. The distance between Madras and Delhi is around 2000 km which can be covered either by air or train or even by a bicycle. A person who wants to do it within the shortest possible time goes by air by paying for the air travel. A person who cannot afford airfare takes a train and covers the distance in a day and a half. On the contrary, a person who has lot of time and who wants to make news chooses to cover the distance by bicycle spending probably two weeks. The preferences for the three modes of travel were ability to pay, economy in travel, and probably, publicity. If, for example, air travel were made free, all the three persons would prefer it. Similarly, in the present context if one chooses three trial vectors, to start with, and goes on iterating without imposing any restrictions, all the three vectors would converge to the first. If, on the contrary, the second one is prevented from converging to the first and the third from converging to the first and the second, obviously we would achieve our aim. This is done by the Gram Schmidt method. Let

$$[K]\{u\} = \omega^2[M]\{u\}$$

be the given problem. Let us assume that the eigenpair λ_1, $\{u_1\}$ is known. Let us choose a trial vector $\{X_1\}$ and subtract from it $\beta_1\{u_1\}$ such that

$$\{\bar{X}_1\} = \{X_1\} - \beta_1\{u_1\} \tag{3.81}$$

where

$$\beta_1 = \{u_1\}^T[M]\{X_1\} \tag{3.82}$$

If $[K]$ and $[M]$ are of size $(n \times n)$, $\{u_1\}$ of size $(n \times 1)$, β_1 of size (1×1) and $\{\bar{X}_1\}$ of $(n \times 1)$. It is easy to prove that $\{\bar{X}_1\}$ is free of the influence of the first mode.

Proof: Let $\{X_1\} = \alpha_1\{u_1\} + \alpha_2\{u_2\} + \dots + \alpha_n\{u_n\}$.
Premultiplying by $\{u_1\}^T[M]$, we have

$$\{u_1\}^T[M]\{X_1\} = \alpha_1\{u_1\}^T[M]\{u_1\} + \alpha_2\{u_1\}^T[M]\{u_2\}$$

$$+ \dots \alpha_n\{u_1\}^T[M]\{u_n\}$$

$$= \alpha_1\{u_1\}^T[M]\{u_1\} = \alpha_1$$

since $\{u_1\}^T[M]\{u_1\} = [I]\{u_1\}^T[M]\{u_1\} = 0 \ (i \neq 1)$

$\therefore \quad \{\bar{X}_1\} = \{X_1\} - \beta_1\{u_1\} = \alpha_1\{u_1\} + \alpha_2\{u_2\} + \dots \alpha_n\{u_n\} - \alpha_1\{u_1\}$

$$= \alpha_2\{u_2\} + \alpha_3\{u_3\} + \dots + \alpha_n\{u_n\} \tag{3.83}$$

Equation (3.83) indicates that $\{\bar{X}_1\}$ is free of the influence of the first. Once a trial vector is chosen, it is modified using Eq. (3.81). It is thus prevented from converging to the first mode. This process of purification can be done at the end of every iteration. When one wants to compute the mth mode, the first $(m - 1)$ modes can be eliminated by modifying the trial vector thus

$$\{\bar{X}_1\} = \{X_1\} - \sum_{i=1}^{m-1} \beta_i\{u_i\} \tag{3.84}$$

where $\beta_i = \{u_i\}^T[M]\{X_1\} \qquad (i = 1, \dots (m-1)) \tag{3.85}$

Example 3.19
Compute the first two natural frequencies of Example 3.14 by inverse iteration in conjunction with the Gram Schmidt deflation technique.

First Frequency

$$\begin{bmatrix} 15 & -10 & 0 & 0 \\ -10 & 25 & -15 & 0 \\ 0 & -15 & 35 & -20 \\ 0 & 0 & -20 & 20 \end{bmatrix} \{u\}$$

$$= \omega^2 \begin{bmatrix} 10 & & & \\ & 20 & & \\ & & 30 & \\ & & & 40 \end{bmatrix} \{u\}$$

Let

$$\{u_1\} = \begin{Bmatrix} \sqrt{0.1} \\ 0 \\ 0 \\ 0 \end{Bmatrix}$$ (The subscript represents the vector and the superscript the iteration cycle)

This leads to

$$\{u_1\}^T [M]\{u_1\} = 1$$

$$[M]\{u_1^1\} = \begin{Bmatrix} 10\sqrt{0.1} \\ 0 \\ 0 \\ 0 \end{Bmatrix}$$

Assuming $[K]\{v^1\} = \begin{Bmatrix} 10\sqrt{0.1} \\ 0 \\ 0 \\ 0 \end{Bmatrix}$

$\{v_1^1\}$ is computed by Potter's method. $\{u_1^2\}$ is obtained by normalising $\{v_1^1\}$ with respect to $[M]$. This is given by

$$\{u_1^2\} = \begin{Bmatrix} 0.1 \\ 0.1 \\ 0.1 \\ 0.1 \end{Bmatrix}$$

For subsequent iterations, proceeding on similar lines

$$\{u_1^3\} = \begin{Bmatrix} 0.0615 \\ 0.0892 \\ 0.1035 \\ 0.1057 \end{Bmatrix}$$

$$\{u_1^4\} = \begin{Bmatrix} 0.06015 \\ 0.08836 \\ 0.1035 \\ 0.1102 \end{Bmatrix}$$

$$\{u_1^5\} = \begin{Bmatrix} 0.06009 \\ 0.08831 \\ 0.1035 \\ 0.1102 \end{Bmatrix}$$

Since $\{u_1^4\}$ and $\{u_1^5\}$ have converged, we have

$$\lambda_1 = \{u_1^4\}^T \{v_1^4\} = 32.89$$

Hence $$\omega_1^2 = 0.0304$$

Second Frequency

Let us assume
$$\{X_1\} = \begin{Bmatrix} \sqrt{0.1} \\ 0 \\ 0 \\ 0 \end{Bmatrix}$$

$$\beta_1 = \{u_1\}^T [M]\{X_1\} = \{0.06, \ 0.088, \ 0.1035, \ 0.1102\} \begin{Bmatrix} 10\sqrt{0.1} \\ 0 \\ 0 \\ 0 \end{Bmatrix}$$

$$= 0.6\sqrt{0.1}$$

$$\therefore \quad \{\bar{X}_1\} = \left\{ \begin{array}{c} \sqrt{0.1} \\ 0 \\ 0 \\ 0 \end{array} \right\} - 0.6\sqrt{0.1} \left\{ \begin{array}{c} 0.06 \\ 0.088 \\ 0.1035 \\ 0.1102 \end{array} \right\}$$

$$= \left\{ \begin{array}{c} 0.964\sqrt{0.1} \\ -0.0528\sqrt{0.1} \\ -0.0618\sqrt{0.1} \\ -0.066\sqrt{0.1} \end{array} \right\} = \{u_2^1\}$$

Proceeding on lines similar to the first mode

$$\{u_2^2\} = \left\{ \begin{array}{c} 2.67 \\ 0.0836 \\ -0.0154 \\ -0.059 \end{array} \right\}$$

$$\{u_2^3\} = \left\{ \begin{array}{c} 0.205 \\ 0.127 \\ 0.0003 \\ -0.079 \end{array} \right\}$$

$$\{u_2^4\} = \left\{ \begin{array}{c} 0.176 \\ 0.138 \\ -0.0101 \\ -0.086 \end{array} \right\}$$

$$\{u_2^5\} = \left\{ \begin{array}{c} 0.165 \\ 0.142 \\ 0.015 \\ -0.089 \end{array} \right\}$$

As before $\lambda_2 = \{u_2^5\}^T \{v_2^5\} = 1.666;$ $\quad \therefore \omega_2^2 = 0.6$
Improving the rate of convergence.

Under Sec. 3.4, we have seen how the Sturm sequence can be used to identify the zones in which the eigen frequencies lie. In

Sec. 3.2.4 we have seen that modifying $[K]$ by $[K - \mu M]$ results in a shift of the natural frequencies by μ. This property can easily be combined with inverse iteration to improve the rate of convergence of the eigenvalues. This can be explained through an example.

Let us assume that approximately the first four natural frequencies for a twisted plate are 5, 21, 22 and 27 units in a nondimensional form obtained by bisection (Section 3.4).

The accuracy of the frequency can be improved by inverse iteration. But improvement inconvergence for the second frequency, 21, will take lot of time since 21 and 22 are close to each other. By using a shift μ of about (20^2) one can make the square of the two natural frequencies modified as (441-400), (484-400) (41 and 84 respectively) which are no more close. This modification results in faster convergence to the second frequency if iteration is used at this juncture. The same procedure can be adopted for the higher order ones.

Summary—Inverse Iteration and Sturm Sequence

Computing the first few natural frequencies and associated vectors by inverse iteration and the Gram Schmidt deflation technique after approximately locating the eigenvalues by Sturm sequence is a very powerful tool. Let us now estimate the multiplications involved in this procedure. Let us assume that on an average six trials of natural frequencies are needed to compute each natural frequency approximately by the Sturm sequence (Sec. 3.4). Let us then use the inverse iteration scheme for computing the eigenpair.

Let us estimate the labour involved in computing $\bar{X}_1$ in Eq. (3.86).

$$\beta_1 = \{u_1\}^T[M]\{X_1\}$$

and
$$\{\bar{X}\} = \{X\} - \beta_1\{u_1\}$$

These two together involve $n + n(2b - 1)$ multiplications where b is the semi-bandwidth of the mass matrix. If the mass matrix is diagonal, it is $2n$. For subsequent β's, it is $4n$, $6n$ and so on. If i vectors are computed, this can be approximately taken as $i\,(i + 1)_n$.

Number of Operations
Sturm sequence:

$$[\bar{K}] = [K - \mu M] \qquad\qquad nb$$

$$[\bar{K}] = [L][D][L]^T \qquad\qquad \frac{1}{2}nb^2 + nb$$

(Triangularisation)
Det $[K - \mu M]$ $\qquad\qquad n$

Inverse Iteration:

$[M]\{u\}$	$n(2b - 1)$
$[K]\{v\} = [M]\{u\}$	$2bn$
Rayleigh quotient	$2n$
(including the	
labour involved in	
normalisation)	
Deflation	$n(2b - 1)(i - 1) + 2n(i - 1)$

Assuming six approximate trials are needed by the Sturm sequence and six iterations subsequently, for p eigenpairs to be computed,
Number of operations involved $= 6pn[b^2/2 + 7b + (p - 1)(2b + 1)$

When $\qquad n = 1000$, $b = 50$, $p = 6$, we have

Number of operations $= 208 \times 8 \times 10^6$
If this is compared with Householder's scheme,
Time involved in tridiagonalisation when mass matrix is diagonal works out to

$$\frac{2}{3}n^3 + \frac{3}{2}n^2 = \frac{2}{3} \times 10^9 + \frac{3}{2} 10^6$$

$$\approx \frac{2}{3} \times 10^9 \text{ operations}$$

which alone is more than three times the inverse iteration approach.

➤ **3.8 SIMULTANEOUS ITERATION METHOD** _____

3.8.1 Standard Eigenvalue Form

The inverse iteration technique in conjunction with the deflation scheme can be used for finding out the first few eigenpairs [3.2]. They are obtained one after the other. But the simultaneous iteration method [3.7] computes them simultaneously thereby substantially improving the computing time.

Let us consider the symmetric eigenvalue problem

$$[K]\{q\} = \omega^2[M]\{q\} \tag{3.86}$$

Since $[K]$ is symmetric it can be expressed as

$$[K] = [L][L]^T \tag{3.87}$$

Let $\qquad\qquad \{q\} = [L]^{-T}\{u\} \tag{3.88}$

Substituting Eqs (3.87) and (3.88) in Eq. (3.86), we have

$$[L][L]^T[L]^{-T}\{u\} = \omega^2[M][L]^{-T}\{u\}$$

or

$$\lambda\{u\} = [L]^{-1}[M][L]^{-T}\{u\}$$

If $\qquad\qquad [A] = [L]^{-1}[M][L]^{-T}, \text{ then} \qquad\qquad (3.89)$

$$[A]\{u\} = \lambda\{u\} \qquad\qquad (3.90)$$

where $\qquad\qquad \lambda = 1/\omega^2 \qquad\qquad (3.91)$

Equation (3.90) is a standard symmetric eigenvalue problem.

If $\{u\}$ is an assumed vector, $(n \times 1)$ the inverse iteration process discussed in Sec. 3.6.2 is continued till the largest λ (lowest ω^2) and its associated eigenvector are obtained. If $\{u\}$ is assumed as a vector of size $(n \times m)$, in order to get m vectors simultaneously the different columns of $\{u\}$ being mutually independent of each other, $\{v\} = [A]\{u\}$ can be obtained. This premultiplication involves backward substitution, straightforward multiplication and forward substitution as discussed below.

$$[A]\{u\} = [L]^{-1}[M][L]^{-T}\{u\}$$

Let $\qquad\qquad [L]^{-T}\{u\} = \{x\}$

Then $\qquad\qquad \{u\} = [L]^{T}\{x\} \qquad\qquad (3.92)$

Since $\{u\}$ and $[L]$ are known, $\{x\}$ is obtained by backward substitution.

$$[M]\{x\} = \{y\} \qquad\qquad (3.93)$$

Since $\{x\}$ is known, computing $\{y\}$ is straight forward multiplication

$$[L]^{-1}\{y\} = \{v\}$$

or $\qquad\qquad \{y\} = [L]\{v\} \qquad\qquad (3.94)$

Since $\{y\}$ is known, computing $\{v\}$ is by forward substitution.

Had the trial vector $\{u\}$ been the correct eigenvector, columns of $\{v\}$ would have been orthogonal. For a general case neither would the columns of $\{v\}$ be orthogonal nor $\{u\}^T\{v\}$ be diagonal, the diagonal elements being $1/\omega^2$. These two conditions would be satisfied when the trial vector $\{u\}$ converges to the correct value. Let us assume that the premultiplied vectors $\{v\}$ of size $(n \times m)$ be multiplied by a vectors of size $(m \times m)$ such that the resultant vector is orthogonal.

Let $\qquad\qquad \{r\} = \{v\}[s] \qquad\qquad (3.95)$

If $\{r\}$ is orthogonal

$$\{r\}^T\{r\} = [I] \qquad\qquad (3.96)$$

or $\qquad\qquad [s]^T\{v\}^T\{v\}[s] = [I] \qquad\qquad (3.97)$

If
$$[v]^T[v] = [L]_0 \cdot [L_0]^T = [D] \qquad (3.98)$$

it is easy to prove that
$$[s] = [L_0]^{-T} \qquad (3.99)$$

If
$$[s] = [L_0]^{-T},$$

then
$$[s]^{-1} = [L_0]^T$$

or
$$[s]^{-T} = [L_0]$$

The coefficients of the matrix $[s]$ can also be directly obtained. [3.6] Substituting in Eq. (3.97), we find

$$[L_0]^{-1}[L_0][L_0]^T[L_0]^{-T} = [I]$$

The new vector $\{r\}$ is used as the new trial vector $\{u\}$. The process of iteration is continued till

$$[B] = \{u\}^T\{v\} = [\Lambda] \qquad (3.100)$$

converges to a diagonal form. The diagonal elements represent largest eigenvalues (λ) and consequently the lowest natural frequencies. Once $\{u\}$ has been obtained the original eigenvector $\{q\}$ can be obtained from Eq. (3.88).

3.8.2 Non-Standard Eigenvalue Form

Another approach to solve the problem is to consider the equation

$$[K]\{u\} = \omega^2[M]\{u\} \text{ from Eq. (3.86)}$$

and solving it by retaining the non-standard form. Let us assume a trial vector $\{u^1\}$ of size ($n \times m$) whose columns are mutually independent. Once can compute $[M]\{u^1\}$ which is straight forward multiplication. It is better to orthonormalise the columns of $\{u'\}$ with respect to $[M]$.

Let
$$[M]\{u^1\} = \{x\} \qquad (3.101)$$

Then one can estimate $\{v\}$ by solving the simultaneous equations

$$[K]\{v\} = \{x\} \qquad (3.102)$$

This can be done by Gaussian elimination, Cholesky's factorisation or by Potters' method (if the core is to be reduced). If $\{u^2\}$ is the desired eigenvector,

then
$$[B] = \{u^2\}^T[K]\{u^2\} = [\Omega^2] \qquad (3.103)$$

If $\{u^2\}$ is the trial vector at the second stage, we know $\{u^2\}$ is to be orthonormalised with respect to $[M]$.

$$\{u^2\}^T[M]\{u^2\} = [I] \tag{3.104}$$

Let $\{v^1\}$ be modified such that

$$\{v^1\}[s] = \{r\} \tag{3.105}$$

$$\times m)(m \times m)(n \times m)$$

Let $\{r\}$ replace $\{u^2\}$. Then substituting in Eq. (3.104)

$$[s]^T\{v^1\}^T[M]\{v^1\}[s] = [I]$$

Let $$\{v^1\}^T[M]\{v^1\} = [L_0][L_0]^T \tag{3.106}$$

then $$[s] = [L_0]^{-T} \tag{3.107}$$

The iteration is terminated when matrix $[B]$ in Eq. (3.103) fully converges.

Example 3.20
Compute the first three natural frequencies of Example 3.14 by the simultaneous iteration method using the standard eigenvalue form.

$$[K] = \begin{bmatrix} 15 & -10 & & \\ -10 & 25 & -15 & \\ & -15 & 35 & -20 \\ & & -20 & +20 \end{bmatrix} = [L][L]^T$$

Hence $$[L] = \begin{bmatrix} 3.873 & & & 0 \\ -2.581 & 4.28 & & \\ 0 & -3.5 & 4.768 & \\ 0 & 0 & -4.195 & 1.55 \end{bmatrix}$$

Let $$\{u^1\} = \begin{Bmatrix} 1 & 0 & 0 \\ 0 & 1 & 0 \\ 0 & 0 & 1 \\ 0 & 0 & 0 \end{Bmatrix}$$

Since $[L]^T\{x\} = \{u_1\}$, we have

$$\{x\} = \begin{Bmatrix} 0.2582 & 0.1557 & 0.1144 \\ 0 & 0.2350 & 0.1755 \\ 0 & 0 & 0.2097 \\ 0 & 0 & 0 \end{Bmatrix}$$

Since $[M]\{x\} = \{y\}$

$$\{y\} = \begin{bmatrix} 10 & & & \\ & 20 & & \\ & & 30 & \\ & & & 40 \end{bmatrix} \{x\} = \begin{bmatrix} 2.582 & 1.557 & 1.144 \\ 0 & 4.671 & 3.431 \\ 0 & 0 & 6.290 \\ 0 & 0 & 0 \end{bmatrix}$$

Since $\{y\} = [L]\{v^1\}$

$$\{v^1\} = \begin{Bmatrix} 0.667 & 0.402 & 0.2953 \\ 0.402 & 1.333 & 0.9797 \\ 0.2975 & 9.797 & 2.039 \\ 0.80 & 2.653 & 5.524 \end{Bmatrix}$$

$$\{v^1\}^T\{v^1\} = [D] = \begin{Bmatrix} 1.33 & 3.216 & 5.60 \\ 3.216 & 9.937 & 18.08 \\ 5.60 & 18.08 & 35.72 \end{Bmatrix}$$

Hence the second trial vector

$$\{u^2\} = \{v^1\}[s] = \{v^1\}[L_0]^{-T}$$

where

$$[L_0][L_0]^T = [D]$$

Hence

$$\{u^2\} = \begin{Bmatrix} 0.5776 & -0.8182 & 0.500 \\ 0.3481 & 0.2459 & -0.9012 \\ 0.2559 & 0.1807 & 0.1448 \\ 0.6925 & 0.489 & 0.3935 \end{Bmatrix}$$

Proceeding on similar lines

$$\{v^2\} = \begin{Bmatrix} 1.155 & 0.2414 & 0.6482 \\ 2.787 & 0.9711 & 1.264 \\ 4.861 & 1.629 & 2.526 \\ 2.705 & 7.572 & 13.4 \end{Bmatrix}$$

$$\{v^2\}^T\{v^2\} = \begin{Bmatrix} 643.55 & 198.05 & 347.59 \\ 197.972 & 60.99 & 106.97 \\ 347.593 & 106.97 & 187.9 \end{Bmatrix}$$

$$\{u^3\} = \begin{Bmatrix} 0.02779 & -0.076 & 0.149 \\ 0.08944 & 0.6135 & 0.117 \\ 0.17925 & 0.685 & 0.1505 \\ 0.9793 & -0.493 & 0.2390 \end{Bmatrix}$$

Similarly

$$\{u^4\} = \begin{Bmatrix} 0.026 & 0.175 & 0.236 \\ 0.087 & 0.678 & 0.628 \\ 0.178 & 0.691 & 0.981 \\ 0.979 & -0.191 & 0.064 \end{Bmatrix}$$

$$\{u^5\} = \begin{Bmatrix} 0.027 & 0.266 & 0.378 \\ 0.088 & 0.691 & 0.591 \\ 0.178 & 0.646 & -0.677 \\ 0.979 & -0.187 & 0.243 \end{Bmatrix}$$

$$\{u^6\} = \begin{bmatrix} 0.027 & 0.298 & 0.448 \\ 0.088 & 0.699 & 0.490 \\ 0.178 & 0.623 & -0.744 \\ 0.979 & -0.184 & 0.078 \end{bmatrix}$$

$$\{v^6\} = \begin{bmatrix} 0.89 & 0.517 & 0.337 \\ 2.905 & 1.179 & 0.313 \\ 5.87 & 1.031 & -0.485 \\ 32.20 & -0.286 & 0.194 \end{bmatrix}$$

$$\therefore \quad [B] = \{u^6\}^T [A] \{u^6\} = \begin{bmatrix} 32.57 & & \text{Small} \\ & & \text{values} \\ & 1.670 & \\ \text{Small} & & 0.653 \\ \text{values} & & \end{bmatrix}$$

Hence

$$\omega_1^2 = \frac{1}{32.57} = 0.0307$$

$$\omega_2^2 = \frac{1}{1.67} = 0.598$$

$$\omega_3^2 = \frac{1}{0.653} = 1.531$$

General remarks

The total number of multiplications needed considering the Eqs (3.87) to (3.99) (if the number of degrees of freedom is n, bandwidth b, trial vectors m) is given

$$= 3nbm + 2\frac{1}{2}nm^2 + \frac{m^3}{6}$$

for every iteration cycle, excluding Cholesky factorisation ($= nb^2/2$) which is needed only once.

It has been found from practice, that if p eigenvectors are needed, it is advisable to choose m trial vectors where m is $\geq (p + 4)$. Convergence of the p vectors is obtained within m cycles, irrespective of the size of n.

Jennings [3.2] has compared the performance of the simultaneous iteration method with Householder's scheme and Sturm sequence. He reports that for $n = 1000$, $b = 40$, for 10 trial vectors, 740×10^6 multiplications are needed when House-holder's scheme is used, 280×10^6 multiplications for Gupta's Sturm sequence [3.5] and 9.5×10^6 multiplication for the simultaneous iteration approach.

It may be observed that when the bandwidth exceeds 10, the simultaneous iteration method has a clear edge over the other two methods. For problems having substantial bandwidth, the simultaneous iteration method is the best.

➤ 3.9 SUBSPACE ITERATION

This method is not essentially different from the simultaneous iteration scheme except for a different procedure for computing the orthogonal vector $\{r\}$ from the premultiplied vector $\{v\}$ [3.3]. Let the premultiplied vector $\{v\}$ of size $(n \times m)$ be known. (referred to non-standard eigenvalue form).

Let us compute

$$[\bar{K}] = \{v^T\} [K] \{v\} \tag{3.108}$$

and

$$[\bar{M}] = \{v^T\} [M] \{v\} \tag{3.109}$$

Size of $[\bar{K}]$ and $[\bar{M}]$ will be $(m \times m)$.

If the columns of $\{v\}$ are orthogonal, $[\bar{K}]$ and $[\bar{M}]$ will be diagonal, but the off-diagonal elements will be present otherwise. Let the eigenvalues and eigenvectors of the eigenvalue problem given by

$$[\bar{K}]\{\phi_i\} = \omega_i^2 [\bar{M}]\{\phi_i\} \qquad (3.110)$$

be $\qquad \lambda_i$ and $\phi_i(i = 1, 2, \ldots 3, \ldots m)$ $\qquad (3.111)$

There are m eigenvalues and m associated eigenvectors.

Let $\qquad\qquad [\phi] = [\phi_1, \phi_2, \ldots, \phi_m] \qquad (3.112)$

$$(m \times m) \qquad (m \times m)$$

If $\{v\}$ is post multiplied by $[\phi]$, it is easy to prove that $\{r\}$, the resulting vector, is orthonormal to $[M]$

$$\{r^T\} [M]\{r\} = [I] \qquad (3.113)$$

where $\qquad\qquad\qquad \{r\} = \{v\}[\phi] \qquad (3.114)$

Proof:

$$[r] = [v][\phi]$$

$$(n \times m) \qquad (n \times m)(m \times m)$$

$$\{r\}^T[M]\{r\} = [\phi]^T\{v\}^T[M]\{v\}[\phi]$$

$$= [\phi]^T [\bar{M}] [\phi]$$

$$= [I]$$

Subsequent operations are exactly similar to the ones in the simultaneous iteration procedure.

Example 3.21

Compute the first two natural frequencies of Example 3.19 by the subspace iteration technique using non-standard eigenvalue form.

The given eigenvalue problem is

$$[K]\{u\} = \omega^2[M]\{u\}$$

$$[K] = \begin{bmatrix} 15 & -10 & & \\ -10 & 25 & -15 & \\ & -15 & 35 & -20 \\ & & -20 & +20 \end{bmatrix}$$

$$[M] = \begin{bmatrix} 10 & & & \\ & 20 & & \\ & & 30 & \\ & & & 40 \end{bmatrix}$$

Let $\qquad \{u^1\} = \begin{Bmatrix} 1 & 0 \\ 0 & 1 \\ 0 & 0 \\ 0 & 0 \end{Bmatrix}$

Normalising $\{u\}^1$ with respect to $[M]$, $\{u\}^1$ can be written as

$$\begin{Bmatrix} 1/\sqrt{10} & 0 \\ 0 & 1/\sqrt{20} \\ 0 & 0 \\ 0 & 0 \end{Bmatrix}$$

$$[M]\{u^1\} = \begin{Bmatrix} \sqrt{10} & 0 \\ 0 & \sqrt{20} \\ 0 & 0 \\ 0 & 0 \end{Bmatrix}$$

If $[K]\{v^1\} = [M]\{u^1\}$, $\{v^1\}$ can be found out, then

$$\{v^1\} = \begin{Bmatrix} 2/\sqrt{10} & 4/\sqrt{20} \\ 2/\sqrt{10} & 6/\sqrt{20} \\ 2/\sqrt{10} & 6/\sqrt{20} \\ 2/\sqrt{10} & 6/\sqrt{20} \end{Bmatrix}$$

Let us now compute $\{r\}$ by the subspace iteration procedure:

$$[\bar{M}] = \{v^1\}^T [M]\{v^1\} \begin{bmatrix} 40 & 116/\sqrt{2} \\ 116/\sqrt{2} & 170 \end{bmatrix}$$

$$[\bar{K}] = \{v^1\}^T [K]\{v^1\} \begin{bmatrix} 2 & 4/\sqrt{2} \\ 4/\sqrt{2} & 6 \end{bmatrix}$$

The eigenvalue problem of the 2×2 subspace is given by

$$[\bar{K}]\{\phi\} = \bar{\omega}^2 [\bar{M}]\{\phi\}$$

Its characteristic equation is of the form

$$\begin{bmatrix} (2 - 40\bar{\omega}^2) & (4/\sqrt{2} - 116/\sqrt{2}\bar{\omega}^2) \\ (4/\sqrt{2} - 116/\sqrt{2}\bar{\omega}^2) & (6 - 170\bar{\omega}^2) \end{bmatrix} = 0$$

$$\therefore \qquad (2 - 40\bar{\omega}^2)(6 - 170\bar{\omega}^2) - \left(\frac{4}{\sqrt{2}} - \frac{116}{\sqrt{2}}\bar{\omega}^2\right) = 0$$

$$\therefore \qquad \bar{\omega}^2 = 0.035 \quad \text{or} \quad 1.57$$

The exact value of the fundamental frequency is 0.03 which is close to 0.035.

$$[\phi] = \begin{bmatrix} 0.158 & 0.0246 \\ 0.016 & -0.0724 \end{bmatrix}$$

Hence

$$\{r\} = \{v^1\}[\phi] \begin{bmatrix} 2/\sqrt{10} & 4/\sqrt{20} \\ 2/\sqrt{10} & 6/\sqrt{20} \\ 2/\sqrt{10} & 6/\sqrt{20} \\ 2/\sqrt{10} & 6/\sqrt{20} \end{bmatrix} \begin{bmatrix} 0.158 & 0.0246 \\ 0.016 & -0.0724 \end{bmatrix}$$

$$= \begin{bmatrix} 0.120 & -0.0537 \\ 0.129 & -0.0862 \\ 0.129 & -0.0862 \\ 0.129 & -0.0862 \end{bmatrix}$$

These can be used for the second trial.

As the iteration progresses, $\bar{\omega}^2$ will converge to actual ω^2 and $\{v\}$ to the actual vector.

Total number of multiplications involved

Let n be the number of degrees of freedom, b the semi-bandwidth of $[K]$ and q the number of vectors needed. For the eigenvalue problem $[K]\{u\} = \omega^2[M]\{u\}$, assuming $\{u\}$ and computing $[M]\{u\}$ needs nq multiplications. Computing $\{v\}$ from $[K]\{v\} = [M]\{u\}$ needs $2nq$ multiplications (in forward and backward substitutions excluding Cholesky's factorisation). Forming $[\bar{K}]$ and $[\bar{M}]$ [Eqs. (3.108) and (3.109) needs $nq(q + 2)$ additional multiplications.

Solving $(q \times q)$ the eigenvalue problem of the subspace needs $(3q^3 + 6q^2)$ multiplications when the Generalised Jacobi method is used [3.3]. Instead, if the non-standard eigenvalue problem is reduced to the standard eigenvalue problem and then the resulting matrix is diagonalised, the number of operations involved is $(q^3 + q^3/6 + 2q^3)$. Forming the new vector needs an additional $nq(q + 1)$ multiplications. Hence, for each iteration the multiplications involved are $= nq$ $(2b + 2q + 4 + (3q^2/n))$. When this is replaced by the simultaneous iteration method $[L_0]$ of Eq. (3.98) needs only $q^3/6$ multiplications in place of $(3q^3 + 6q^2)$. For subspace iteration, Bathe [3.3] suggests the size of vector q which is the minimum $[2p, p + 8]$, where p are the desired eigenvectors. It is seen that the number of multiplications for every iteration in the simultaneous iteration scheme is less than the one in the subspace scheme. But no comparison is available on the rate of convergence using both the schemes for a large size problem.

➤ 3.10 LANCZOS' METHOD

Lanczos pointed out that his method could be used for finding out the first few eigenvectors of a large size symmetric matrix [3.2]. However, the method was used for a long time for tridiagonalisation only. But it was Paige who established that it is an effective tool for eigenvalue extraction [3.8].

This matrix reduction method is based on Crandall's tailoring of the eigenvalue routine created by Lanczos' [3.9]. Starting with a single trial vector, the algorithm generates a sequence of mutually orthogonal vectors. This process in effect transforms the original matrix $[A]$ into tridiagonal form. The eigenvalues of the tridiagonal matrix (which are the same as those of the original matrix) can be computed either by Sturm sequence and inverse iteration or by any other method.

Let
$$[A] \{u\} = \lambda\{u\} \tag{3.115}$$

be the standard eigenvalue problem

Let
$$[P]\{\bar{u}\} = \{u\} \tag{3.116}$$

where $[P]$ is an orthogonal matrix of size (n, n)
Then we know,
$$[\bar{A}]\{\bar{u}\} = \lambda \{\bar{u}\} \tag{3.117}$$

where
$$[\bar{A}] = [P^T][A][P] \tag{3.118}$$

Recasting,
$$[A][P] = [P^{-T}][\bar{A}]$$

or,
$$[A][P] = [P][\bar{A}] \tag{3.119}$$

since $[P]$ is an orthogonal matrix of size (n, n)
Let the columns of the matrix $[P]$ be $\{y_1\}, \{y_2\}, \{y_3\}$ each of size $(n, 1)$

Since $[P]$ is orthogonal $\{y_i\}^T\{y_i\} = 1$

and $\qquad\qquad\qquad \{y_i\}^T\{y_j\} = 0 \quad$ for $\ i \neq j$

Let us aim at $[\bar{A}]$ becoming a tridiagonal matrix. Then Eq. (3.119) can be written as

$$[A][y_1, y_2, y_3 \ldots y_n] = [y_1, y_2 \ldots y_n] \begin{bmatrix} \alpha_1, \beta_1 & 0 & 0 & 0 & 0 \\ \beta_1 \alpha_2 \beta_2 & & 0 & 0 & 0 \\ & \ldots\ldots & & & \alpha_{n-1}\beta_{n-1} \\ & & & \ldots & \beta_{n-1}, \alpha_n \end{bmatrix} \tag{3.120}$$

Equations (3.120) when expanded give n equations

$$[A]\{y_1\} = \alpha_1\{y_1\} + \beta_1\{y_2\}$$

$$[A]\{y_2\} = \beta_1\{y_1\} + \alpha_2\{y_2\} + \beta_2\{y_3\}) \tag{3.121}$$

$$\ldots \quad \ldots \quad \ldots$$

$$[A]\{y_n\} = \beta_{n-1}\{y_{n-1}\} + \alpha_n\{y_n\}$$

$[A]$ is the given matrix of size $(n \times n)$, $\{y_1\}$, $\{y_2\} \ldots \{y_n\}$ are vectors of size $(n \times 1)$, and the elements of tridiagonal matrix α's and β's are constants.

Let the starting vector $\{y_1\}$ be a non-null vector such that $\{y_1\}^T\{y_1\} = 1$. If the first equation of (3.121) is premultiplied by $\{y_1\}^T$, it can be proved that when $\alpha_1 = \{y_1\}^T [A]\{y_1\}$, $\{y_1\}^T\{y_2\} = 0$. In other words, choice of $\alpha_1 = \{y_1\}^T [A]\{y_1\}$ makes $\{y_1\}$ orthogonal to $\{y_2\}$. From the first equation

$$\beta_1\{y_2\} = [A]\{y_1\} - \alpha_1\{y_1\} \tag{3.122}$$

Let this be equal to $\{z_1\}$
If $\{y_2\}$ is such that $\{y_2\}^T\{y_2\} = 1$, then

$$\{y_2\}^T\{y_2\} = \frac{\{z_1\}^T}{\beta_1} \cdot \frac{\{z_1\}}{\beta_1} = 1$$

$$\therefore \qquad\qquad \beta_1 = [\{z_1\}^T \cdot z_1]^{1/2} \tag{3.123}$$

The same procedure can be adopted for the other terms. The summary is as follows:

$$\{V_j\} = [A]\{y_j\} - \beta_{j-1}\{y_{j-1}\} \qquad (\beta_0 = 0)$$

$$\alpha_j = \{y_j\}^T\{V_j\}$$

$$\{z_j\} = \{V_j\} - \alpha_j\{y_j\} \tag{3.124}$$

$$\beta_j = [\{z_j\}^T\{z_j\}]^{1/2}$$

$$\{y_{j+1}\} = (1/\beta_j)\{z_j\}$$

When n is very large, the orthogonality condition $\{y_i\}^T\{y_j\} = 0$ may be satisfied for adjacent i and j, but not for $j > (i + 1)$, due to round-off errors. This difficulty may be overcome by orthogonalising $\{y_{j+1}\}$ with respect to $y_1, y_2 \dots y_j$. Then the performance will be satisfactory. But this involves additional labour.

For a matrix $[A]$ of size $n \times n$ and bandwidth b the tridiagonalisation (Eq. 3.121) needs $(b + 5)n^2$ multiplications. Extra orthogonalisation for improved performance needs n^3 multiplications approximately. Hence the total labour involved in tridiagonalisation alone is $n^2(n + b + 5)$. This is only $2/3\ n^3 + 3/2\ n^2$ for Householder's scheme. Hence for a full matrix there is no saving in labour. If the tridiagonalisation is halted when only m rows have been formed ($m \ll n$) and then the eigenvalues extracted, there is a considerable saving in labour. It is only $nm(b + m + 4)$. The eigenvectors $\{u\}$ of $[A]$ corresponding to eigenvectors $\{q\}$ of the tridiagonal matrix are given by

$$\{u\} = [p]\{q\} \tag{3.125}$$

Hence an additional nmc multiplications are required to compute c vectors of $[A][3.10]$ to $[3.13]$ describe the special features of this approach. For a specific problem $[3.11]$ compares the subspace iteration scheme with Lanczos' method.

If the trial vector $\{y_1\}$ is iterated several times till its individual rows converge, this process is similar to Gram Schmidt deflation of the first vector. Hence α_1 will be close to λ_1. Use of the first equation in (3.121) indicates that the value of β_1 will be close to zero. If this argument is extended to $\{y_2\}$, α_2 will be close to λ_2, β_2, as a consequence, will be small. The same argument can be extended to the other columns. Computing the eigenvalues involves negligible labour since $[\bar{A}]$ is tridiagonal. This method turns out to be extremely efficient since it combines in itself the advantages of the Gram Schmidt deflation and simultaneous iteration.

Example 3.22

Tridiagonalise $\quad [A] = \begin{bmatrix} 27 & 14 & 8 & 4 \\ 14 & 16 & 9 & 3 \\ 8 & 9 & 8 & 2 \\ 4 & 3 & 2 & 1 \end{bmatrix}$

I Trial

Let $\quad \{y_1\} = \begin{Bmatrix} 1 \\ 0 \\ 0 \\ 0 \end{Bmatrix}$

Compute $\qquad [A]\{y_1\} = \begin{Bmatrix} 27 \\ 14 \\ 8 \\ 4 \end{Bmatrix} = \{x_1\}$

$$\{x_1\}^T\{x_1\} = 1005$$

II Trial

Orthonormalising $\quad \{y_1\} = \begin{Bmatrix} 27/31.7 \\ 14/31.7 \\ 8/31.7 \\ 4/31.7 \end{Bmatrix} = \begin{Bmatrix} 0.852 \\ 0.44 \\ 0.215 \\ 0.125 \end{Bmatrix}$ (As $\sqrt{1005} = 31.7$)

$$[A]\{y_1\} = \begin{Bmatrix} 31.67 \\ 21.60 \\ 13.03 \\ 5.48 \end{Bmatrix} = \{x_1\}$$

$$\{x_1\}^T\{x_1\} = 1669$$

III Trial

Orthonormalising $\quad \{y_1\} = \{x_1\}/\sqrt{1669} = \begin{Bmatrix} 0.775 \\ 0.528 \\ 0.319 \\ 0.132 \end{Bmatrix}$

$$[A]\{y_1\} = \begin{Bmatrix} 3.41 \\ 22.57 \\ 13.12 \\ 5.45 \end{Bmatrix} = \{x_1\}$$

IV Trial

Orthonormalising $\qquad \{y_1\} = \begin{Bmatrix} 0.762 \\ 0.548 \\ 0.328 \\ 0.132 \end{Bmatrix}$

Since $\{y_1\}$ has fairly converged,

$$\alpha_1 = \{y_1\}^T[A]\{y_1\} = 41.20$$

Using Eq. (3.122)

$$\beta_1\{y_2\} = z_1 = [A]\{y_1\} - \alpha_1\{y_1\} = \left\{ \begin{array}{c} -0.085 \\ 0.12 \\ 0.08 \\ 0.07 \end{array} \right\}$$

Therefore $$\beta_1 = \sqrt{z_1^T z_1} = 0.18$$

It is observed that β_1 is very small in comparison with α_1. The extra effort put in, for establishing convergence of $\{y_1\}$, pays dividends in the form of faster convergence of α's and hence eigenpairs.

Hence, $$\{y_2\} = \{z_1\}/\beta_1 = \left\{ \begin{array}{c} -0.472 \\ 0.66 \\ 0.44 \\ 0.39 \end{array} \right\}$$

Let us check whether $\{y_2\}^T\{y_1\} = 1$

$\{y_2\}^T\{y_1\} = (-0.472 \times 0.762) + (0.66 \times 0.548) + (0.44 \times 0.318)$
$$+ (0.39 \times 0.132) = 0.19$$

Let us use $\{y_1\}$ to correct $\{y_2\}$ such that $\{y_2\}^T\{y_1\} = 0$

$$\text{Hence corrected } \{y_2\} = \left\{ \begin{array}{c} -0.472 - (0.19 \times 0.762) \\ 0.66 - (0.19 \times 0.548) \\ 0.44 - (0.19 \times 0.318) \\ 0.39 - (0.19 \times 0.132) \end{array} \right\} = \left\{ \begin{array}{c} -0.616 \\ 0.56 \\ 0.38 \\ 0.36 \end{array} \right\}$$

Now $$\{y_2\}^T\{y_1\} = 0$$

$$\{V_2\} = [A]\{y_2\} - \beta_1\{y_1\} = \left\{ \begin{array}{c} -4.425 \\ 4.738 \\ 8.554 \\ 8.867 \end{array} \right\}$$

Therefore $\alpha_2 = \{y_2\}^T\{V_2\} = 12.44$

This procedure can now be extended to α_3 and α_4.
Tridiagonal matrix truncated up to 2 terms is now equal to

$$\begin{bmatrix} 41.20 & 0.18 \\ 0.18 & 12.44 \end{bmatrix}$$

For more number of terms one may observe the principal diagonal dominant and the off-diagonal terms very small.
Eigenvalue problem in non-standard form is given by

$$[K]\{q\} = \omega^2[M]\{q\} \qquad (3.126)$$

This can be brought to the form

$$[A]\{u\} = \lambda\{u\}$$

by the following transformation. Cholesky decomposition of $[K]$ yields

$$[K] = [L][L]^T$$

Substituting in Eq. (3.126), we have

$$[A]\{u\} = \lambda\{u\} \qquad (3.127)$$

where
$$\{q\} = [L]^{-T}\{u\} \qquad (3.128)$$

$$[A] = [L]^{-1}[M][L]^{-T} \qquad (3.129)$$

and
$$\lambda = \frac{1}{\omega^2} \qquad (3.130)$$

Comparing with standard eigenvalue problem, for any given $\{y_i\}$, $[A]\{y_i\}$ is done in three storages as given by Eq. (3.91).

For the starting vector $\{y_1\}$, one can make any choice, iterate a few times by premultiplying by $[A]$ till convergence is obtained.

Example 3.24
Calculate the first four natural frequencies of the system given in Example 3.15. (All I's in N mm sec^2 and q's in N/mm).

$$I_1 = I_2 = I_3 = I_4 = 1, \quad I_5 = 2, \quad I_6 = I_7 = 5, \quad q = 10$$

Since the first natural frequency will be zero for this free-free system, let us use a shift μ of -2 (Sec. 3.2.4).

$$[K] = \begin{bmatrix} 10 & -10 & & & & & \\ & 20 & 0 & 0 & -10 & & \\ & & 10 & -10 & & & \\ & & & 20 & -10 & & \\ & & & & 30 & -10 & \\ \text{(Sym)} & & & & & 20 & -10 \\ & & & & & & 10 \end{bmatrix}$$

$$[M] = \begin{bmatrix} 1 & & & & & & 0. \\ & 1 & & & & & \\ & & 1 & & & & \\ & & & 1 & & & \\ & & & & 2 & & \\ 0 & & & & & 3 & \\ & & & & & & 5 \end{bmatrix}$$

Transformed tridiagonal matrix [T] is computed up to 5 terms and is given by

$$\begin{bmatrix} 0.500 & 0.00027 & & & \\ & 0.283 & & 0.00023 & & \\ & & 0.1698 & & 0.0074 & \\ & \text{Symmetric} & & 0.145 & & 0.00007 \\ & & & & 0.065 \end{bmatrix}$$

The computed natural frequencies in rad/sec. are

$$1.241, \quad 1.954 \quad \text{and} \quad 2.236$$

Associated vectors are

$$\omega_1 = 1.241 \begin{Bmatrix} -0.1968 \\ -0.1665 \\ -0.1968 \\ -0.1665 \\ -0.1105 \\ 0.0354 \\ 0.1541 \end{Bmatrix}; \ \omega_2 = 1.954 \begin{Bmatrix} 0.2492 \\ 0.1541 \\ -0.2494 \\ -0.1541 \\ 0.1572 \\ 0.000 \\ 0.000 \end{Bmatrix}; \ \omega_3 = 2.236 \begin{Bmatrix} 0.0871 \\ 0.0435 \\ 0.0871 \\ 0.0436 \\ -0.0219 \\ -0.1304 \\ 0.0869 \end{Bmatrix}$$

It is to be observed that the first mode has one node, the second two and the third three (locations of zero twist). Off diagonal terms of the tri-diagonal matrix are negligible in comparison with the diagonal terms.

The computer program for this problem is given in Program 3.8.

Labour involved

Let the given problem be of size n and bandwidth b. Let us compute the first m vectors. Let us compare the labour involved in the simultaneous iteration method and in Lanczos scheme. For simultaneous iteration scheme, the labour (excluding Cholesky factorisation) is given by

$$c\left(3nmb + 2\frac{1}{2}nm^2 + \frac{m^3}{6}\right)$$

for c iterations.

Let s iterations be involved in improving the starting vector. Then this process involves $s(3nb + 2n)$ operations. Labour involved in computing α_i is $m(3nb + n)$ and in computing β_i is $m \times 4n$.

Labour involved in orthonormalising

$$\text{Lanczos vectors} = 2n + 4n + \ldots + 2nm \simeq nm^2$$

Computing the eigenvalues of
tridiagonal matrix of size n $\left.\begin{array}{l} \\ \\ \\ \end{array}\right\}$ $m\left(6m^2 + 2\frac{1}{2}m^3 + \frac{m^3}{6}\right)$
doing the iteration m times

Computing the eigenvectors
subsequently $\qquad nm^2 + nb$

Hence, the total labour $= nb(3m + 3s + 1)$

$$+ n(2s + 2m^2 + 5)$$

$$+ m^3(6 + 2.66\ m)$$

For a specific problem of $n = 1000$, $b = 50$, $c = 10$, $m = 10$, and $s = 5$.

Labour involved in the s.i. method $= 10^6 \times 17.5$

Labour involved in the Lanczos scheme $= 10^6 \times 2.43$

For a bandwidth more than 50 and when more than ten eigenpairs are to be computed, the Lanczos scheme has considerable saving in labour. This is specially true when one computes closely spaced eigenvalues as in ship hulls and automobile bodies.

➤ 3.11   COMPONENT MODE SYNTHESIS

3.11.1   Rayleigh-Ritz Method

It is well-known that the energy method of computing the natural frequency of a system yields a result which is higher than the exact value. Consider the system shown in Fig. 3.9. In Rayleigh's method one can assume a dynamic deflection curve which does not violate the boundary conditions. The maximum kinetic energy of the vibrating system is equated to the strain energy of the springs. If the maximum displacements are X_1 and X_2,

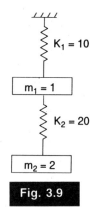

Fig. 3.9

Two mass systems

$$\text{Maximum kinetic energy} = \frac{1}{2}(m_1 X_1^2 + m_2 X_1^2)\omega^2$$

$$\text{Maximum potential energy} = \frac{1}{2}[k_1 X_1^2 + k_2(X_2 - X_1)^2]$$

(3.131)

Hence
$$\omega^2 = \frac{k_1 X_1^2 + k_2(X_2 - X_1)^2}{(m_1 X_1^2 + m_2 X_2^2)}$$

Let us assume $X_1 = 1$ and $X_2 = 2$ represent the fundamental mode.

Then
$$\omega^2 = \frac{(10 + 20(2-1)^2)}{(1^2 + 2.2^2)} = \frac{30}{9}$$

This can also be represented in another fashion.
Let the first eigenvector be $\{u_1\}$.
Then we know

$$\omega_1^2 = \frac{\{u_1\}^T [K]\{u_1\}}{\{u_1\}^T [M]\{u_1\}}.$$

where $\{u_1\}$ is of size $(n \times 1)$, $[K]$ and $[M]$ of size $(n \times n)$. In this problem let us assume the first mode $\{u_1\}$ to be given by X_1 and X_2.

Here
$$[K] = \begin{bmatrix} 30 & -20 \\ -20 & 20 \end{bmatrix}$$

$$[M] = \begin{bmatrix} 1 & 0 \\ 0 & 2 \end{bmatrix}$$

$$\{u_1\} = \begin{Bmatrix} 1 \\ 2 \end{Bmatrix}$$

$$\therefore \quad \omega_1^2 = \frac{[1 \ 2]\begin{bmatrix} 30 & -20 \\ -20 & 20 \end{bmatrix}\begin{Bmatrix} 1 \\ 2 \end{Bmatrix}}{[1 \ 2]\begin{bmatrix} 1 & 0 \\ 0 & 2 \end{bmatrix}\begin{Bmatrix} 1 \\ 2 \end{Bmatrix}} = \frac{30}{9}$$

Extending this to the Rayleigh-Ritz approach if one is in a position to get the first two vectors ψ approximately the 2×2 subspace will have the stiffness and mass matrices $[\bar{K}]$ and $[\bar{M}]$ given by

$$[\bar{K}] = \psi^T [K] \psi$$

$$[\bar{M}] = \psi^T [M] \psi$$

If the eigenvalue problem connected with subspace $[\bar{K}]\phi = \bar{\omega}^2[\bar{M}]\phi$ is solved one can compute the first two frequencies and vectors. In this specific case let

$$\psi = \begin{bmatrix} 1 & 1 \\ 2 & -1 \end{bmatrix}$$

(The second assumed mode shape has one mode between the two masses.)

Then

$$[\bar{K}] = \psi^T[K]\psi = \begin{bmatrix} 30 & -30 \\ -30 & 90 \end{bmatrix} \quad [\bar{M}] = \psi^T[M]\psi = \begin{bmatrix} 9 & -3 \\ -3 & 3 \end{bmatrix}$$

Hence for the eigenvalue problem,

$$[\bar{K}] = \bar{\omega}^2[\bar{M}]\phi$$

$\bar{\omega}^2$ values are 2.78 and 37.32.

For the problem given in Fig. 3.9, the exact values of $\bar{\omega}^2$ are 2.42 and 77.78. It may be observed, 1×1 approximation gave ω^2 of 3.33, 2×2 gave ω^2 of 2.78 whereas the actual value is 2.42 (which is lower than the other two).

3.11.2   Mode Synthesis

This method is suitable for computing the first few eigenpairs of large size problems by treating the problem as if it is made up of different groups. For example, if one is interested in analysing a turbomachine, consisting of a rotor, disc and blades, each one of these groups can be separately analysed by arresting the degrees of

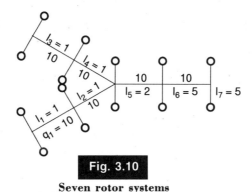

Fig. 3.10

Seven rotor systems

freedom of nodes connecting different groups. In the second stage, these can be released (one at a time) and the unarrested ones can be made zero. Since these two steps involve modes which are mutually independent, these together can be used as trial vectors in the Rayleigh-Ritz approach. This is explained by working out the following example.

Example 3.25
Solve the branched torsional vibration problem of Example 3.14. Let us arrest rotor 5 and solve the three branches separately (Fig. 3.10).

Let the first eigenvector be $\{u_1\}$.
Then we know

$$\omega_1^2 = \frac{\{u_1\}^T[K]\{u_1\}}{\{u_1\}^T[M]\{u_1\}}$$

The natural frequencies for the branch given by Fig. 3.11 are

$$\begin{bmatrix} (10 - \omega^2) & -10 \\ -10 & (20 - \omega^2) \end{bmatrix} = 0$$

Hence $\quad \omega^2 = 5(3 \pm \sqrt{5}) = 3.820 \quad$ or $\quad 26.18$

Ratio of displacements $X_1 : X_2$ is given by $1 : 0.618$ and $1 : -1.618$

The natural frequencies for the system given by Fig. 3.12 are

$$\begin{bmatrix} (20 - 5\omega^2) & -10 \\ -10 & (10 - 5\omega^2) \end{bmatrix} = 0$$

Hence $\quad \omega^2 = 3 \pm \sqrt{5} = 0.761 \quad$ or $\quad 5.239$

Ratio of X_1 and X_2 is given by $1 : 1.618$ or $1 : -0.618$.

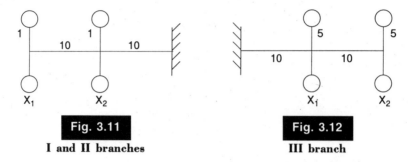

Fig. 3.11

I and II branches

Fig. 3.12

III branch

The eigenvalue problem for the entire system is given by

$$
\begin{bmatrix}
10 & -10 & 0 & & & & \\
-10 & 20 & 0 & 0 & -10 & & \\
 & & 10 & -10 & & & \\
 & & -10 & 20 & -10 & & \\
 & -10 & & -10 & 30 & -10 & \\
 & & & & -10 & 20 & -10 \\
 & & & & & -10 & 10
\end{bmatrix}
\begin{Bmatrix}
X_1 \\ \cdot \\ \cdot \\ \cdot \\ \cdot \\ \cdot \\ X_7
\end{Bmatrix}
$$

$$
= \omega^2
\begin{bmatrix}
1 & & & & & & \\
 & 1 & & & & & \\
 & & 1 & & & & \\
 & & & 1 & & & \\
 & & & & 2 & & \\
 & & & & & 5 & \\
 & & & & & & 5
\end{bmatrix}
\{X\}
$$

Rearranging, this can be written as

$$
\begin{bmatrix}
10 & & -10 & & & & \\
0 & 10 & & -10 & & & \\
-10 & 0 & 20 & 0 & -10 & & \\
 & -10 & 0 & 20 & -10 & & \\
 & & -10 & -10 & 30 & -10 & \\
 & & & & -10 & 20 & -10 \\
 & & & & & -10 & 10
\end{bmatrix}
\{X\}
$$

$$= \omega^2 \begin{bmatrix} 1 & & & & & & \\ & 1 & & & & & \\ & & 1 & & & & \\ & & & 1 & & & \\ & & & & 2 & & \\ & & & & & 5 & \\ & & & & & & 5 \end{bmatrix} \{X\}$$

Since this problem has the lowest natural frequency zero and since we are interested in the nonzero frequency, we will use a shift μ of 1.0 and rewrite $[K]$ as $[K - \mu M]$.

$$\begin{bmatrix} 9 & -10 & & & & & \\ 9 & & -10 & & & & \\ -10 & 19 & & -10 & & & \\ & -10 & 19 & -10 & & & \\ & -10 & -10 & 28 & -10 & & \\ & & & -10 & 15 & -10 \\ & & & & -10 & 5 \end{bmatrix}$$

The first two trial vectors in the Ritz approach can be written as

$$\begin{Bmatrix} 1.00 & & 1.00 \\ 1.00 & & -1.618 \\ 0.618 & & 1.000 \\ 0.618 & & -1.618 \\ 0 & & 0 \\ -1.00 & & 1.00 \\ -1.618 & & -0.618 \end{Bmatrix}$$

They are the first two vectors of the isolated system when junction is fixed. The third vector can be assumed as

$$\begin{Bmatrix} 0 \\ 0 \\ 0 \\ 0 \\ 1 \\ 0 \\ 0 \end{Bmatrix}$$

In this case the junction is having unit displacement, the rest zero. Hence

$$\psi = \begin{Bmatrix} 1.00 & 1.00 & 0 \\ 1.00 & 1.00 & 0 \\ +0.618 & -1.618 & 0 \\ +0.618 & -1.618 & 0 \\ 0.00 & 0.00 & 1.00 \\ -1.00 & 1.00 & 0 \\ -1.618 & -0.618 & 0 \end{Bmatrix}$$

$$[K]\psi = \begin{Bmatrix} 2.82 & 25.18 & 0 \\ 2.82 & 25.18 & 0 \\ 1.74 & -40.742 & -10 \\ 1.74 & -40.742 & -10 \\ -02.36 & 16.180 & 28 \\ +1.18 & 34.27 & -10 \\ +1.91 & -21.18 & 0 \end{Bmatrix}$$

$$[\bar{K}] = \psi^T[K]\psi = \begin{Bmatrix} 3.51 & 0 & -2.36 \\ 0 & 258.83 & 16.18 \\ -2.36 & 16.18 & 28 \end{Bmatrix}$$

Similarly

$$[\overline{M}] = \psi^T[M]\psi = \begin{Bmatrix} 20.86 & 0 & 0 \\ 0 & 25.32 & 0 \\ 0 & 0 & 2.00 \end{Bmatrix}$$

The fundamental frequency of the eigenvalue problem $[\overline{K}]\{\phi\} = \overline{\omega}^2[\overline{M}]\{\phi\}$ is given by $\overline{\omega}^2 = 0.17$.

Since a shift μ of 1.00 was used the fundamental frequency ω^2 is given by $\omega^2 = 1.17$. The exact value is 1.54 rad/sec.

➤ 3.12 FIGURES OF SIGNIFICANCE

We have discussed methods of computing the natural frequencies, in particular, the first few lower ones of a large system. It is good to have an idea of the magnitudes of these frequencies. The lowest natural frequency of a car in bouncing mode is around lcps, the torsional critical speed of I.C. engines around 150 cps and the lateral critical speed of steam turbines as low as 50 cps. The blades lowest natural frequency may be around 20 cps, and impellers around 15 cps. In general, the first few natural frequencies of practical machine elements can vary from around 0.5 cps to 1000 cps, in rpm (revolutions per minute) from 30 to 60,000, in radians per second (ω) from around 3 to 6000. As a consequence, ω^2 varies from around 10 to 3.6×10^7 hence λ, which is ($1/\omega^2$), can vary from 10^{-8} to 10^{-1}. It is advisable to use appropriate units for K and M when computing eigenvalues and not to non-dimensionalise them. These will help in ascertaining that the values of λ or ω^2 computed subsequently are in the meaningful range.

➤ 3.13 INDUSTRIAL APPLICATIONS

3.13.1 Determination of Critical Speeds Including Gyroscopic Effect

Let us consider a typical beam element bending in one plane as discussed in Section 1.3.1. Let us suppress the axial degrees of freedom. A typical element will now have only two degrees of freedom v and θ at each node, refer to Fig. 3.13. Let us consider a disc as shown in Fig. 3.14 rotating uniformly with angular velocity μ about the axis OQ which is stationary in space (precessional rotation), and rotates uniformly with angular velocity γ relative to the plane of the diagram

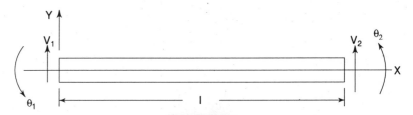

Fig. 3.13

The beam element

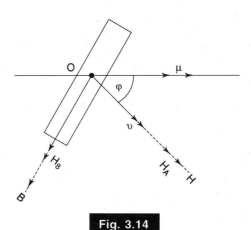

Fig. 3.14

Spin and precessional rotation of the disc

(which itself rotates uniformly with angular velocity μ) about axis OA, fixed to the disc and at right angles to the plane of the disc (OA and OB are the two principal axes of inertia).

The components of angular velocity along OA and OB respectively are [3.14]

$$\omega_A = \mu \cos \phi + v, \qquad \omega_B = -\mu \sin \phi \qquad (3.132)$$

and similarly, the components of angular moment are

$$H_A = I_A\omega_A = I_A (\mu \cos \phi + v), \quad H_B = I_B\omega_B = I_B(-\mu \sin \phi) \quad (3.133)$$

where I_A and I_B are mass moments of inertia along OA and OB.

The rate of change of $\vec{H}_A$ is given by

$$(d/dt)\vec{H}_A = \vec{\mu} \times \vec{H}_A = \mu I_A(\mu \cos \phi + v) \sin \phi \qquad (3.134)$$

(away from the observer). The rate of change of $\vec{H}_B$ is

$$(d/dt)\vec{H}_B = \vec{\mu} \times \vec{H}_B = -\mu^2 I_B \sin \cos \phi \qquad (3.135)$$

(away from the observer). Therefore, the accelerating torque is

$$\vec{\mu} \times (\vec{H}_A + \vec{H}_B) = \mu I_A(\mu \cos \phi + v) \sin \phi - \mu^2 I_B \sin \phi \cos \phi \quad (3.136)$$

(away from the observer). For small ϕ this torque is

$$\mu I_A \phi(\mu + v) - \mu^2 I_B \phi \quad (3.137)$$

(away from the observer). Therefore, the inertia torque T, the negative of the accelerating torque, is

$$T = \mu^2 I_B \phi - \mu I_A \phi(\mu + v) \quad (3.138)$$

(away from the observer).

In the case of forward precession $\omega_A = \Omega$ and $\mu = \Omega$. From Eq. (3.129) it follows that $v = 0$. From Eq. (3.138)

$$T = \Omega^2 I_B \phi - \Omega^2 I_A \phi = (I_B - I_A)\Omega^2 \phi = C\Omega^2 \phi \quad (3.139)$$

(away from the observer), where $C = (I_B - I_A)$.

In the case of backward precession $\omega_A = \Omega$ and $\mu = -\Omega$. From Eq. (1.132) it follows that $v = 2\Omega$. From Eq. (3.138)

$$T = \Omega^2 I_B \phi + \Omega^2 I_A \phi = (I_B + I_A)\Omega^2 \phi = C\Omega^2 \phi \quad (3.140)$$

(away from the observer), where $C = (I_B + I_A)$.

Therefore, T can be expressed as

$$T = C\Omega^2 \phi \quad (3.141)$$

(away from the observer), where

$$C = I_B \pm I_A , \quad (3.142)$$

with the + sign for backward precession and the − sign for forward precession.

The stiffness matrix of any given element is given by

$$[K] = \frac{EI}{l^3} \begin{bmatrix} 12 & 61 & -12 & -61 \\ & 41^2 & -61 & 21^2 \\ & & 12 & -61 \\ (\text{Sym}) & & & 41^2 \end{bmatrix} \quad (3.143)$$

By using the lumped mass approach, the mass matrix can be expressed as

$$[M] = \begin{bmatrix} m_1 & & & 0 \\ & C_1 & & \\ & & m_2 & \\ 0 & & & C_2 \end{bmatrix} \quad (3.144)$$

where m_1 and m_2 are the masses at the two nodes and C_1 and C_2 are the values of C as obtained from Eq. (3.142) at the nodes.

The eigenvalues (critical speed of rotation) are computed after assembling the element stiffness and mass matrices and incorporating the boundary conditions.

Two examples are presented. The results are obtained for an overhanging shaft with a flat disc (Fig. 3.15), and for a shaft with three discs (Fig. 3.16). For flat discs, $I_B = I_A/2$ and from Eq. (3.142).

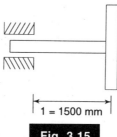

Fig. 3.15

Overhanging shaft with single disc. Disc weight 100 N; shaft diameter 40 mm

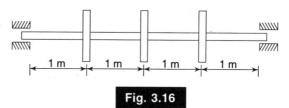

Fig. 3.16

Shaft with three discs. Disc weights 100 N each; shaft diameter 40 mm

$$C = \begin{cases} -I_B \text{ (forward precession)} \\ 3I_B \text{ (backward precession)} \end{cases} \qquad (3.145)$$

If $I_B = mC^2$ then

$$C = \begin{cases} -mC^2 \text{ (forward precession)} \\ 3mC^2 \text{ (backward precession)} \end{cases} \qquad (3.146)$$

For an overhanging shaft (Fig. 3.15)

$$\frac{EI}{l^2} = \begin{bmatrix} 12 & -61 \\ -61 & 41^2 \end{bmatrix} \begin{Bmatrix} v_2 \\ \theta_2 \end{Bmatrix} = \omega^2 \begin{bmatrix} m & 0 \\ 0 & -mC^2 \end{bmatrix} \begin{Bmatrix} v_2 \\ \theta_2 \end{Bmatrix} \quad (3.147)$$

for forward precession·

$$\frac{EI}{l^3} = \begin{bmatrix} 12 & -61 \\ -61 & 41^2 \end{bmatrix} \begin{Bmatrix} v_2 \\ \theta_2 \end{Bmatrix} = \omega^2 \begin{bmatrix} m & 0 \\ 0 & 3mC^2 \end{bmatrix} \begin{Bmatrix} v_2 \\ \theta_2 \end{Bmatrix} \quad (3.148)$$

for backward precession

ω_C^2 (critical speed ignoring gyroscopic effect) $= \dfrac{3EI}{ml^3}$

when $\dfrac{C}{l} = 1$, for case   (i) $\dfrac{\omega}{\omega_C} = 1.74$

and for case   (ii) $\dfrac{\omega}{\omega_C} = 0.31$ or 2.07

The results are similar to those given in Reference [3.15]. The results are shown in Figs 3.17 and 3.18 respectively.

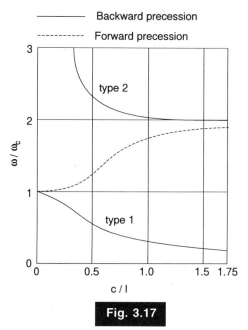

Fig. 3.17

Overhanging shaft with single disc

Advantages of the method

1. Shafts of varying cross-sectional area, i.e. stepped shafts can be treated easily.
2. Boundary conditions can be taken care of easily (for example, intermediate supports).

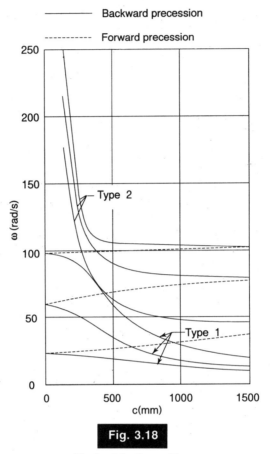

Fig. 3.18

Shaft with three discs

3. Shafts with continuously distributed mass can be considered by taking more elements or reducing the length of the element.
4. The influence of non-rigidity supported bearings and oil cushion in bearings can be included.
5. The dynamic stresses in the shaft can be computed at any given rotational speed.

3.13.2   Eigenpairs of Boiler Frame

Let us compute the eigenpairs of the boiler frame given in Example 2.11. Referring to Fig 2.5, we have 24 effective degrees of freedom when the nodes 1 and 10 are assumed to be fixed. Using the simultaneous iteration scheme and standard eigenvalue form, the first six eigenpairs of this steel structure have been calculated. The actual frequencies are 167, 484, 855, 1012, 1528 and 1926 rad./sec. The mode shapes are plotted in Fig. 3.19. It may be observed that

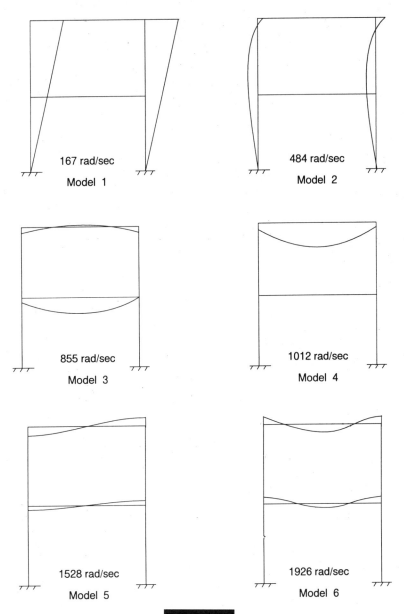

Fig. 3.19

Mode shapes of boiler frame

the first mode is close to the vertical members behaving like cantilever beams in the fundamental mode and the second mode corresponding to the second cantilever mode. The third, the fourth, the fifth and the sixth mode shapes essentially correspond to the beam modes of the horizontal members. It may be observed that the fundamental frequency of 167 rad./sec. corresponds to 1565 rpm. If an induction motor running at 1440 rpm is located on the frame, its running speed may be close to resonance.

3.13.3   Eigenvalues of a Compressor Disc

Half-sectional elevation of one-half of the compressor disc taken up for investigation is shown in Fig. 3.20. The disc is assumed to be fixed at the inner radius. For the triangular elements chosen for analysis, three degrees of freedom are allowed at all other nodes. They are expressed as follows:

$$U = (a_1 z + a_4 r^2 z + a_7) \cos \theta$$
$$V = (a_2 z + a_5 r^2 z + a_8) \sin \theta \qquad (3.149)$$
$$W = (a_3 z + a_6 r^2 z^2 + a_9) \cos \theta$$

Here U, V and W are radial, circumferential and axial displacement, r, θ and z being radial, circumferential and axial coordinates of the node. These terms represent approximately the displacements of a beam circular in cross-section subjected to bending. Total number of degrees of freedom for the problem shown in Fig. 3.20 is more than 500. Eight trial vectors are chosen and after eight iterations the first three natural frequencies converge up to the first four significant digits and the fourth up to two digits. To verify the results, the natural frequencies are determined experimentally. The experimental values are compared with the theoretical results in Table 3.1. The deformed shape of the compressor disc in all the four modes is shown in Figs 3.21 to 3.24. It may be observed that the buckling of the disc is predominant in the third and the fourth modes.

3.13.4   Critical Speeds of a Sugar Centrifugal

Sugar cane juice, after being subjected to chemical processing, is known as massecuite. Centrifugal basket (Fig. 3.25) is used to separate sugar and molasses from massecuite. The basket has a number of holes around the circumference to allow the molasses to escape and is provided with circumferential hoops to withstand the centrifugal forces. The basket is lined with metal gauge on its inner surface,

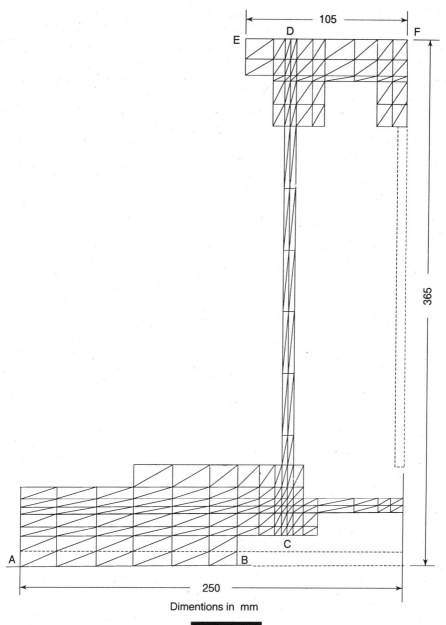

Dimentions in mm

Fig. 3.20

Half sectional elevation of the compressor disc

Table 3.1

Comparison of Frequencies (c/s)

Mode	Experiment	Computation	% difference
1	126	149	+ 16.4
2	842	750	− 10.9
3	1490	1622	+ 8.86
4	1746	2070	+ 18.55

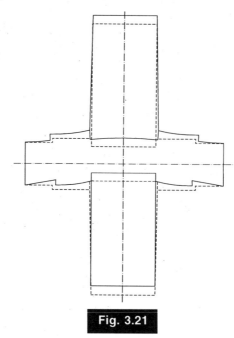

Fig. 3.21

Response of the compressor disc (I Mode)

which serves to retain the sugar while allowing the molasses to pass through. The centrifugal is driven by an electric motor. A mechanical coupling is used to couple the motor to the centrifugal basket. Two bearings are used to support the rotor of the electric motor and another two for the shaft of centrifugal basket. All the bearings are antifriction bearings.

Centrifugal basket receives the massecuite as charge at 200 rpm. After charging, its speed increases to 1200 rpm in about 45 seconds. At this speed, the molasses and sugar get separated, and the molasses leaves the basket through the holes. Once the entire molasses has

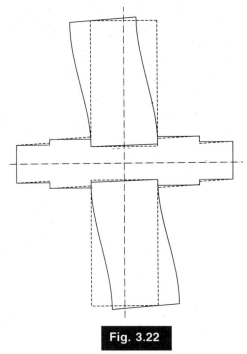

Fig. 3.22

Response of the compressor disc (II Mode)

left the basket, its speed is reduced to 200 rpm, exact determination of lateral critical speeds is warranted. Since these should not coincide with either the idling speed of 200 rpm or the operating speed of 1200 rpm and sugar is scooped out after which the cycle gets completed.

On scrutiny, it is observed that the drive shaft along with the heavy basket with the rotating fluid inside, is the most flexible member. Its lateral critical speeds are computed using finite element method, treating the driving shaft along with the basket as a beam. Beam element with 2 degrees of freedom (dof) per node is used for modelling the system.

For simplicity, the bearing supports can be considered to be rigid. In order to study the influence of flexibility of the bearings, the critical speeds can be found by assuming the bearings as flexible. This is done by taking into account the radial stiffness of bearings.

The stiffness of ball bearings is calculated from the Ref. [3.17]

$$K = 1.3z^{2/3}d^{1/3}F^{1/3} \qquad (3.150)$$

where z is the number of balls, d is the diameter of ball in mm, F is the force on bearing in Newton.

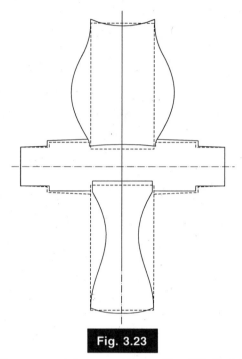

Fig. 3.23

Response of the compressor disc (III Mode)

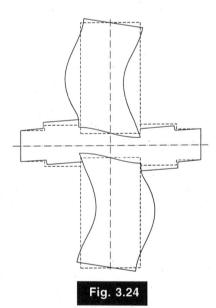

Fig. 3.24

Response of the compressor disc (IV Mode)

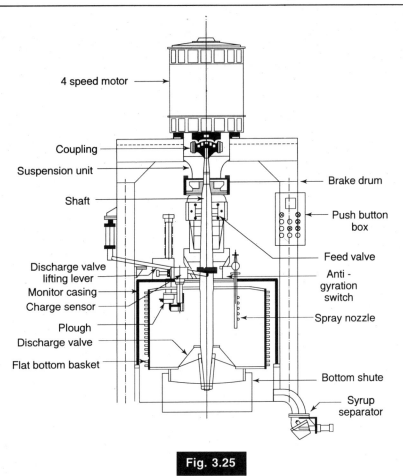

4 speed motor ⟶

Coupling ⟶

Suspension unit ⟶

Shaft ⟶

⟵ Brake drum

⟵ Push button box

Feed valve

Discharge valve lifting lever
Monitor casing
Charge sensor
Plough
Discharge valve
Flat bottom basket

Anti - gyration switch

Spray nozzle

⟵ Bottom shute

Syrup separator

Fig. 3.25

Sugar centrifugal

Similarly, the stiffness of roller bearings is calculated by the equation

$$K = 4z^{0.9}l^{0.8}F^{0.1} \qquad (3.151)$$

where l is length of roller in mm.

Lanczos method is used to determine the eigenvalues and eigenvectors.

Since the basket has to rotate when it is empty and when it is charged with massecuite, critical speeds are computed for both the cases [3.18]. The effect of the rotating liquid is taken into account only in the elemental mass matrix. The stiffness matrix will be same as given earlier. The mass matrix due to the additional liquid mass will be as follows

$$[M] = \frac{\rho + \rho_1}{2} \begin{bmatrix} 1 & & & \\ & 0 & & \\ & & 1 & \\ & & & 0 \end{bmatrix} \qquad (3.152)$$

ρ_L is liquid mass density.

If the coupling connecting the motor shaft and the centrifugal is treated as fully flexible while determining the lateral critical speeds, the centrifugal gets delinked from the motor and gets supported on two bearings with the basket overhanging. However, in the present analysis, the motor and the basket are taken together and are assumed to be supported on four bearings, since a close look at the flexible coupling construction justifies this assumption. (Slope and displacement continuity existing at the node corresponding to the coupling). The first and the third natural frequencies of the system with the motor and the basket taken together agrees very well with the first and second frequencies corresponding to the fully flexible case, since the cantilevering action of the basket is dominant.

The model assumes that the basket is connected to the central shaft through the flat bottom only. The total unit is discritized into 190 elements with 191 nodes. In the basket portion, first the nodes 119 to 155 are numbered from top to bottom along the shaft axis where each element is a part of shaft only and renumbered backwards (viz. from 155 to 191), where each element is a part of basket only (Fig. 3.26).

Critical speeds in rpm when the gyroscopic effect is neglected from lumped mass matrix approach and consistent mass matrix approach are tabulated in Table 3.2. These two sets of values are close to each other. Consistent mass matrix approach yielded the first few eigenpairs with excellent convergence within four iterations whereas the lumped mass approach needed more than ten iterations.

The critical speeds are also determined considering gyroscopic effect. The forward and backward critical speeds in rpm are tabulated in Table 3.3. Out of these two, the forward critical speeds are normally considered important. The first three modes are shown in Fig. 3.27.

Now let us consider the effect of liquid mass on the critical speeds. With the addition of liquid, fundamental critical speed has come down from 541 rpm to 400 rpm. Since this drop is significant, one can realize the importance of liquid mass on critical speeds.

The critical speeds are also determined by introducing the finite stiffness of bearings instead of assuming them as rigid. With this fundamental critical speed has decreased to 394 rpm, where as its

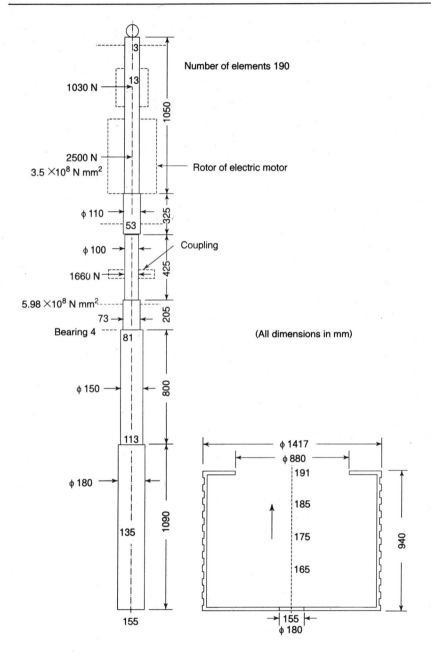

Number of elements 190

1030 N

2500 N
3.5 ×10⁸ N mm²

Rotor of electric motor

φ 110

Coupling

φ 100

1660 N

5.98 ×10⁸ N mm²

Bearing 4

(All dimensions in mm)

φ 150

φ 180

φ 1417
φ 880

φ 180

Shaft and basket numbered separately

Fig. 3.26

FEM model

Table 3.2

Critical Speeds

	Lumped mass matrix	Consistent mass matrix
1	511.6	511.6
2	2776.9	2777.9
3	3807.8	3754.4

Table 3.3

Critical Speeds Considering Gyroscopic Effect

	Forward critical speeds	Backward critical speeds
1	541.6	436.4
2	3857.6	1448.7
3	4428.0	3638.1

value with rigid bearing assumption is 541 rpm. This is also a significant decrease.

The fundamental critical speed is also determined by conducting free vibration test on the centrifugal. This is done using an impulse hammer and a spectrum analyser. The experimental frequency is 405 rpm. The value obtained using FEM considering flexibility of bearings is 400 rpm. It can be concluded that the chosen FEM model is fairly accurate.

The fundamental critical speed obtained by considering both the liquid mass and finite stiffness of bearings, is 289 rpm. But the lowest speed of operation (idling speed) is 200 rpm. As sufficient margin is there between these two speeds, there is no danger of resonance while the basket is idling.

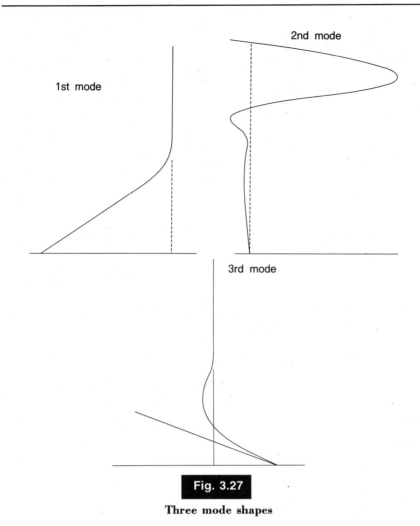

1st mode

2nd mode

3rd mode

Fig. 3.27

Three mode shapes

REFERENCES

3.1 Pestel, E.C. and F.A. Leekie, *Matrix Methods in Electromechanics,* McGraw-Hill, London 1963.
3.2 Jennings, A., *Matrix Computation for Engineers and Scientists,* John Wiley and Sons, New York, 1977.
3.3 Bathe, K.J. and E.L. Wilson, *Numerical Methods in Finite Element Analysis,* Prentice-Hall of India, New Delhi, 1978.
3.4 Meirovitch, L., *Computational Methods in Structural Dynamics,* Sijthoff and Noordhoff, Rockville, USA, 1980.
3.5 Gupta, K.K., Eigen problem solution of banded damped structural systems, *Int. J. Num. Methods in Engg.*, vol. 18, pp 877–911, 1974.

3.6 Wilkinson, J.H., *Algebraic Eigenvalue Problem*, Clarandon University, Oxford, 1965.

3.7 Jennings, A., A direct iteration method of obtaining latent roots and vectors of a symmetric matrix, *Proc. Camb. Phi. Soc.*, vol. 63, pp 755–765, 1967.

3.8 Parlett, B.N., *The Symmetric Eigenvalue Problem*, Prentice-Hall, Englewood Cliffs, N.J., 1980.

3.9 Crandall, S.H., *Engineering Analysis*, McGraw-Hill, New York, 1956.

3.10 Weaver, W. and D.M. Yoshida, The eigenvalue problem for banded matrices, *Comp. Struct.*, vol. 1, pp 651–664, 1971.

3.11 Noorr-Omid, B., B.N. Parlett and R.L. Taylor, Lanczos versus subspace iteration for solution of eigenvalue problems, *Int. J. Num. Methods in Engg.*, vol. 19, pp 859–871, 1983.

3.12 Ojalvo, I.U. and M. Newman, Vibration modes of large structures by an automatic matrix reduction method, *AIAAJ*, vol. 8, pp 1234–39, 1970.

3.13 Chowdhury, P.C. The truncated Lanczos algorithm for partial solution of the symmetric eigenproblem, *Comp. Struct.*, vol. 8, pp 439–446, 1976.

3.14 Ramamurti, V. and A.R. Pradeep Simha, Finite element calculations of critical speeds of rotation of shafts with gyroscopic action of discs, *J. Sound and Vibration*, vol. 117, no. 3, pp 578–582, 1987.

3.15 Biezeno, C.B. and R. Grammel, *Engineering dynamics*, vol. 3, *Steam Turbines*, Blackie and Sons, London, 1954.

3.16 Ramamurti, V. and O. Mahrenholtz, Application of simultaneous iteration method to flexural vibration problems, *Int. J. Mech. Sci.*, vol. 16, pp 269–283, 1974.

3.17 Krämer, E., Dynamics of Rotors and Foundations, Springer Verlag, Berlin, 1993.

3.18 C.V. Satyanarayana, 'Critical speeds of sugar centrifugals', M.Tech. Thesis, I.I.T. Madras, 1995.

EXERCISES

3.1 For a given system

$$[K] = \begin{bmatrix} 5 & -4 & 1 & 0 & 0 \\ -4 & 6 & -4 & 1 & 0 \\ 1 & -4 & 6 & -4 & 1 \\ 0 & 1 & -4 & 6 & -4 \\ 0 & 0 & 1 & -4 & 5 \end{bmatrix}$$

$$[M] = \begin{bmatrix} 1 & 0 & 0 & 0 & 0 \\ 0 & 2 & 0 & 0 & 0 \\ 0 & 0 & 3 & 0 & 0 \\ 0 & 0 & 0 & 2 & 0 \\ 0 & 0 & 0 & 0 & 1 \end{bmatrix}$$

compute the first two natural frequencies by Sturm sequence and bisection technique.

3.2 Check the results of Exercise 3.1 by inverse iteration scheme and Gram Schmidt deflation.

3.3 Compute the highest frequency of Exercise 3.1 by forward iteration scheme using the following two approaches.
(i) converting into a standard eigenvalue problem

$$[A] = [M]^{-1/2}[K][M]^{-1/2}$$

(ii) using the logic $[M]^{-1}[K]\{u\}$.

3.4 Compute the eigenvalues and eigenvectors of the system shown in Fig. 3.28 by forming the characteristic equation.
Normalise the eigenvectors and prove
$\{\delta_i\}^T[K]\{\delta_i\} = \omega_i^2$ $(i = 1, 2, 3)$
(Ans: $\omega_1 \approx 4.9$ rad/sec)

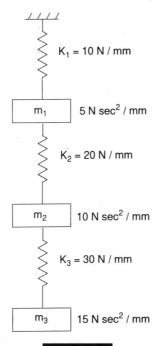

Fig. 3.28

Three mass system

3.5 Compute the lowest eigenpair by inverse iteration and the highest eigenpair by forward iteration of the system shown in Fig. 3.28. Hint for trial vector:
Inverse Iteration:

$$\{u^1\} = \begin{Bmatrix} 1 \\ 2 \\ 3 \end{Bmatrix} \quad \text{or} \quad \begin{Bmatrix} 1 \\ 0 \\ 0 \end{Bmatrix} \quad \text{or} \quad \begin{Bmatrix} 1 \\ 1 \\ 1 \end{Bmatrix}$$

Forward Iteration:

$$\{u^1\} = \begin{Bmatrix} 1 \\ -1 \\ 1 \end{Bmatrix} \quad \text{or} \quad \begin{Bmatrix} 1 \\ 0 \\ 0 \end{Bmatrix}$$

3.6 Compute the first two natural frequencies and associated vectors of Exercise 3.1 by the simultaneous iteration method by choosing three simultaneous trial vectors.

3.7 Attempt Exercise 3.4 by subspace iteration technique.

3.8 Solve the Exercise 3.4 by Lanczos method.

3.9 Shown in Fig. 3.29 is a branched torsional system. Reduce the system to a standard eigenvalue problem. Using Lanczos' method, tridiagonalise it. Compute the first natural frequency and associated vector by inverse iteration. Check tridiagonalisation by Householder's scheme. All the stiffnesses are in Nmm/rad and inertias in Nmm sec^2.

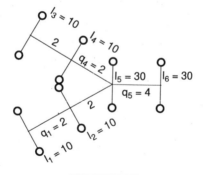

Fig. 3.29

Branched systems with 6 rotors

3.10 Figure 3.30 shows two steel discs each weighing 160 N with a radius of gyration of 150 mm located at one-third span of a shaft simply supported at the ends. Compute the critical speeds of this steam turbine shaft (i) including gyroscopic effect; and (ii) excluding gyroscopic effect of the discs ($E = 2.1 \times 10^5$ N/mm^2).

Note on computer programs related to Chapter 3.

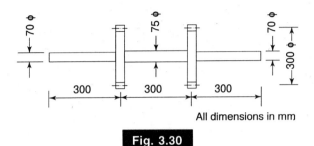

All dimensions in mm

Fig. 3.30

Critical speed determination including gyroscopic effect

Programs 3.1 to 3.8 deal with the following:

3.1 Sturm sequence to have the approximate locations of the natural frequencies.

3.2 Secant iteration scheme to determine the natural frequencies with a specified accuracy using the information given by Prog. 3.1.

3.3 Inverse iteration scheme with Gram Schmidt deflation technique to find the eigen frequencies and vectors of the system.

3.4 Simultaneous iteration scheme (nonstandard symmetric paranthesis) eigenvalue problem is transformed to a standard eigenvalue problem).

3.5 Simultaneous iteration scheme (nonstandard symmetric eigenvalue problem is retained as a nonstandard eigenvalue problem).

3.6 Subspace iteration scheme.

3.7 Simultaneous iteration scheme (nonstandard unsymmetric eigenvalue problem).

3.8 Lanczos' method.

In all these programs mass and stiffness matrices are supplied in banded form. Programs 3.1, 3.2, 3.4, 3.5 and 3.6 use Example 3.9 for demonstration. Example 3.19 is used for the Program 3.3, Example 3.12 and 3.16 are chosen to demonstrate Progs 3.7 and 3.8 respectively.

CHAPTER 4

Cyclic Symmetric Structures

Many engineering problems have identical substructures in the circumferential direction. It is in the field of mechanical engineering that one finds a variety of such applications. Some examples are— turbomachinery blades and discs, centrifugal fan impellers, rotary pumps, gear wheels, cooling towers, pressure vessels with ribs, grinding mills, turbochargers, rotor windings of electrical machines, milling cutters, reamers, gear hobbers, power saws and melting furnaces. These units can be analysed by the methods presented so far but at a cost which is prohibitive. It is good to exploit the properties associated with cyclic symmetry and reduce the labour involved in handling them. Problems of this type are discussed in References [4.1 to 4.7]. We will present in this chapter mathematical refinements one can think of in attacking these problems. The problems themselves will be grouped under three heads: (i) static or steady-state analysis of cyclic symmetric objects subjected to cyclic symmetric loading; (ii) static or steady-state analysis of cyclic symmetric objects subjected to general loading; and (iii) eigenvalue problems. Both in-core and out-of-core solutions will be presented. Methods of solution will be logical extensions of the method presented in Chapters 2 and 3.

The reason why these need special treatment is quite obvious. If one is interested in discretising the whole structure, the core and the time needed would be enormous. Let us take a very simple example of a cylindrical pressure vessel with 12 ribs along the circumference. Let us allow for 5 nodes along the axis and 5 nodes along the radial direction for each rib and allow for the same number of nodes for the sector of the pressure vessel between two successive ribs. This will demand a total of 600 (50×12) nodes for the whole problem if thin shell elements are used. Since each node has 6 degrees of freedom, this will involve 3600 unknowns with a bandwidth of 606 (101×6). If one contemplates on using Cholesky's scheme for solving a static problem, the core needed for the stiffness matrix alone would be (3600, 606) $\not\subset 2.2 \times 10^6$ dimensions (or around 18 mega bytes of computer core). The, time needed to solve the problem would be substantial. Hence, it is essential to think of a more elegant approach to handle such problems.

The ideal finite elements for solving problems under this category are triangular shell elements with six degrees of freedom at each node and eight-noded brick elements with three degrees of freedom at each node. In both the cases, stiffness and mass matrices formulated in Cartesian co-ordinates are to be transformed to the polar form resulting in displacements along the radial, circumferential and axial directions. The shell element will adequately model problems associated with bladed discs, fabricated gear wheels and pressure vessels. Brick elements can be used to solve problems of milling cutters, gear hobbers, drills and reamers, bevel, helical and hypoid gears, and thick shells with periodicity.

➤ 4.1 STATIC ANALYSIS UNDER CYCLIC SYMMETRIC LOADING

Let us consider a structure which has geometric periodicity in the circumferential direction as shown in Fig. 4.1. Let us assume that this has N repeated structures. Each repeated structure is connected to only two adjacent structures—one on either side. Let us also assume that each repeated structure is subjected to the same type of load. Examples of such loading are internal pressure and centrifugal loading. Obviously, all the repeated structures would have identical response though within a structure, the response can vary from point to point. In other words, we can consider just one repeated structure for analysis. Let the first of such structures consist of substructures (as distinct from repeated structure), as shown in Fig. 4.1.

The behaviour of the first substructure of the repeated structure (1) would be identical to the first substructure of repeated structure (2).

4.1.1 Solution Using Skyline Approach

Let us assume that the number of nodes to be handled in the circumferential direction is less than that in the radial direction. (Examples are cases wherein the number of repeated structures are very large and consequently, sector angle small). In such cases, numbering can be done along the circumferential direction commencing from the innermost circle and proceeding towards the outermost circle. For the configuration shown in Fig. 4.2, there are ten nodes from the start to the end of the first circumferential line. It may be noticed that the first node of repeated structure (1) is connected to

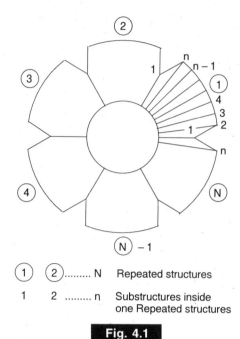

(1) (2) .... N   Repeated structures

1   2 .... n   Substructures inside
one Repeated structures

Fig. 4.1

A rotationally periodic structure

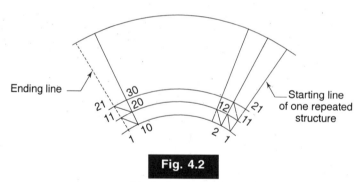

Fig. 4.2

Typical numbering

the tenth node at the repeated structure (N) and the tenth node of repeated structure (1) connected to the first node of repeated structure (2). The same will be true of other circumferential lines. If each node has 6 degrees of freedom (as in the case of shell elements), the bandwidth at node 1 will be 66 and at node 10 will also be 66. The starting line and ending line nodes like, 1, 11, 21, ..., and 10, 20, 30, ... will also have contributions from elements occupying the boundaries. This problem can easily be solved using the skyline approach and in-core solution techniques discussed in Chapter 2.

Illustrative Example

Consider the static problem as given below:

$$\begin{bmatrix} 25 & 9 & -3.5 & -5.0 & 0 & 0 & 0 & 0 & -3.5 & -4.0 \\ & 17 & -4.0 & -4.5 & 0 & 0 & 0 & 0 & -5.0 & -4.5 \\ & & 25 & 9 & -3.5 & -5.0 & 0 & 0 & 0 & 0 \\ & & & 17 & -4.0 & -4.5 & 0 & 0 & 0 & 0 \\ & & & & 25 & 9 & -3.5 & -5.0 & 0 & 0 \\ & & & & & 17 & -4.0 & -4.5 & 0 & 0 \\ & & & & & & 25 & -4.0 & -4.5 & -5.0 \\ & & & & & & & 17 & -4.0 & -4.5 \\ & & & & & & & & 25 & 9 \\ & & & & & & & & & 17 \end{bmatrix} \begin{Bmatrix} u_1 \\ v_1 \\ u_2 \\ v_2 \\ u_3 \\ v_3 \\ u_4 \\ v_4 \\ u_5 \\ v_5 \end{Bmatrix} = \begin{Bmatrix} fu_1 \\ fv_1 \\ \\ \\ \\ \\ \\ \\ fu_5 \\ fv_5 \end{Bmatrix}$$

This is a problem in 10 unknowns. This can be pictorially represented as shown in Fig. 4.3. If the firm lines represent stiffness, matrix terms k_{ij} which are non-zero and u and v at any i (varying from 1 to 5) represent two degrees of freedom at the node). k_{11}, k_{22}, k_{33}, k_{44} and k_{55} are square matrices of size (2×2) and they are identical. It may also be seen that k_{12}, k_{23}, k_{34}, k_{45}, and k_{51} are also identical. This can be considered as a cyclic symmetric problem with 5 repeated sectors. Let us now consider a set of force vectors as given below:

$$\{f_I\} = \begin{Bmatrix} 10 \\ 0 \\ 10 \\ 0 \\ 10 \\ 0 \\ 10 \\ 0 \\ 10 \\ 0 \end{Bmatrix}$$

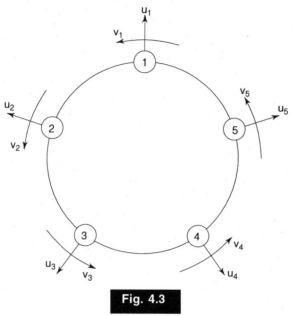

Fig. 4.3

Pictorial representation

If this problem is solved by the skyline approach (the profile being just 43) the results obtained can be summarised as follows:

$$
\{\delta_{\mathrm{I}}\} = \left\{ \begin{array}{c} .555 \\ 0 \\ .555 \\ 0 \\ .555 \\ 0 \\ .555 \\ 0 \\ .555 \\ 0 \end{array} \right\}
$$

Since the set of five force vectors in the five repeated structures are equal and are in phase, so are the displacements. The same problem could have been tackled in a much more simplified fashion thus:

$$[B_1]\{z_1\} + [A_1]\{z_2\} + [C_1]\{z_5\} = \{f_1\}$$
$$[C_2]\{z_1\} + [B_2]\{z_2\} + [A_2]\{z_3\} = \{f_2\}$$

$$\cdots \qquad \cdots \qquad \cdots \qquad \cdots$$
$$\cdots \qquad \cdots \qquad \cdots \qquad \cdots$$

$$[C_5]\{z_4\} + [B_5]\{z_5\} + [A_5]\{z_1\} = \{f_5\}$$

where
$$[B_1] = [B_2] = [B_5] = \begin{bmatrix} 25 & 9 \\ 9 & 17 \end{bmatrix}$$

$$[A_1] = [A_2] = [A_5] = \begin{bmatrix} -3.5 & -5.0 \\ -4.0 & -4.5 \end{bmatrix}$$

$$[C_1] = [C_2] = [C_5] = [A_1]^T$$

$$\{f_1\}, \ldots, \{f_5\} = \left\{ \begin{array}{c} 10 \\ 0 \end{array} \right\}$$

The problem can be solved by using only the first set of equations (or by restricting the analysis to one repeated structure).

I Set:

$$\begin{bmatrix} 25 & 9 \\ 9 & 17 \end{bmatrix} \{z_1\} + \begin{bmatrix} -3.5 & -5.0 \\ -4.0 & -4.5 \end{bmatrix} \{z_1\} + \begin{bmatrix} -3.5 & -4.0 \\ -5.0 & -4.5 \end{bmatrix} \{z_1\} = \left\{ \begin{array}{c} 10 \\ 0 \end{array} \right\}$$

Therefore

$$\begin{bmatrix} 18 & 0 \\ 0 & 8 \end{bmatrix} \left\{ \begin{array}{c} u_1 \\ v_1 \end{array} \right\} = \left\{ \begin{array}{c} 10 \\ 0 \end{array} \right\}$$

$$u_1 = 10/8 = 0.555, \qquad v_1 = 0$$

$$\{z_1\} = \{z_2\} = \{z_3\} = \{z_4\} = \{z_5\} = \left\{ \begin{array}{c} 0.555 \\ 0 \end{array} \right\}$$

It is seen from this example that the analysis can be restricted to one repeated structure if the loads on all the repeated structures are identical. When one repeated structure is, by itself, big, out-of-core solution for this structure is to be resorted to.

It is not very difficult to visualise cases where the in-core solution becomes very difficult even though the analysis is restricted to only one repeated structure. To cite an example, let us assume that there are 20 nodes along each circumferential line and there are about 30 such circumferential lines starting from the inner radius to the outer radius. For six degrees of freedom at each node, this problem will have a total 3600 unknowns with a band width of $(21 \times 6) = 126$. The in-core needed for storing the stiffness matrix alone would be more than 0.5×10^6. This obviously is a fairly large core. For such problems, it is better to use Potters' scheme and out-of-core solution.

4.1.2   Out-of-Core Solution

A structure having N repeated structures is shown in Fig. 4.1. Each repeated structure is assumed to be made up of n substructures. Loading on each repeated structure is assumed to be identical. The static equalibrium for any of the repeated structures can be written as follows [4.8]:

$$
\begin{bmatrix}
B_1 & A_1 & & & C_1 \\
C_2 & B_2 & A_2 & & \\
A_n & & & C_n & B_n
\end{bmatrix}
\begin{Bmatrix}
z_1 \\
z_2 \\
z_n
\end{Bmatrix}
=
\begin{Bmatrix}
g_1 \\
g_2 \\
g_n
\end{Bmatrix}
\tag{4.1}
$$

Submatrices elimination scheme

When the finite element method is employed, the basic stiffness matrix is symmetric. In Eq. (4.1) $[C_i] = [A_{i-1}^T]$. Hence Eq. (4.1) can be written as follows:

$$
\begin{bmatrix}
[B_1] & [A_1] & 0 & 0 & & & & & [A_n^T] \\
[A_1^T] & [B_2] & [A_2] & 0 & & & & & 0 \\
0 & [A_2^T] & [B_3] & [A_3] & & & & & 0 \\
\cdots & \cdots & \cdots & \cdots & \cdots & & & & \cdots \\
& & & & [A_{n-3}^T] & [B_{n-2}] & [A_{n-2}] & & 0 \\
& & & & & [A_{n-2}^T] & [B_{n-1}] & [A_{n-1}] \\
[A_n] & & & & & & [A_{n-1}^T] & [B_n]
\end{bmatrix}
$$

$$
\begin{bmatrix}
\{z_1\} \\
\{z_2\} \\
\{z_3\} \\
\cdots \\
\{z_{n-2}\} \\
\{z_{n-1}\} \\
\{z_n\}
\end{bmatrix}
=
\begin{bmatrix}
\{g_1\} \\
\{g_2\} \\
\{g_3\} \\
\cdots \\
\{g_{n-2}\} \\
\{g_{n-1}\} \\
\{g_n\}
\end{bmatrix}
\tag{4.2}
$$

In the first cycle of operation, the above equation can be simplified to the following form:

$$
\begin{bmatrix}
[L_1^T] & [P_1] & & & & & [Q_1] \\
& [L_2^T] & [P_2] & & & & [Q_2] \\
& & [L_3^T] & [P_3] & & & [Q_3] \\
\dots & \dots & \dots & \dots & \dots & & \dots \\
& & & [L_{n-2}^T] & [P_{n-2}] & & [Q_{n-2}] \\
& & & & [L_{n-1}^T] & & [Q_{n-1}] \\
A_n & & & & & [A_{n-1}^T] & [B_n]
\end{bmatrix}
$$

$$
\begin{bmatrix}
\{z_1\} \\
\{z_2\} \\
\{z_3\} \\
\dots \\
\{z_{n-2}\} \\
\{z_{n-1}\} \\
\{z_n\}
\end{bmatrix}
=
\begin{bmatrix}
\{q_1\} \\
\{q_2\} \\
\{q_3\} \\
\dots \\
\{q_{n-2}\} \\
\{q_{n-1}\} \\
\{q_n\}
\end{bmatrix}
\tag{4.3}
$$

where $[L_1]$, $[P_1]$, $[Q_1]$ are given by

$$[L_1][L_1^T] = [B_1]$$

$$[L_1][P_1] = [A_1]$$

$$[L_1][Q_1] = [A_n^T]$$

$$[L_1]\{q_1\} = \{g_1\} \tag{4.4}$$

Similarly $[L_i]$, $[P_i]$, $[Q_i]$ can be calculated from,

$$[L_i][L_i^T] = [[B_i] - [P_{i-1}^T][P_{i-1}]],$$

$$[L_i][P_i] = [A_i]$$

$$[L_i][Q_i] = -[[P_{i-1}^T]\cdot[Q_{i-1}]]$$

$$[L_i]\{q_i\} = [\{g_i\} - [P_{i-1}^T]\{q_{i-1}\}], \quad i = 1, 2, 3, \dots, n-1 \tag{4.5}$$

and

$$[\overline{Q}_{n-1}] = [P_{n-1}] + [Q_{n-1}]] \tag{4.6}$$

These operations are similar to the ones in Eq. (2.65).

Cholesky factorisation of the relevant square matrices is possible since all of them are real and symmetric.

The last row in Eq. [4.2] is given by

$$[A_n]\{z_1\} + [B_n]\{z_n\} + [A_{n-1}^T]\{z_{n-1}\} = \{g_n\} \tag{4.7}$$

Pre-multiplying the first row in Eq. (4.3) by $[Q_1^T]$ gives,

$$[Q_1^T][L_1^T]\{z_1\} + [Q_1^T][P_1]\{z_2\} + [Q_1^T][Q_1]\{z_n\} = [Q_1^T]\{q_1\} \tag{4.8}$$

$$[Q_1^T][L_1^T] = ([L_1][Q_1])^T, \text{ which from Eq. (4.4)} = [A_n] \tag{4.9}$$

Substituting Eq. (4.9) in Eq. (4.8) and subtracting from Eq. (4.7) gives,

$$([B_n] - [Q_1^T][Q_1])\{z_n\} - [Q_1^T][P_1]\{z_2\} + [A_{n-1}^T]\{z_{n-1}\}$$

$$= (\{g_n\} - [Q_1^T]\{q_1\}) \tag{4.10}$$

Operations on the second row similar to the ones in Eqs (4.8), (4.9) and (4.10) give

$$[Q_2^T][L_2^T]\{z_2\} + [Q_2^T][P_2]\{z_3\} + [Q_2^T][Q_2]\{z_n\} = [Q_2^T]\{q_2\} \tag{4.11}$$

$$[Q_2^T][L_2^T] = ([L_2][Q_2])^T = -([P_1^T][Q_1])^T = -[Q_1^T][P_1] \tag{4.12}$$

$$([B_n] - [Q_1^T][Q_1] - [Q_2^T][Q_2]\{z_n\} - [Q_2^T][P_2]\{z_3\} + [A_{n-1}^T]\{z_{n-1}\}]$$

$$= (\{g_n\} - [Q_1^T]\{q_1\} - [Q_2^T]\{q_2\} \tag{4.13}$$

This can be continued for the first $(n-2)$ rows, resulting in an equation similar to Eq. (4.13) given by

$$([B_n] - [Q_1^T][Q_1] - [Q_2^T][Q_2] - \ldots - [Q_{n-1}^T][Q_{n-2}]\{z_n\}$$

$$- [Q_{n-2}^T][P_{n-2}]\{z_{n-1}\} + [A_{n-1}^T]\{z_{n-1}\}$$

$$= \{g_n\} - [Q_1^T]\{q_1\} - [Q_2^T]\{q_2\} - \ldots - [Q_{n-2}^T]\{q_{n-2}\} \tag{4.14}$$

Pre-multiplying both sides of the $(n-1)$th row of Eq. (4.3) by $[\bar{Q}_{n-1}^T]$, we have

$$[\bar{Q}_{n-1}^T][L_{n-1}^T]\{z_{n-1}\} + [\bar{Q}_{n-1}^T][\bar{Q}_{n-1}]\{z_n\} = [\bar{Q}_{n-1}^T]\{q_{n-1}\} \tag{4.15}$$

Here,

$$[\bar{Q}_{n-1}^T][L_{n-1}^T] = ([L_{n-1}][\bar{Q}_{n-1}])^T \tag{4.16}$$

Equation (4.16) can be simplified and written in the following form

$$([\bar{Q}_{n-1}])^T[L_{n-1}^T] = ([L_{n-1}]\cdot([P_{n-1}] + [Q_{n-1}]))^T \text{ from Eq. (4.6)}$$

$$= [[A_{n-1}] - [P_{n-2}^T]\cdot[Q_{n-2}]]^T \text{ from Eq. (4.5)}$$

$$= [A_{n-1}^T] - [Q_{n-2}^T]\cdot[P_{n-2}] \tag{4.17}$$

Combining Eqs (4.14), (4.15) and (4.17), we have

$$[\bar{B}_n]\{z_n\} = \{\bar{g}_n\} \tag{4.18}$$

where

$$[\bar{B}_n] = [[B_n] - [Q_1^T][Q_1] - [Q_2^T][Q_2] - \ldots$$

$$- [Q_{n-2}^T]\{Q_{n-2}\} - [\bar{Q}_{n-1}^T]\{\bar{Q}_{n-1}\}]$$

and

$$\{\bar{g}_n\} = [\{g_n\} - [Q_1^T]\{q_1\} - [Q_2^T]\{q_2\} - \ldots$$

$$-[Q_{n-2}^T]\{q_{n-2}\} - [\bar{Q}_{n-1}^T]\{q_{n-1}\}] \tag{4.19}$$

Hence Eq. (4.3) can be written as follows:

$$\begin{bmatrix} [L_1^T] & [P_1] & & & & & [Q_1] \\ & [L_2^T] & [P_2] & & & & [Q_2] \\ & & [L_3^T] & [P_3] & & & [Q_3] \\ \ldots & \ldots & \ldots & \ldots & & \ldots & \ldots \\ & & & [L_{n-2}^T] & [P_{n-2}] & [Q_{n-2}] \\ & & & & [L_{n-1}^T] & [\bar{Q}_{n-1}] \\ & & & & & [L_n^T] \end{bmatrix}$$

$$\begin{Bmatrix} \{z_1\} \\ \{z_2\} \\ \{z_3\} \\ \ldots \\ \{z_{n-2}\} \\ \{z_{n-1}\} \\ \{z_n\} \end{Bmatrix} = \begin{Bmatrix} \{q_1\} \\ \{q_2\} \\ \{q_3\} \\ \ldots \\ \{q_{n-2}\} \\ \{q_{n-1}\} \\ \{q_n\} \end{Bmatrix} \tag{4.20}$$

Here, $[L_n]$ and $\{q_n\}$ are estimated from

$$[L_n][L_n^T] = [\bar{B}_n]$$

$$[L_n]\{q_n\} = \{\bar{g}_n\} \tag{4.21}$$

Thus, for each partition, values of $[L]^T$, $[Q]$, $[P]$ and $\{q\}$ are calculated. A simple backward substitution in the last row of Eq. (4.20) gives $\{z_n\}$. Then a series of similar backward substitutions in Eq. (4.20) gives the values of $\{z_{n-1}\}$, $\{z_{n-2}\}$, $\{z_{n-3}\}$, ..., $\{z_1\}$.

A flowchart explaining scheme II is given in Fig. 4.4.

4.1.3   Number of Multiplications Involved

Table 4.1 gives the labour involved. For a specific case of s (number of degrees of freedom in each substructure) = 8 and n (number of substructures) = 8, total multiplications are = 10.9×10^6.

Example 4.1

A square plate with line load at the central circular hole [4.9] has been selected and is shown in Fig. 4.5. Its outer edges are clamped. To get a line load, a uniformly distributed load is assumed to be acting on a very narrow annular space of length (0.5×10^{-4} times the side of the square at the inner periphery of the hole). Deflections at different points of the circular hole for various inner radii are calculated. They are compared with closed form solutions in Fig. 4.6. The agreement is very fair.

In Fig. 4.7 deflections at the central hole for ring load have been estimated and they are checked with classical solutions in Table 4.2. The agreement is found to be good.

➤ 4.2   STATIC ANALYSIS UNDER GENERALISED LOADING _____

4.2.1   Electrical Analogy

A cyclic symmetric structure subjected to general loading can be best explained by an electrical analogy. Let us consider the electrical circuit shown in Fig. 4.8. This circuit has four branches. The resistances of the four branches are identical. But the valtage in these four branches are not identical. Hence the currents flowing in these four branches will not be the same. This problem is analogous to the behaviour of a cyclic symmetric structure subjected to asymmetric loading. Here we have four repeated structures.

Let the currents flowing in these branches be I_1, I_2, I_3 and I_4.

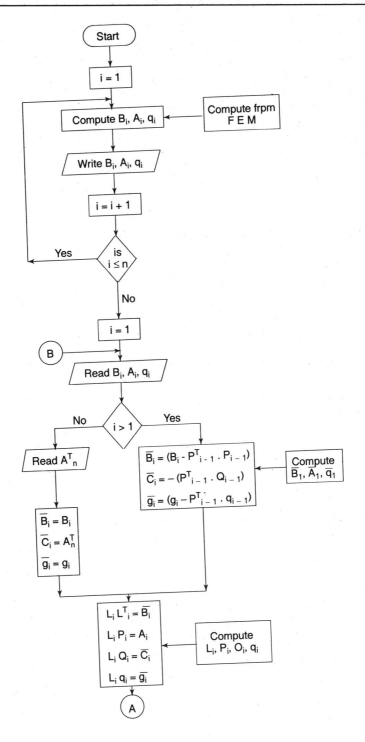

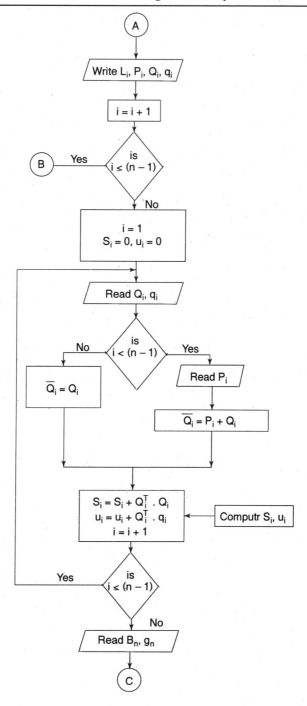

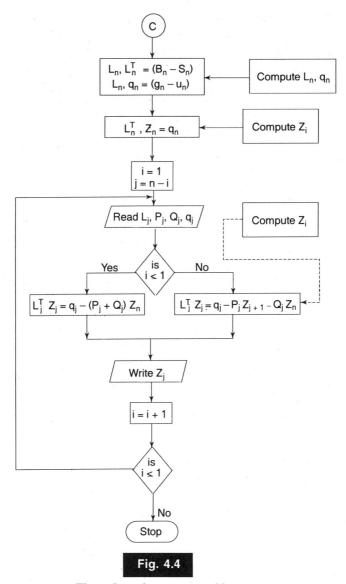

Fig. 4.4

Flow chart for static problem

Let us solve this problem by using Kirchhoffs law. The resulting equations for the four branches are as follows:

$$
\left.
\begin{aligned}
10I_1 + 5(I_1 - I_2) + 5(I_1 - I_4) &= -1.5 \\
10I_2 + 5(I_2 - I_3) + 5(I_2 - I_1) &= 0.5 \\
10I_3 + 5(I_3 - I_2) + 5(I_3 - I_2) &= 0.5 \\
10I_4 + 5(I_4 - I_3) + 5(I_4 - I_3) &= 0.5
\end{aligned}
\right\}
\tag{4.22}
$$

Table 4.1

Total Number of Multiplications Involved in Cyclic Symmetry (Static)

1. *Reduction stage:*
   First partition $[s^3(7/6) + s^2(1/2)]$
   Second to $(n - 1)$th $[s^3(16/6)(n - 2)$
   partition $+ s^3(3/2)(n - 2)]$
   nth partition $[s^3(3n - 2/6)$
                                                     $+ s^2(2n - 1/2)]$

2. *Back substitution:*
   nth partition $[s^2(1/2)]$
   $(n - 1)$th partition $[s^2(3/2)]$
   $(n - 1)$th partition $[s^2(5/2)(n - 2)]$
   first partition
   Total $[s^3(19n - 27/6)$
                                                     $+ s^2(5n - 6)]$

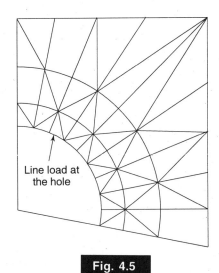

Fig. 4.5

Square plate with line load at the hole

Line load at
the hole

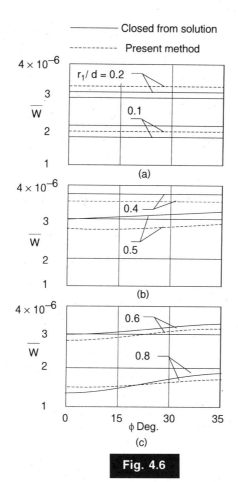

——————— Closed from solution

----------- Present method

Fig. 4.6

Comparison of non-dimensional deflection at circular hole for line load at the hole

Rearranging,

$$
\left.
\begin{aligned}
20I_1 - 5I_2 - 5I_4 &= -1.5 \\
20I_2 - 5I_3 - 5I_1 &= 0.5 \\
20I_3 - 5I_4 - 5I_2 &= 0.5 \\
20I_4 - 5I_1 - 5I_3 &= 0.5
\end{aligned}
\right\}
\tag{4.23}
$$

Solving, we have $I_1 = -0.066$ A, $I_2 = 0.0166$ A, $I_3 = 0.033$ A, $I_4 = 0.0166$ A. Let us redo this problem by another approach. The variation in voltage around the circumference (Fig. 4.8) can be

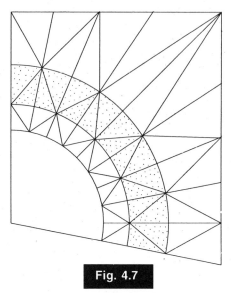

Fig. 4.7

Square plate with uniform ring load

Table 4.2

Square Plate Having Ring Loading
$r_2/d = 0.6$ and $r_3/d = 08$
Boundary conditions: Outer edges clamped and inner circle free

r_1/d	w by Reference [4.29]	w by method of cyclic symmetry
0.2	– 0.00464	– 0.005102
0.4	– 0.00446 to – 0.00456	– 0.004195 to –0.004264
0.5	– 0.00388 to – 0.00413	– 0.00355 to –0.00417

considered as a sum of a finite Fourier series with four terms expressed as follows

$$V_k = a_0 + \sum_{m=1}^{3} a_m e^{i\,m(k-1)\psi} = 0 + \sum_{m=1}^{3} - 0{,}5e^{i\,m(k-1)\psi} \quad (4.24)$$

where $\psi = \pi/2$ and k represents the branch.

a_0, a_1, a_2, a_3 can be considered as the magnitudes of the Fourier harmonics of the voltage for $m = 0, 1, 2$ and 3, their values being 0, – .5, – .5 and – .5 respectively and V_1, V_2, V_3, V_4 being the voltages in each branch.

Similarly i_1, i_2, i_3, i_4 are the Fourier harmonics ($m = 0, 1, 2, 3$) of the current and I_1, I_2, I_3, I_4 are the currents in each branch.

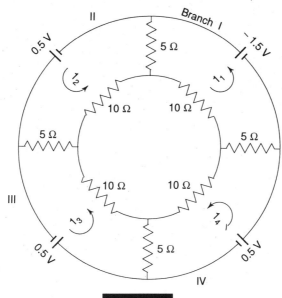

Fig. 4.8

Electrical analogy

$V_1 = 0 + (3 \times - 0.5) = - 1.5$ V,

$V_2 = 0 - 0.5(e^{i\pi/2} + e^{i\pi} + e^{i3\pi/2}) = - 0.5$ V,

$V_3 = 0 - 0.5(e^{i\pi} + e^{i2\pi} + e^{i3\pi}) = - 0.5$ V,

$V_4 = 0 - 0.5(e^{i3\pi/2} + e^{i3\pi} + e^{i9\pi/2}) = - 0.5$ V.

For any harmonic m, let us assume that the current in any branch bears a constant ratio to the adjacent branch $(k - 1), e^{im\psi}$. This is valid since the voltage in any branch k, for any harmonic m, bears a constant ratio to the voltage in the $(k - 1)$th branch, $e^{im\psi}$. Then considering the first of the four equations for any harmonic, the Fourier components of the current can be inferred.

When $m = 0$, $(20i_1 - 5i_1e^0 - 5i_1e^0) = a_0$ from the first of Eq. (4.23)

Hence, $i_1(20 - 5 - 5) = 0.0$, $\therefore i_1 = 0.0$ A.

When $m = 1$, $i_2 (20 + 5 (e^{i\pi/2} + e^{-i\pi/2}) = 0.5$, $\therefore i_2 = 0.025$ A.

Similarly,

When $m = 2$, $i_3 = - 0.0166$ A.

When $m = 3$, $i_4 = - 0.025$ A.

Hence,

$$I_1 = i_1 + i_2 + i_3 + i_4 = -.066 \text{ A}.$$
$$I_2 = i_1 + i_3 (e^{-2\pi}) = 0.166 \text{ A}.$$
$$I_3 = i_1 + i_2(e^{-3i\pi/2}) + i_3(e^{-i3\pi}) + i_4(e^{-i9\pi/2}) = 0.033 \text{ A}.$$

The same values as the ones obtained earlier.

These values I_1 to I_4 were obtained by using the first of Eq. (4.23) (using one branch only) without using all the four equations. (Solving for all the branches together).

The following inferences can be made from this study:

(i) The general variation in loading (here voltage) along the circumference can be expressed as a finite Fourier series.

(ii) The resulting displacement vector (here the current) for every harmonic m bears a constant ratio $e^{m(k-1)\psi}$ between successive repeated structures (here branches) k and $(k + 1)$.

(iii) The number of distinct harmonics are $(N/2 + 1)$ when N is even and $(N + 1)/2$ when N is odd, where N is the number of repeated structures (here branches).

For the problem considered here, the number of unknowns n in each repeated structure is only one (current). In industrial problems these will be the total number of degrees of freedom in each repeated structure, which will be very high (around 6000 to 10000); and the number of repeated structures, N may vary from as low as 6 (in the case of fabricated gears) to as high as 100 (in the case of blades along the circumference of a disc in turbo machines). For this problem connected with the electrical circuit, N is 4, $n = 1$ and hence the classical solution (approach 1) is easy. But when n is very large, the classical approach requires $(N \times n)$ simultaneous equations to be solved, which is impossible. Hence the second approach is adopted wherein we have N decoupled problems, each of size n. But these decoupled problems of size n are themselves banded and partly complex. They are the ones discussed in this chapter.

If we consider a ribbed pressure vessel under internal pressure and a pair of mating gear wheels, we see the difference—though in both the cases the geometry is periodic. In the former case, the loading is axisymmetric whereas in the other case, typically a single pair of gear teeth is assumed to take the load. In the previous section, we consider the analysis of cyclic symmetric structures under cyclic symmetric loading. Since the geometry is periodic and the loading symmetric, the deformations of all repeating structures are identical. In the more general case of non-cyclic symmetric loading, this, obviously, is not true. We shall discuss such problems here. The concepts developed in this section will also be useful in the subsequent discussion on the eigenvalue problems of cyclic symmetric structures.

4.2.2 Illustrative Example

Let us consider the problem discussed in Section 4.1.1.

$$\begin{bmatrix} 25 & 9 & -3.5 & -5 & 0 & 0 & 0 & 0 & -3.5 & -4 \\ & 17 & -4 & -4.5 & 0 & 0 & 0 & 0 & -5 & -4.5 \\ & & 25 & 9 & -3.5 & -5 & 0 & 0 & 0 & 0 \\ & & & 17 & -4 & -4.5 & 0 & 0 & 0 & 0 \\ & & & & 25 & 9 & -3.5 & -5 & 0 & 0 \\ & & & & & 17 & -4 & -4.5 & 0 & 0 \\ & & & & & & 25 & 9 & -3.5 & -5 \\ & & & & & & & 17 & -4 & -4.5 \\ & & & & & & & & 25 & 9 \\ \text{Sym.} & & & & & & & & & 17 \end{bmatrix} \begin{Bmatrix} u_1 \\ v_1 \\ u_2 \\ v_2 \\ u_3 \\ v_3 \\ u_4 \\ v_4 \\ u_5 \\ v_5 \end{Bmatrix} = \begin{Bmatrix} f_{u1} \\ f_{v1} \\ . \\ . \\ . \\ . \\ . \\ . \\ f_{u5} \\ f_{v5} \end{Bmatrix}$$

4.26

Consider three sets of force vectors as given below:

$$f_{\rm I} = \{10 \quad 0 \quad 10 \quad 0 \quad 10 \quad 0 \quad 10 \quad 0 \quad 10 \quad 0\}^T$$
$$f_{\rm II} = \{10 \quad e^{i2\pi/5} \quad 0 \quad 10\,e^{i4\pi/5} \quad 0 \quad 10 \quad e^{i6\pi/5} \quad 0 \quad 10\,e^{i8\pi/5} \quad 0 \quad 10 \quad 0\}^T$$
$$f_{\rm III} = \{10 \quad e^{i4\pi/5} \quad 0 \quad 10\,e^{i8\pi/5} \quad 0 \quad 10 \quad e^{i12\pi/5} \quad 0 \quad 10^{ei16\pi/5} \quad 0 \quad 10 \quad 0\}^T$$

(4.27)

On solving equation (1) by Skyline approach [4.10] the corresponding deformations are obtained as:

$$\delta_{\rm I} = \{ \; 0.555 \quad 0 \quad 0.555 \quad\quad 0 \quad\quad ... \quad\quad 0.555 \quad\quad 0 \; \}^T$$

$$\delta_{\rm II} = \{ \; u_1 e^{i2\pi/5} \quad v_1 e^{i2\pi/5} \quad u_1 e^{i4\pi/5} \quad\quad v_1 e^{i4\pi/5} \quad ... \quad u_1 v_1 \; \}^T$$

where

$$u_1 = 0.4987 + i0.0 \text{ and } v_1 = -0.2181 - i0.033,$$

$$\delta_{\rm III} = \{u_1 \, e^{i4\pi/5} \quad v_1 \, e^{i4\pi/5} \quad u_1 \, e^{i8\pi/5} \quad v_1 \, e^{i8\pi/5} \quad ... \quad u_1 v_1\}^T \qquad (4.28)$$

where

$$u_1 = 0.5068 + i0.0 \quad \text{and} \quad v_1 = -0.3398 - i0.0123$$

The force-deformation relationship is depicted in Fig. (4.9). It is observed that the phase relationship of the forces is directly reflected in the resulting deformations. It implies that in each case we could

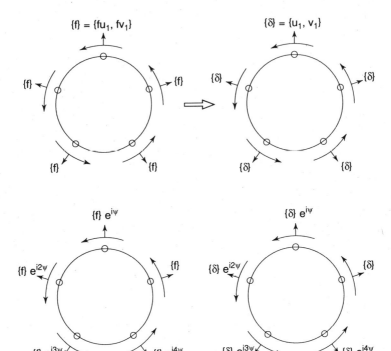

Fig. 4.9

Response of a cyclic symmetric structure

have simply considered just one repeating sector, the displacements in other sectors being elegantly related to those in this.

The equilibrium equation for the fifth nodal point (repeated structure) from Eq. (4.26) is

$$B_1 z_1 + A_1 z_2 + C_1 z_5 = f_f \qquad (4.29)$$

where

$$\mathbf{B} = \begin{bmatrix} 25 & 9 \\ 9 & 17 \end{bmatrix}, \qquad \mathbf{A} = \begin{bmatrix} -3.5 & -5 \\ -4 & -4.5 \end{bmatrix}$$

$$\mathbf{C} = \begin{bmatrix} -3.5 & -4 \\ -5 & -4.5 \end{bmatrix}, \qquad \mathbf{f} = \begin{Bmatrix} 10 \\ 0 \end{Bmatrix}$$

Also

$$z_2 = z_1\, e^{i\psi} \text{ and } z_5 = z_1\, e^{-\lambda\psi} \qquad (4.30)$$

where

$$\psi = 0,\ 2\pi/5,\ 4\pi/5 \text{ for } \mathbf{f}_{\mathrm{I,II,III}}.$$

Substituting Eq. (4.30) in Eq. (4.29)

$$\begin{bmatrix} (25 - 3.5e^{i\psi} - 3.5e^{-i\psi}) & (9 - 5e^{i\psi} - 4e^{-i\psi}) \\ (9 - 4e^{i\psi} - 5e^{-\psi}) & (17 - 4.5e^{i\psi} - 4.5e^{-i\psi}) \end{bmatrix} \begin{Bmatrix} u_1 \\ v_1 \end{Bmatrix} = \begin{Bmatrix} 10 \\ 0 \end{Bmatrix}$$
(4.31)

On solving Eq. (4.31)

$$\begin{Bmatrix} u_1 \\ v_1 \end{Bmatrix}_{\text{I}} = \begin{Bmatrix} 0.555 \\ 0 \end{Bmatrix}, \quad \begin{Bmatrix} u_1 \\ v_1 \end{Bmatrix}_{\text{II}} = \begin{Bmatrix} 0.498 \\ -0.2181 - i0.033 \end{Bmatrix}$$

$$\begin{Bmatrix} u_1 \\ v_1 \end{Bmatrix}_{\text{III}} = \begin{Bmatrix} 0.5068 \\ -0.3398 - i0.0123 \end{Bmatrix}$$

These are the same as those obtained upon direct solution of Eq. (4.26). To summarize:

(i) If the forces in the repeating sectors are related as in Fig. (4.9), the corresponding displacements in the substructures are so related.

(ii) The structure behaviour can then be obtained from the analysis of a single repeating sector, stipulating the above variation as a precondition.

Thus, if a generalized external force acting on the cyclic symmetric structure (e.g. load carried by a pair of meshing gears) is represented as a sum of Fourier harmonics, then for each such harmonic, analysis of a single repeated sector is possible. It is worth mentioning here that this is, in fact, an extension of the standard practice for axisymmetric structures under arbitrary external loading. The forces and corresponding displacements are expressed in terms of Fourier series (orthogonal trigonometric series expressions), and analysis is carried out for each harmonic individually. The total response is obtained as a sum of the individual harmonic responses.

4.2.3 Finite Fourier Series

The Fourier series representation of an external force on one of the n substructures of the kth repeated structure can be expressed in the form

$$a_k\,(r, z) = f_1 + f_2 e^{i(k-1)\psi} + f_3 e^{i2(k-1)\psi} + f_4 e^{i3(k-1)\psi} + f_5 e^{i4(k-1)\psi} +$$
$$\cdots\cdots + f_N e^{i(N-1)(k-1)\psi}$$
(4.32)

where $f_j\,(r, z)$ are the coeficients corresponding to the jth harmonic for any load vector in a substructure and ψ is the angle between any two repeated structures. There will be n such sets for the load on all the n substructures inside one repeated structure. In matrix notation

$$\begin{bmatrix} 1 & 1 & 1 & \cdots \\ 1 & e^{i\psi} & e^{2i\psi} & \cdots \\ \vdots & \vdots & \vdots & \\ 1 & e^{i(N-1)i\psi} & e^{i2(N-1)i\psi} & e^{i(N-1)^2 i\psi} \end{bmatrix} \begin{Bmatrix} f_1 \\ f_2 \\ \vdots \\ f_N \end{Bmatrix} = \begin{Bmatrix} a_0 \\ a_1 \\ \vdots \\ a_{N-1} \end{Bmatrix} \quad (4.33)$$

where $\psi = 2\pi/N$. Such Fourier coefficients are to be evaluated for each substructure. The choice of $e^{i\psi}$ in Eq. (4.33) results in $e^{-i\psi}$ in evaluating the Fourier coefficients. Such an elegant simplification would not be possible for the choice of simple trigonometric functions like $\cos \psi$ and $\sin \psi$ used separately.

Example 4.2
Given in Fig. 4.10 is a cyclic symmetric structure with four identical substructures and subjected to loads as shown. Compute the Fourier coefficients. Load on the first structure is 1, the loads on other substructures are zero.

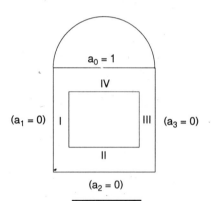

Fig. 4.10

Cyclic symmetric structure with general loading

Hence $a_0 = 1$, $a_1 = a_2 = a_3 = 0$,
Hence

$$\begin{Bmatrix} f_1 \\ f_2 \\ f_3 \\ f_4 \end{Bmatrix} = \frac{1}{4} \begin{bmatrix} 1 & 1 & 1 & 1 \\ 1 & e^{-\pi i/2} & e^{-2\pi i/2} & e^{-2\pi i/2} \\ 1 & e^{-\pi i} & e^{-2\pi i} & e^{-3\pi i} \\ 1 & e^{-3\pi i/2} & e^{-3\pi i} & e^{-4.5\pi i} \end{bmatrix} \begin{Bmatrix} 1 \\ 0 \\ 0 \\ 0 \end{Bmatrix}$$

$$= \frac{1}{4} \begin{Bmatrix} 1 \\ 1 \\ 1 \\ 1 \end{Bmatrix} = \begin{Bmatrix} 0.25 \\ 0.25 \\ 0.25 \\ 0.25 \end{Bmatrix}$$

All the four Fourier harmonics have equal intensity. This can easily be verified.

$$a_0 = f_0 + f_1 + f_2 + f_3 = 1$$
$$a_1 = (f_3 e^{i3\pi/2} + f_0 + f_1 e^{i\pi/2} + f_2 e^{i\pi}) = 0$$
$$a_2 = (f_3 e^{i3\pi} + f_0 + f_1 e^{i\pi} + f_2 e^{i2\pi}) = 0$$
$$a_3 = (f_3 e^{i9\pi/2} + f_0 + f_1 e^{i3\pi/2} + f_2 e^{i3\pi}) = 0$$

Typical general loading

Let us reconsider Eq. (4.26) but with the right-hand-side force vector as given below:

$$\mathbf{f} = \{10 \quad 0 \quad 8 \quad 0 \quad 6 \quad 0 \quad 4 \quad 0 \quad 2 \quad 0\}^T \tag{4.34}$$

This problem can easily be solved directly, obtaining

$$\{\delta\} = \{0.5344 \quad -0.1215 \quad ... \quad 0.1322 \quad 0.1016\}^T \tag{4.35}$$

Let us now apply the cyclic symmetry approach of the preceding sections. The Fourier coefficients following Eq. (4.33) are given as

$$f_{2,5} = -1 \mp i1.376, f_{3,4} = -1 \mp i0.31684$$

$$f_1 = 6 \tag{4.36}$$

It is to be observed that f_1, the symmetric load, is the average of the five loads $a_1, a_2, ..., a_5$. Solving the five cases independently for the fifth substructure alone,

$$\begin{bmatrix} (25 - 7\cos\psi_j) & (9 - 9\cos\psi_j - i\sin\psi_j) \\ (9 - 9\cos\psi_j + i\sin\psi_j) & (17 - 9\cos\psi_j) \end{bmatrix} \begin{Bmatrix} u_5 \\ v_5 \end{Bmatrix} = \begin{Bmatrix} f_j \\ 0 \end{Bmatrix} \tag{4.37}$$

where $j = 1, 2, 3, 4, 5$ and $\psi_j = 2\pi(j-1)/5$

Solving the Hermitian matrix problem of Eq. (4.37) with the corresponding values from Eq. (4.36), the following results are obtained:

$$z_{11} = \begin{Bmatrix} -0.0498 - 0.068i \\ 0.0257 + 0.034i \end{Bmatrix}, \quad z_{12} = \begin{Bmatrix} -0.0508 - 0.0162i \\ 0.0339 + 0.012i \end{Bmatrix}$$

$$z_{13} = \begin{Bmatrix} -0.0508 + 0.0162i \\ 0.0339 - 0.012i \end{Bmatrix}, \quad z_{14} = \begin{Bmatrix} -0.0498 + 0.068i \\ 0.0257 - 0.034i \end{Bmatrix} \tag{4.38}$$

$$z_{15} = \begin{Bmatrix} 0.33 \\ 0 \end{Bmatrix}$$

For the first substructure, the net displacement

$$\begin{Bmatrix} u_1 \\ v_1 \end{Bmatrix} = \sum_{j=1}^{5} z_{1j} = \begin{Bmatrix} 0.1322 \\ 0.1016 \end{Bmatrix} \qquad (4.39)$$

For the second repeated structure

$$\begin{Bmatrix} u_2 \\ v_2 \end{Bmatrix} = \sum_{j=1}^{5} z_{2j} e^{i\psi} = \begin{Bmatrix} 0.5344 \\ -0.1215 \end{Bmatrix}$$

These are the same as those given in Eq. (4.35).
We have

$$a_1 = f_1 + f_2 e^{i\psi} + f_3 e^{2i\psi} +, \; ...$$

Since $\psi = 2\pi/5$, $e^{i\psi}$ and $e^{4i\psi}$ are complex conjugate, likewise are $e^{2i\psi}$ and $e^{3i\psi}$. Since a_2 and a_5 are real, f_2 and f_5 should be complex conjugate, so should f_3 and f_4 be. It may be observed that there are only five unknowns in f_1 to f_5 (real and imaginary parts put together) and there are only five arbitrary quantities a_1 to a_5. Hence, f_1 to f_5 will be unique for a given set of a's. Associated displacement components in any repeated structure due to harmonics 2 and 5 are complex conjugate, so are the ones due to 3 and 4.

In general, for a problem having N repeated structures, for values of $\psi = 2\pi i/N$ and $\psi = (2\pi(N - i))/N$ the imaginary parts of the displacement values will be equal and opposite. Hence, the cumulative sum will always be real. For a complete problem, it is enough to compute only $(N/2) + 1$ harmonic, when N is even and $(N + 1)/2$ harmonics when N is odd. All subsequent stress calculations can be restricted to the real part alone.

For cyclic symmetric objects, the generalised loading can be expressed as the sum of finite number of Fourier components of load, this number being equal to the number of periodic structures. The displacement vector corresponding to any sector j is connected to the adjacent sector by the relationship

$$\{u_j\} = \{u_{j-1}\} \, e^{i\psi} \qquad (4.40)$$

where

$$\psi = \frac{2\pi}{N} p \quad (p = 1, 2, ..., m)$$

N—number of repeated structures and m—number of harmonics to be computed. The governing equation takes the following form for any one repeated structure:

$$\begin{bmatrix} B_1 & A_1 & & & A_n^T e^{-i\psi} \\ A_1 & B_2 & A_2 & & \\ \vdots & \vdots & & & \\ A_n e^{i\psi} & & & A_{n-1}^T & B_n \end{bmatrix} \begin{Bmatrix} z_1 \\ z_2 \\ \vdots \\ z_n \end{Bmatrix} = \begin{Bmatrix} f_1 \\ f_2 \\ \vdots \\ f_n \end{Bmatrix} \qquad (4.41)$$

Here $z_1, z_2, ..., z_n$ represent the displacements of n substructures within the first repeated structure as shown in Fig. 4.2. $f_1, f_2, ..., f_n$ represent the load vector for the corresponding substructures for a given Fourier harmonic p. The square matrix on the L.H.S. will be Hermitian for any Fourier harmonic. Sum of the displacements computed for any repeated structure due to all the harmonics will be real. After the net displacements, $z_1, z_2, ..., z_n$ in any repeated structure are computed (which are real), stresses are calculated. Stress calculations will not involve any complex number.

4.2.4 Submatrices Elimination Scheme

The submatrices elimination scheme explained in Chapter 2 is extended to analyse the Hermitian complex simultaneous problem. Equation (4.41) is expanded as

$$
\begin{bmatrix}
[B_1] & [A_1] & 0 & \cdots & \cdots & & 0 & [C_1 k_1] \\
[C_2] & [B_2] & [A_2] & & & & & 0 \\
\cdots & \cdots & \cdots & & & & & \\
& & & [C_{n-1}] & [B_{n-1}] & [A_{n-1}] & & \\
[A_n k_2] & & & 0 & [C_n] & [B_n] &
\end{bmatrix}
\begin{Bmatrix}
z_1 \\ z_2 \\ \cdots \\ z_{n-1} \\ z_n
\end{Bmatrix}
=
\begin{Bmatrix}
g_1 \\ g_2 \\ \cdots \\ g_{n-1} \\ g_n
\end{Bmatrix}
\quad (4.42)
$$

where

$$ k_1 = e^{-i\psi}, \quad \text{and} \quad k_2 = 1/k_1, \quad C_2 = A_1^T, \text{ and so on} \quad (4.43) $$

Equation (4.42) can be written as

$$
\begin{bmatrix}
[B_1] & [A_1] & 0 & \cdots & \cdots & \cdots & \cdots & [A_n^T k_1] \\
[A_1^T] & [B_2] & [A_2] & \cdots & \cdots & \cdots & \cdots & 0 \\
\cdots & \cdots & \cdots & \cdots & \cdots & \cdots & \cdots & \cdots \\
& & & & [A_{n-2}^T] & [B_{n-1}] & [A_{n-1}] \\
[A_n k_2] & & & & & [A_{n-1}^T] & [B_n]
\end{bmatrix}
$$

$$
\begin{Bmatrix}
z_1 \\ z_2 \\ \cdots \\ z_{n-1} \\ z_n
\end{Bmatrix}
=
\begin{Bmatrix}
g_1 \\ g_2 \\ \cdots \\ g_{n-1} \\ g_n
\end{Bmatrix}
\quad (4.44)
$$

$$
\begin{bmatrix}
[L_1^T] & [P_1] & & & & & [Q_1 k_1] \\
0 & [L_2^T] & [P_2] & & & & [Q_2 k_1] \\
\cdots & \cdots & \cdots & & & & \cdots \\
& & & [L_{n-2}^T] & [P_{n-2}] & [Q_{n-2} k_1] \\
& & & & [L_{n-1}^T] & [\overline{Q}_{n-1}] \\
[A_n k_2] & & & & [A_{n-1}^T] & [B_n]
\end{bmatrix}
$$

$$
\begin{Bmatrix}
z_1 \\
z_2 \\
\cdots \\
z_{n-2} \\
z_{n-1} \\
z_n
\end{Bmatrix}
=
\begin{Bmatrix}
q_1 \\
q_2 \\
\cdots \\
q_{n-2} \\
q_{n-1} \\
g_n
\end{Bmatrix}
\tag{4.45}
$$

where $\{L_i\}$, $[P_i]$ and $[Q_i]$ values are obtained from Eqs (4.2) and (4.3)

$$
[\overline{Q}_{n-1}] = ([P_{n-1}] + [Q_{n-1}]k_1) \tag{4.46}
$$

The first row of Eq. (4.45) is premultiplied by $[Q_1]^T k_2$, and hence

$$
k_2[Q_1^T][L_1^T]\{z_1\} + k_2[Q_1^T][P_1]\{z_2\} + [Q_1^T][Q_1]\{z_n\} = k_2[Q_1^T]\{q_1\} \tag{4.47}
$$

Using Eq. (4.9).

$$
k_2[A_n]\{z_1\} + k_2[Q_1^T][P_1]\{z_2\} + [Q_1^T][Q_1]\{z_n\} = k_2[Q_1^T]\{q_1\} \tag{4.48}
$$

Subtracting the above equation from the last row of Eq. (4.64) gives

$$
([B_n] - [Q_1^T][Q_1])\{z_n\} + [A_{n-1}^T]\{z_{n-1}\} - k_2[Q_1^T][P_1]\{z_2\} = \{g_n - k_2[Q_1^T]\{g_1\} \tag{4.49}
$$

Premultiplying the second row of Eq. (4.45) by $k_2 Q_2^T$ provides,

$$
k_2[Q_2^T][L_2^T]\{z_2\} + k_2[Q_2^T][P_2]\{z_3\} + [Q_2^T][Q_2]\{z_n\} = k_2[Q_2^T]\{q_2\} \tag{4.50}
$$

Substituting for $Q_2^T L_2^T$ in Eq. (4.50) from Eq. (4.20) and combining with Eq. (4.49) simplifies

$$
([B_n] - [Q_1^T][Q_1] - [Q_2^T][Q_2])\{z_n\} + [A_{n-1}^T]\{z_{n-1}\} - k_2[Q_2^T][P_2]\{z_3\}
$$
$$
= \{g_n\} - k_2[Q_1^T]\{q_1\} - k_2[Q_2^T]\{q_2\} \tag{4.51}
$$

The same procedure is extended up to the $(n-2)$th partition and the resulting equation is given by

$$
([B_n] - [Q_1^T][Q_1] - [Q_2^T][Q_2] - \ldots - Q_{n-2}^T[Q_{n-2}])\{z_n\} + [A_{n-1}^T]\{z_{n-1}\}
$$
$$
- k_2[Q_{n-2}^T][P_{n-2}]\{z_{n-1}\} = [\{g_n\} - k_2[Q_1^T]\{q_1\} - k_2[Q_2^T]\{q_2\}
$$
$$
- \ldots - [Q_{n-2}^T]\{q_{n-2}\} \tag{4.52}
$$

The $(n-1)$th row of Eq. (4.45) is premultiplied by $\bar{Q}_{n-1}^H$ and hence,

$$[\bar{Q}_{n-1}^H][L_{n-1}^T]\{z_{n-1}\} + [\bar{Q}_{n-1}^H][Q_{n-1}] = [\bar{Q}_{n-1}^H]\{q_{n-1}\} \qquad (4.53)$$

Using Eqs (4.5) and (4.46)

$$[\bar{Q}_{n-1}^H][L_{n-1}^T] = ([L_{n-1}][\bar{Q}_{n-1}])^H = [L_{n-1}]([P_{n-1}] + [Q_{n-1}]k_1)^H$$

$$= ([L_{n-1}^T][P_{n-1}] + [L_{n-1}^T][Q_{n-1}]k_1)$$

$$= [[A_{n-1}] - k_1[P_{n-1}^T][Q_{n-2}]]^H = [[A_{n-1}^T] - k_2[Q_{n-2}^T][P_{n-2}]] \quad (4.54)$$

Substituting Eq. (4.54) in Eq. (4.53) simplifies

$$[\bar{\bar{B}}_n]z_n = [\bar{g}_n] \qquad (4.55)$$

where

$$[\bar{B}_n] = [[B_n] - [Q_1^T][Q_1] - [Q_2^T][Q_2] - \dots - [Q_{n-2}^T][Q_{n-2}]] \qquad (4.56)$$

$$[B_n] = [[\bar{B}_n] - [\bar{Q}_{n-1}^H][\bar{Q}_{n-1}] \qquad (4.57)$$

$$\{\bar{g}_n\} = [\{g_n\} - [Q_1^T]\{q_1\} - [Q_2^T]\{q_2\} - \dots - [Q_{n-1}^T]\{q_{n-2}\} - [\bar{Q}_{n-1}^H]\{q_{n-1}\}] \qquad (4.58)$$

Hence Eq. (4.45) can be written as

$$\begin{bmatrix} [L_1^T] & [P_1] & & & & & [Q_1 k_1] \\ 0 & [L_2^T] & [P_2] & & & & [Q_2 k_1] \\ 0 & & [L_3^T] & [P_3] & & & [Q_3 k_1] \\ \dots & \dots & \dots & \dots & \dots & & \dots \\ 0 & & & & [L_{n-2}^T] & [P_{n-2}] & [Q_{n-2} k_1] \\ 0 & & & & & [L_{n-1}^T] & [\bar{Q}_{n-1}] \\ 0 & & & & & & [L_n^T] \end{bmatrix}$$

$$\begin{Bmatrix} z_1 \\ z_2 \\ z_3 \\ \dots \\ z_{n-2} \\ z_{n-1} \\ z_n \end{Bmatrix} = \begin{Bmatrix} q_1 \\ q_2 \\ q_3 \\ \dots \\ q_{n-2} \\ q_{n-1} \\ q_n \end{Bmatrix} \qquad (4.59)$$

Here $[L_n^T]$ and $\{q_n\}$ are estimated from,

$$[L_n][L_n^H] = [B_n] \quad \text{and} \quad [L_n]\{q_n\} = \{\bar{g}_n\} \qquad (4.60)$$

The backward substitution in the last row of Eq. (4.59) gives z_n. Having computed the values of P, Q and $[L^T]$ for each partition, a series of backward substitutions in each row gives all values of $\{z_{n-1}\}$, $\{z_{n-2}\}$ ... $\{z_2\}$ and $\{z_1\}$. Here also calculations of $[\bar{B}_n]$, $[L]$, $[P]$ and $[\bar{Q}]$ are independent of μ value. Hence most of the decomposition operations are performed in real mode but only once.

Comparison of results showing cyclic symmetric approach and semi analytic approach is presented in [4, 11].

4.2.5   Skyline Approach

One of the effective schemes used to store stiffness matrix is the skyline approach [4.10] in which only the non-zero terms in the reduced matrix are stored in a single dimension array.

The basic factorization algorithm for the solution of real simultaneous equations can be extended to a Hermitian simultaneous equation solver as follows.

The equilibrium equation of the repeated structure for the pth harmonic can be written as

$$[K_p]\{\delta_p\} = \{f_p\} \tag{4.61}$$

Factorization of $[K_p]$,

$$[K_p] = [U^H][U] \tag{4.62}$$

$[U]$ is calculated as below

$$u_{11} = \sqrt{k_{11}}$$
$$u_{12} = k_{12}/u_{11}$$
$$u_{13} = k_{13}/u_{11}$$
$$\cdots \qquad \cdots$$
$$u_{22} = \sqrt{k_{22} - u_{12}^H u_{12}}$$
$$u_{23} = (k_{23} - u_{12}^H u_{13}/u_{22} \tag{4.63}$$

Upper triangle matrix is evaluated by the Eq. (4.63). Substituting

Equation (4.61) can be written as

$$[U^H]\{Z\} = \{f_p\} \tag{4.64}$$

where $\qquad \{Z\} = \{U\}\{\delta_p\}$

Hence, {Z} can be evaluated from

$$z_1 = f_{p1}/u_{11}$$

$$z_2 = \{f_{p2} - u_{12}^H z_1\}/u_{22}$$

$$z_3 = \{f_{p3} - (u_{13}^H z_{1+} + u_{23}^H z_2)\}/u_{33} \qquad (4.65)$$

Similarly {δ} can be evaluated from

$$\delta_{pJ} = z_J/u_{JJ}$$

$$\delta_{p(J-1)} = \{z_{J-1} - u_{(J-1)(J)}\,\delta_{pJ}\}/u_{(J-1)(J-1)}$$

$$\delta_{p(J-2)} = \{z_{J-2} - \{u_{(J-2)(J-1)}\,\delta_{p(J-1)} + u_{(J-2)(J)}\,\delta_{pJ}\}\}/u_{(J-2)(J-2)} \qquad (4.66)$$

...   ...

where J = Size of the stiffness matrix.

The out-of-core approach involves a large number of operations and hence organisation of the program becomes difficult. In this aspect, the skyline approach scores over the out-of-core approach, as organisation of the program is simple [4.11].

4.2.5.1 Skyline approach I (numbering of nodes along radial direction)

For a cyclic sysmmetric problem, the stiffness matrix becomes complex and hence the core required to store it completely as a complex array using the skyline approach becomes quite large.
If one classifies the nodes as

(i) **Master nodes and slave boundary nodes** The slave boundary nodes are those boundary nodes connected to the master boundary nodes by a complex constraint.
(ii) **Adjacent nodes**—nodes which are adjacent to the slave boundary nodes.
(iii) **Interior nodes**—nodes which are neither boundary nodes nor adjacent nodes.

Then the stiffness matrix can be partitioned as shown in Fig. 4.11, [4.10].
For a typical case of an annular disk fixed at the inner edge with node numbering as in Fig. 4.12, the form of the stiffness matrix $[K_p]$, is as in Fig. 4.13. It is seen that a majority of the nodes are interior nodes. This is true in the case of most engineering problems where the number of nodes in the circumferential direction is large. The complex constraints affect only the coupling terms of the adjacent

	Interior nodes	Adjacent nodes	Master / slave boundary nodes
Interior nodes	Stiffness coefficients of interior nodes (Real) $[K_{AA}]$	Coupling terms (Real) $[K_{AB}]$	Coupling terms (Real) $[K_{AC}]$
Adjacent nodes	Coupling terms (Real) $[K_{BA}]$	Stiffness coefficients of adjacent nodes (Real) $[K_{BB}]$	Coupling terms (Complex) $[K_{BC}]$
Master / slave boundary nodes	Coupling terms (Real) $[K_{CA}]$	Coupling terms (Complex) $[K_{CB}]$	Stiffness coefficients of Master / slave boundary nodes (Real) $[K_{CC}]$

Fig. 4.11

Stiffness matrix partitions

nodes with the boundary nodes. This portion of the stiffness matrix is relatively small. Thus a large portion of the stiffness matrix is real and unaffected by the complex constraints.

Since the stiffness matrix is symmetric, it is sufficient to store one half only. The terms of the stiffness matrix in the columns of the d.o.f. corresponding to the interior and adjacent nodes are stored in a real array, and the terms in the columns of the d.o.f. corresponding to master/slave bundary nodes are stored in complex array. The real array is factorised only once, while the complex array is modified and factorised for each harmonic. This saves the labour of factorising the complete stiffness matrix for each harmonic. The other computations such as forward and backward substitutions are carried out in such a way that the complex and real parts of the stiffness matrix are used in the same order as the original stiffness matrix. After using the elements of one array for computation, control is passed on to the other, array with no compromise on accuracy of the results.

It is seen that if the complex terms in the stiffness matrix are brought closer to the principal diagonal as in Fig. 4.12, there is a

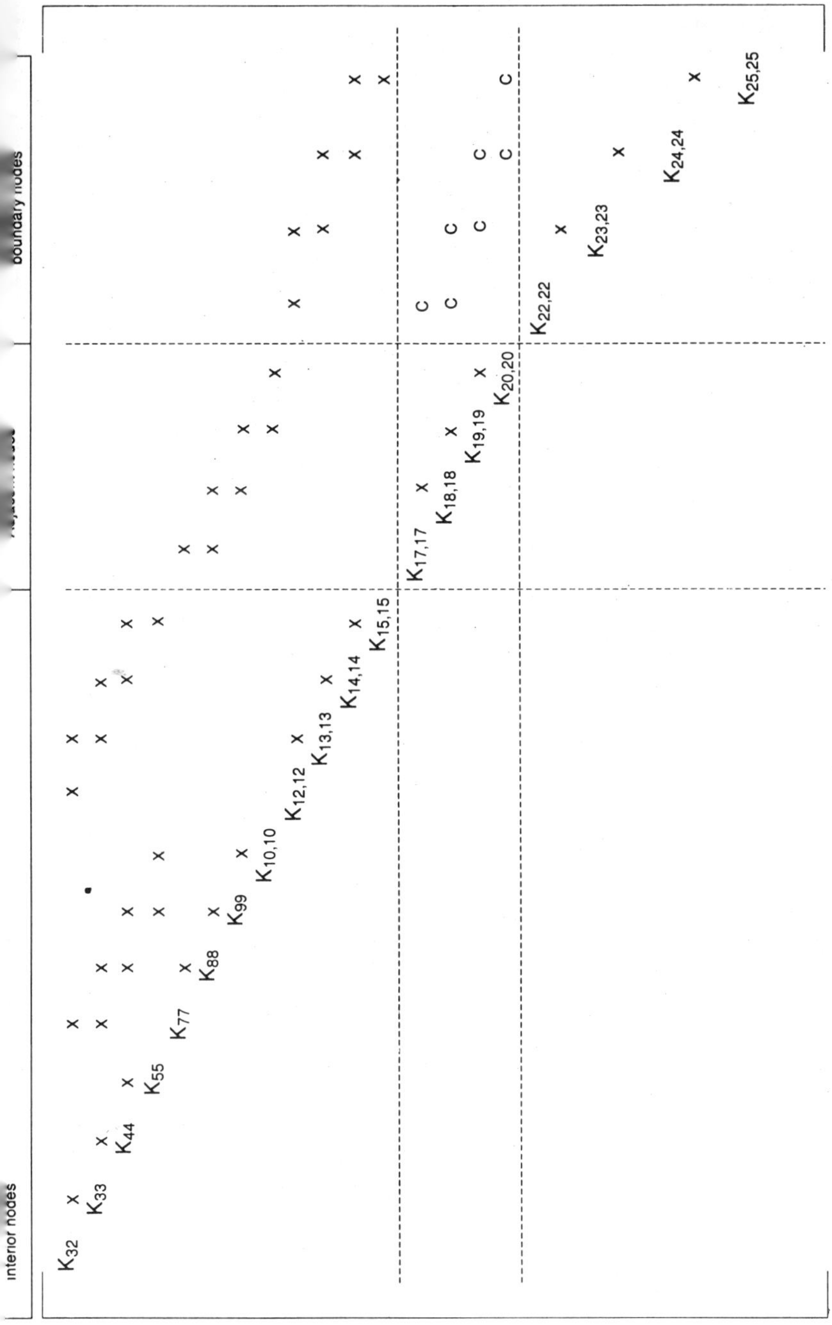

Fig. 4.12

Stiffness matrix for skyline approach I

significant decrease in the number of complex matrix operations. This can be achieved by numbering the nodes in such a way that the difference in the node numbers between the adjacent nodes and master nodes is minimal, as in Fig. 4.13. Although this type of node numbering increases the number of real matrix operations, since complex operations take more time, there is an overall decrease in the time used.

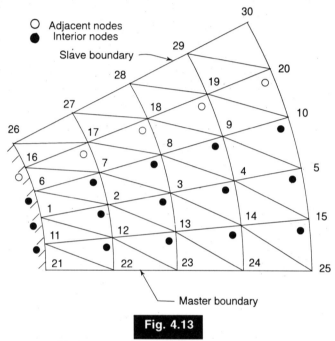

Fig. 4.13

Node numbering for skyline approach I

4.2.5.2 Skyline Approach II (Numbering the nodes in circumferential direction)

When the number of nodes, in any repeated structure, along the circumferential direction is significantly less than the number of nodes in the radial direction, it is advantageous to number the nodes along the circumferential direction as shown in Fig. 4.14. Due to this type of node numbering the band width of the problem reduces significantly, thus reducing the core required for the problem. But this type of node numbering distorts the arrangement of the stiffness matrix, i.e. the complex terms are mixed along with the real terms. This makes it difficult to use either the out-of-core approach which requires the complex terms at the top corner of the stiffness matrix, or the skyline approach I which requires the complex terms to be grouped together at the end. However, one can use the skyline approach to solve cyclic symmetric structures by defining the complete stiffness matrix as complex and performing all the computations in complex arithmetic. For every harmonic, the stiffness matrix is modified, factorised and used for substitutions.

In addition to the above approaches, a hybrid approach, which is a modification of the skyline approach with the use of file operations to reduce core required, can be used to solve the problem.

4.2.6 Core and Labour Required

Core The core required to solve a problem using out-of-core approach, skyline approach-I and skyline approach II is given in Tables 4.3, 4.4 and 4.5 respectively.

Table 4.3

Core Required-Out-of-Core Approach

Matrix	Size	Type	Purpose
[STR]	[s, KB$_1$]	real	Assembly
[A]	[s, s]	real	Store stiffness matrix partition
[B]	[s, s]	real	Store stiffness matrix partition
[P]	[s, s]	real	To store [P] matrix
[QS]	[s, s]	real	To store [Q] matrix
[S]	[s, s]	real	To store $\Sigma\, Q^T Q$
[XC]	[s, s]	complex	To store $[Q]^T[Q]$
[QSC]	[s, s]	complex	To store $[Q]$
{G}	[s, 1]	complex	Force vector
{V}	[s, 1]	complex	To store displacements
{Y}	[s, 1]	complex	To store $[P]^T[q]$
{QC}	[s, 1]	complex	To store $\Sigma\,[Q]^T[q]$

Total core required = $9s^2 + 4s + sKB_1$
(considering complex numbers require twice the core as real numbers)
s = size of each partition
KB_1 = bandwidth of the problem

Table 4.4

Core Required-Skyline Approach I (with radial node numbering)

Matrix	Size	Type	Purpose
[A]	$KB_2 \sum\limits_{i=1}^{n_c-1} n_{ri}$	real	Storing real part of the stiffness matrix
[B]	$KB_2 n_{rm_c}$	complex	Storing complex part of the stiffness matrix
{F}	$\sum\limits_{i=1}^{n_c} n_{ri}$	complex	Force vector
{KDIAG}	$\sum\limits_{i=1}^{n_c} n_{ri}$	real	To store the location of the diagonal element of the stiffness matrix
{DIS}	$\sum\limits_{i=1}^{n_c} n_{ri}$	complex	Array to add up the displacements

Total core required (approximately) $= KB_2 \sum\limits_{i=1}^{n_c} n_{ri}$

KB_2 = bandwidth of the problem

n_{ri} = the number of degrees of freedom along the ith radial line.

n_c = number of nodes along the circumference.

Labour It is assumed that multiplication of a real value by a complex value is equivalent to two real multiplications, and multiplication of two complex values to four real operations. Also, the bandwidth of the problem is assumed to be uniform. The labour (number of multiplications) involved in the approaches for m harmonics is as below:

(a) Out-of-core approach
For each harmonic : $(19n_s - 27/6)s^3 + s^2(5n_s - 6)$
For m harmonics : $m[(19n_s - 27/6)s^3 + s^2(5n_s - 6)]$
n_s number of partitions, s-size of the problem for each partition.

(b) Skyline approach
(i) Skyline approach I (using radial node numbering)
For first harmonic:

Factorization of real part of $[K_p]$ is $[KB_2^2/2] \sum\limits_{i=1}^{n_c-1} n_{ri}$

Factorization of complex part of $[K_p]$ is $[2KB_2^2] \, n_{rn_c}$

Table 4.5

Core Required-Skyline Approach II (with circumferential node numbering)

Matrix	Size	Type	Purpose
[A]	$\sum\limits_{i=1}^{n_r} n_{ci}\, KB_{3i}$	complex	Storing the stiffness matrix
{F}	$\sum\limits_{i=1}^{n_r} n_{ci}$	complex	Force vector
{KDIAG}	$\sum\limits_{i=1}^{n_r} n_{ci}$	real	To store the location of the diagonal element of the stiffness matrix
{DIS}	$\sum\limits_{i=1}^{n_r} n_{ci}$	complex	Array to add up the displacements

KB_{3i} = bandwidth of the nodes on the ith circumferential line
$\quad n_{ci}$ = number of degrees of freedom along the ith circumferential line
$\quad n_r$ = number of nodes along the radial line
e.g. For the case of an annular disk as shown in the Fig. 4.15
KB_{3i} = $(n_c + 1)$ for interior nodes
KB_{3i} = $(2n_c)$ for adjacent nodes

The approximate total core required is $2n_c^2 n_r + 9n_r n_c$.

Substitutions involving real part of $[K_p]$ is $[4KB_2]\sum\limits_{i=1}^{n_c-1} n_{ri}$

Substitutions involving complex part of $[K_p]$ is $[8KB_2]\, n_{rn_c}$

$$\text{Total labour} = (KB_2^2/2)\left(\left(\sum_{i=1}^{n_c} n_{ri}\right) + 3n_{rn_c}\right) + 4KB_2\left(\left(\sum_{i=1}^{n_c} n_{ri}\right) + n_{rn_c}\right)$$

For each of the higher harmonics:

$$\text{Total labour} = \left[(KB_2^2/2)(4n_{rn_c}) + 4KB_2\left(\left(\sum_{i=1}^{n_c} n_{ri}\right) + n_{rn_c}\right)\right]$$

For m harmonics, the labour involved is:

$$(m-1)\left[(KB_2^2/2)(4n_{rn_c}) + 4KB_2\left(\left(\sum_{i=1}^{n_c} n_{ri}\right) + n_{rn_c}\right)\right]$$

$$+ \left[(KB_2^2/2)\left(\left(\sum_{i=1}^{n_c} n_{ri}\right) + 3n_{rn_c}\right) + 4KB_2\left(\left(\sum_{i=1}^{n_c} n_{ri}\right) + n_{rn_c}\right)\right]$$

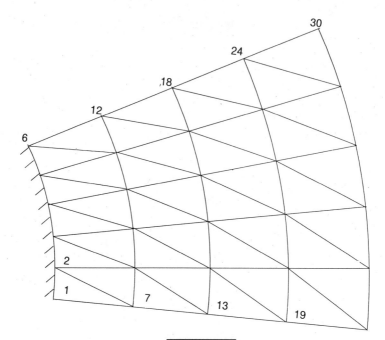

Fig. 4.14

Node numbering for skyline approach II

$$= m\left[2KB_2^2\,n_{rn_c} + 4KB_2\left(\left(\sum_{i=1}^{n_c} n_{ri}\right) + n_{rn_c}\right)\right] + [KB_2^2/2]\sum_{i=1}^{n_c-1} n_{ri}$$

where

KB_2 = bandwidth of the problem

n_{ri} = number of d.o.f. along the ith radial line

n_c = number of nodes along a circumferential line.

(ii) Skyline approach II (using circumferential node numbering)

For each harmonic   : $4[KB_3^2/2 + 2KB_3]\sum\limits_{i=1}^{n_r} n_{ci}$

For m harmonics   : $m\left\{4[KB_3^2/2 + 2KB_3]\sum\limits_{i=1}^{n_r} n_{ci}\right\}$

where

KB_3 = bandwidth of the problem

n_{ci} = number of d.o.f. along the ith circumferential line

n_r = number of nodes along a radial line.

Example 4.3

A problem of an annular disc fixed at the inner edge as shown in Fig. 4.15 was sloved using the approaches (with different node numbering). The particulars are as below:

Inner radius	500 mm
Outer radius	1000 mm
Thickness	5 mm
Sector angle	30°
Number of nodes	30
Number of elements	48

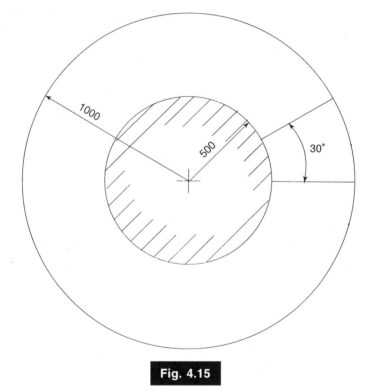

Fig. 4.15

Dimensions of annular disk

The results using the approaches are given in Table 4.6.

4.2.7 Application to Large Industrial Problems

In problems involving cyclic symmetric structures the two types of finite elements used are 2-D triangular plate/shell element with 6 d.o.f. at each node and 3-D brick element with 3 d.o.f. at each node.

Table 4.6

Numerical Results

Case 1:
Uniform loading of 7×10^{-4} N/mm^2 in all the sectors

Parameter	Out-of-core approach	Skyline approach-I	Skyline approach-II	Closed form solution
Maximum displacement	2.46 mm	2.50 mm	2.50 mm	2.50 mm

Case 2:
Uniform loading of 7×10^{-3} N/mm^2 in one sector

Parameter	Out-of-core approach	Skyline approach-I	Skyline approach-II
Maximum displacement	13.16 mm	12.94 mm	12.96 mm

Guidelines for tackling various situations encountered while solving industrial problems are given below.

Let n_a be the number of d.o.f. along an axial line, n_c be the number of nodes along a circumferential line, n_r be the number of nodes along a radial line.

4.2.7.1   In-core Memory Available is Large

Here the method to be selected to solve the problem is based on the CPU time taken. In such a case, the skyline approach I is very effective.

The time taken for solving the problem using out-of-core of approach would be around 15 to 40 times the time taken by skyline approach I.

The time taken for solving the problem using skyline approach II is around 1.5 to 3 times the time taken for skyline approach I, for n_r/n_c greater than 0.75.

If n_r/n_c is less than 0.75, the time required by skyline approach I and skyline approach II are comparable.

e.g. For $n_r = 10$; $n_c = 10$; six degrees of freedom at each node, for seven harmonics,

Assuming a speed of 5×10^{-6} s per multiplication, the approximate time required by skyline approach I is 73.5 s, and the time required by skyline approach II is 151 s, and the time required by out-of-core approach is 1435 s.

Figure 4.16 gives the variation of the CPU time required for skyline

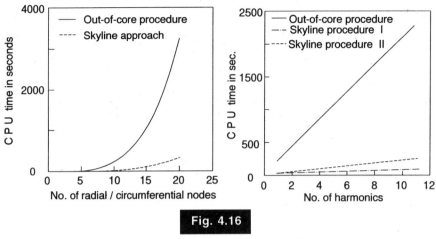

Fig. 4.16

Comparison of time

approach I and out-of-core approach with the number of radial nodes (number of circumferential nodes being equal to the number of radial nodes) and six degrees of freedom at each node.

Figure 4.16 gives the variation of the CPU time required with the number of harmonics, for solving a problem using the three approaches for a problem with $n_r = 10$ and $n_c = 10$, and six degrees of freedom at each node.

4.2.7.2. Core Available is Limited (say the case of a PC 486)

Case (i) n_c is greater than n_r

Considering the core required, skyline approach I is preferable to skyline approach II
e.g. For $n_r = 20$ and $n_c = 35$ and six degrees of freedom at each node,
The approximate core required by skyline approach I is   8 MB
and the core required by skyline approach II is   14 MB

Case (ii) n_c is less than n_r

Considering the core required, skyline approach II is preferred to skyline approach I.
e.g. For $n_r = 35$ and $n_c = 15$ and six degrees of freedom at each node,
The approximate core required by skyline approach I is   11 MB
and the core required by skyline approach II is   5 MB
The core required by skyline approach II is less than that required by out-of-core approach for the following cases:

    (i) problem using 2-D plate element with six d.o.f. at each node and $n_c < $ sqrt $(5n_r)$

   (ii) problem using 3-D brick element with three d.o.f. at each node, the nodes being numbered axially last in skyline approach II and $n_c < $ sqrt $(5n_a)$

 (iii) problem using 3-D brick element with three d.o.f. at each node, the nodes being numbered radially last in skyline approach II and $n_c < $ sqrt $(5n_r)$.

Figure 4.17 (a) gives the variation of core required for skyline approach II and Out-of-core approach, with the number of circumferential nodes (the number of radial nodes being kept constant at 15).

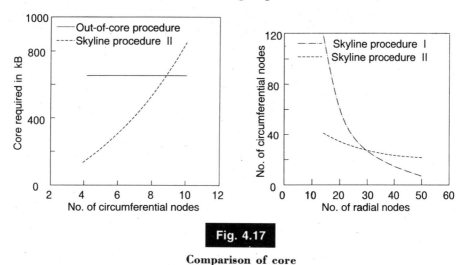

Fig. 4.17

Comparison of core

Case (iii) If the core required by both skyline approach I and skyline approach II cannot be accommodated, then out-of-core approach will have to be resorted to.

e.g. $n_r = 40$ and $n_c = 30$ and six degrees of freedom at each node,

The approximate core required by skyline approach I is	28 MB
the core required by skyline approach II is	22 MB
and the core required by out-of-core approach is	5 MB

    Figure 4.17 gives the upper limit on the number of variables for solving a problem using skyline approach I and skyline approach II.

➤ 4.3 EIGENVALUE PROBLEM

4.3.1 Simple Case of Illustration

Consider the case of the free vibration problem of four rods [4.12]

having concentrated masses m at each corner (Fig. 4.18). Let us presuppose that the four rods can only stretch or compress (truss

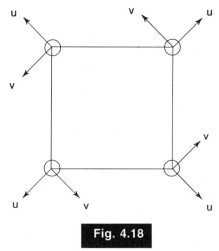

Fig. 4.18

Four rods connected together

elements) and their masses can be ignored. This combination can be considered to be a cyclic symmetric structure with 2 degrees of freedom at each node, u radially outwards and v along the circumference (anti-clockwise positive). Since every rod has stiffness along its length direction AE/l (A—the area, E—Young's modulus and l—the length) it is easy to write the governing equation for free vibration. It takes the following form:

$$[K]\{\delta\} = \omega^2[M]\{\delta\} \qquad (4.67)$$

where

$$\{\delta\} = \begin{Bmatrix} u_1 \\ v_1 \\ u_2 \\ v_2 \\ u_3 \\ v_3 \\ u_4 \\ v_4 \end{Bmatrix}$$

$$[K] = \frac{AE}{2l} \begin{bmatrix} 2 & 0 & 1 & 1 & 0 & 0 & 1 & -1 \\ 0 & 2 & -1 & -1 & 0 & 0 & 1 & -1 \\ 1 & -1 & 2 & 0 & 1 & 1 & 0 & 0 \\ 1 & -1 & 0 & 2 & -1 & -1 & 0 & 0 \\ 0 & 0 & 1 & -1 & 2 & 0 & 1 & 1 \\ 0 & 0 & 1 & -1 & 0 & 2 & -1 & -1 \\ 1 & 1 & 0 & 0 & 1 & -1 & 2 & 0 \\ -1 & -1 & 0 & 0 & 1 & -1 & 0 & 2 \end{bmatrix}$$

M is (8×8) diagonal mass matrix with m in each principal diagonal.

Let us now try to solve Eq. (4.66) by standard procedures. The three non-zero values of ω^2 (repeated roots) and associated eigenvectors are given by

$$\omega^2 = \frac{2AE}{m^2}; \quad \{U_1\}^T = \{1, 0, 1, 0, 1, 0, 1, 0\}^T$$

$$\{U_2\}^T = \{1, 0, 0, +1, -1, 0, 0, -1\}^T$$

$$\{U_3\}^T = \{0, 1, 0, -1, 0, 1, 0, -1\}^T$$

4.3.2   Cyclic Symmetric Approach

Case (i) Axisymmetric mode ($\psi = 0$) (Fig. 4.19)

$$z_i = z_{i-1} \, (e^{i\psi})$$

$$u_1 = u_2 = u_3 = u_4 = u, \, v_1 = v_2 = v_3 = v_4 = 0$$

Then from the first two rows, we have

$$\frac{AE}{2l} \begin{bmatrix} 4 & 0 \\ 0 & 0 \end{bmatrix} \begin{Bmatrix} u \\ v \end{Bmatrix} = 2 \begin{bmatrix} m & 0 \\ 0 & m \end{bmatrix} \begin{Bmatrix} u \\ v \end{Bmatrix}$$

$$\therefore \qquad\qquad \omega^2 = 2AE/ml$$

Case (ii) Second mode shape as shown in Fig. 4.20.

$$z_i = z_{i-1}(e^{i\psi})$$

$$\psi = \pi/2, \, z_2 = z_1(i), \, z_3 = z_1(i^2), \, z_4 = z_1(i^3)$$

$$z_1 = \begin{Bmatrix} u_1 \\ -iv_1 \end{Bmatrix}; \quad z_2 = \begin{Bmatrix} iu_1 \\ +v_1 \end{Bmatrix}; \quad z_4 = \begin{Bmatrix} -iu_1 \\ v_1 \end{Bmatrix}$$

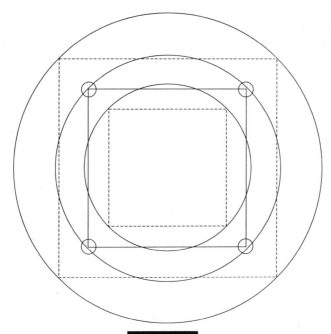

Fig. 4.19

Rods in tension or compression

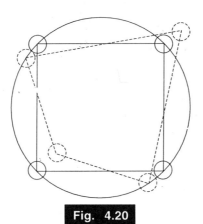

Fig. 4.20

Rods, tension, tension, compression and compression

$$z_3 = \left\{ \begin{array}{c} -u_1 \\ -iv_1 \end{array} \right\}; \quad z_4 = \left\{ \begin{array}{c} -iu_1 \\ v_1 \end{array} \right\}$$

From the first two rows

$$\frac{AE}{2l}\begin{bmatrix} 2 & 2 \\ 2 & 2 \end{bmatrix}\begin{Bmatrix} u_1 \\ v_1 \end{Bmatrix} = \omega^2 \begin{bmatrix} m & 0 \\ 0 & m \end{bmatrix}\begin{Bmatrix} u_1 \\ v_1 \end{Bmatrix}$$

$\therefore$ $\qquad\qquad\qquad \omega^2 = 2AE/lm;$

Case (iii): The response covers two waves along the circumference (Fig. 4.21 ($\psi = \pi$))

$$z_2 = -z_1, z_3 = z_1; z_4 = -z_1$$

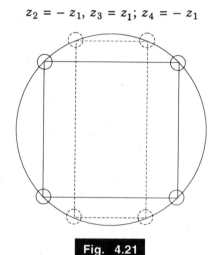

Fig. 4.21

Rods, tension, compression tension compression

From the first two rows

$$\frac{AE}{2l}\begin{bmatrix} 0 & 0 \\ 0 & 4 \end{bmatrix}\begin{Bmatrix} u_1 \\ v_1 \end{Bmatrix} = \omega^2 \begin{bmatrix} m & 0 \\ 0 & m \end{bmatrix}\begin{Bmatrix} u_1 \\ v_1 \end{Bmatrix}$$

$\therefore \omega^2 = 2AE/ml; u_1 = 0, v_1 = v,$ or $u_1 = u, v_1 = 0$

It may be observed in Fig. 4.19, all the four members either experience tension/compression, in Fig. 4.20 in the order tension, tension, compression and compression and in Fig. 4.21 in the order tension, compression, tension, compression.

4.3.3   General Case

Let us now generalise the eigenvalue problem [4.13, 4.14 and 4.15]. For a general case, the eigenvector $\{u_i\}$ of the ith repeated structure will be connected to the $(i + 1)$th by the relationship

$$\{u_{i+1}\} = e^{i\psi}(u_i) \tag{4.68}$$

The stiffness and mass matrices for the entire structure may be assembled and the frequency equation written in the form

$$
\begin{bmatrix}
[K_{1,1}] & [K_{1,2}] & 0 & 0 & 0 & [K_{1,N}] \\
[K_{2,1}] & [K_{2,2}] & [K_{2,3}] & & & 0 \\
0 & \cdots & \cdots & \cdots & \cdots & \cdots \\
\cdots & \cdots & \cdots & \cdots & \cdots & \cdots \\
& & & [K_{N-1,N-2}] & [K_{N-1,N-1}] & [K_{N-1,N}] \\
[K_{N,1}] & 0 & 0 & 0 & [K_{N,N-1}] & [K_{N,N}]
\end{bmatrix}
\begin{Bmatrix}
\{z_1\} \\ \{z_2\} \\ \cdots \\ \cdots \\ \{z_{N-1}\} \\ \{z_N\}
\end{Bmatrix}
$$

$$
= [\omega^2]
\begin{bmatrix}
[M_{1,1}] & [M_{1,2}] & 0 & 0 & 0 & [M_{1,N}] \\
[M_{2,1}] & [M_{2,2}] & [M_{2,3}] & 0 & 0 & 0 \\
0 & \cdots & \cdots & \cdots & \cdots & \cdots \\
0 & 0 & 0 & [M_{N-1,N-2}] & [M_{N-1,N-1}] & [M_{N-1,N}] \\
[M_{N,1}] & 0 & 0 & 0 & [M_{N,N-1}] & [M_{N,N}]
\end{bmatrix}
$$

$$
\begin{Bmatrix}
\{z_1\} \\ \{z_2\} \\ \cdots \\ \cdots \\ \{z_{N-1}\} \\ \{z_N\}
\end{Bmatrix}
\tag{4.69}
$$

Here K and M are all real symmetric positive definite matrices. Using Eq. (4.68) the global matrices may be decoupled for each repeated structure as follows.

$$
\begin{bmatrix}
[K_{1,1}] & [K_{1,2}] & 0 & 0 & 0 & [K_{1,N}] \\
[K_{2,1}] & [K_{2,2}] & [K_{2,3}] & 0 & 0 & 0 \\
\vdots & \cdots & \cdots & \cdots & \cdots & \cdots \\
0 & 0 & 0 & [K_{N-1,N-2}] & [K_{N-1,N-1}] & [K_{N-1,N}] \\
[K_{N,1}] & 0 & 0 & 0 & [K_{N-1,N-1}] & [K_{N,N}]
\end{bmatrix}
$$

$$\begin{Bmatrix} \{z_1\} \\ \{z_1 e^{1i\psi}\} \\ \{z_1 e^{2i\psi}\} \\ \cdots \\ \{z_1 e^{(N-2)i\psi}\} \\ \{z_1 e^{(N-1)i\psi}\} \end{Bmatrix}$$

$$= [\omega^2] \begin{bmatrix} [M_{1,1}] & [M_{1,2}] & 0 & 0 & 0 & [M_{1,N}] \\ [M_{2,1}] & [M_{2,2}] & [M_{2,3}] & 0 & 0 & 0 \\ \vdots & \cdots & \cdots & \cdots & \cdots & \cdots \\ [M_{N,1}] & 0 & 0 & 0 & [M_{N,N-1}] & [M_{N,N}] \end{bmatrix}$$

$$\begin{Bmatrix} \{z_1\} \\ \{z_1 e^{i\psi}\} \\ \{z_1 e^{2i\psi}\} \\ \cdots \\ \{z_1 e^{(N-1)i\psi}\} \end{Bmatrix} \tag{4.70}$$

Let us consider only the stiffness matrix.

Since all the substructures are identical,

$$[K_{1,1}] = [K_{2,2}] = [K_{3,3}] = \ldots = [K_{N,N}]$$
$$[K_{1,2}] = [K_{2,3}] = [K_{3,4}] = \ldots = [K_{N,1}]$$
$$[K_{2,1}] = [K_{1,2}^T]$$
$$[K_{2,1}] = [K_{3,2}] = [K_{4,3}] = \ldots = [K_{N-1,N-1}] = [K_{1,N}] \tag{4.71}$$

If each repeated structure is considered to consist of substructures 1, 2, 3, 4, ..., n, etc. then the global stiffness matrix for each repeated structure is of the form

$$\begin{bmatrix} [B_1] & [A_1] & 0 & & & & & & [C_1] \\ [C_2] & [B_2] & [A_2] & 0 & & & & & \\ 0 & [C_3] & [B_3] & [A_3] & & & & & \\ \cdots & \cdots & \cdots & \cdots & \cdots & \cdots & \cdots & \cdots & \cdots \\ 0 & & [C_n] & [B_n] & [A_n] & & & & \\ & & & [C_1] & [B_1] & [A_1] & & & \\ & & & & [C_2] & [B_2] & [A_2] & & \\ \cdot & & & & & [C_3] & [B_3] & [A_3] & \\ \cdot & & & & & & & & \\ 0 & & & & & & & & \\ [A_n] & 0 & \cdots & \cdots & \cdots & \cdots & 0 & [C_n] & [B_n] \end{bmatrix} \tag{4.72}$$

Similarly, for the mass matrix.

But $[C_2] = [A_1^T]$ and so on.

The global eigenvector is,

$$\{z\} = [\{z_1\}\ \{z_2\}\ \{z_3\}\ \ldots\ \{z_n\}\ \{z_1 e^{i\psi}\}\ \{z_2 e^{i\psi}\}\ \ldots\ \{z_n e^{i\psi}\}$$

$$\{z_1 e^{i2\psi}\}\ \{z_2 e^{i2\psi}\}\ \{z_3 e^{i2\psi}\}\ \ldots\ \{z_n e^{i2\psi}\}\ \ldots\ \{z_n e^{-i\psi}\}]^T \quad (4.73)$$

Substituting Eqs (4.71) and (4.72) in Eq. (4.70), the obtained decoupled equations for the first repeated structure are [4.13]:

$$
\begin{bmatrix}
[B_1] & [A_1] & & & & [A_n^T e^{-i\psi}] \\
[A_1^T] & [B_2] & [A_2] & & & \\
 & [A_2^T] & [B_3] & [A_3] & & \\
 & & \cdots & \cdots & \cdots & \cdots \\
[A_n e^{i\psi}] & & & & [A_{n-1}^T] & [B_n]
\end{bmatrix}
\begin{Bmatrix}
z_1 \\ z_2 \\ z_3 \\ \cdots \\ z_n
\end{Bmatrix}
$$

$$
= [\omega^2]
\begin{bmatrix}
M_1 & & & & \\
 & M_2 & & & \\
 & & M_3 & & \\
 & \cdots & \cdots & \cdots & \\
 & & & & M_n
\end{bmatrix}
\begin{Bmatrix}
z_1 \\ z_2 \\ z_3 \\ \cdots \\ z_n
\end{Bmatrix}
\quad (4.74)
$$

Here the lumped mass approach is employed.

Similarly, for the second repeated structure,

$$
\begin{bmatrix}
[B_1] & [A_1] & & & & [A_n^T e^{-i\psi}] \\
[A_1^T] & [B_2] & [A_2] & & & \\
 & [A_2^T] & [B_3] & [A_3] & & \\
 & & \cdots & \cdots & \cdots & \cdots \\
[A_n e^{i\psi}] & & & & [A_{n-1}^T] & [B_n]
\end{bmatrix}
\begin{Bmatrix}
z_1 \\ z_2 \\ z_3 \\ \cdots \\ z_n
\end{Bmatrix} e^{i\psi}
$$

$$
= [\omega^2]
\begin{bmatrix}
[M_1] & & & & \\
 & [M_2] & & & \\
 & & [M_3] & & \\
 & \cdots & \cdots & \cdots & \\
 & & & & [M_n]
\end{bmatrix}
\begin{Bmatrix}
z_1 \\ z_2 \\ z_3 \\ \cdots \\ z_n
\end{Bmatrix} e^{i\psi}
\quad (4.75)
$$

On dividing both sides of Eq. (4.75) by $e^{i\psi}$ one gets Eq. (4.74). Hence, we have to solve Eq. (4.74) for $(N/2 + 1)$ values of ψ if N is even (or $(N + 1)/2$ values of if N is odd) to get all the eigenvalues of the entire structure. Equation (4.74) has complex stiffness and real mass matrices. If consistent mass is used then the mass matrix is also complex. The real parts are symmetric and imaginary parts skew-symmetric. Hence, it is an eigenvalue problem of Hermitian matrix. Its eigenvalues are real but eigenvectors are complex. However, this eigenvalue problem becomes real for $\psi = 0$ and $\psi = \pi$ (if N is even). For all cases, the mode shape for other substructures are obtained from Eq. (4.68).

4.3.4   Simultaneous Iteration Scheme

Simultaneous iteration scheme [3.8] is modified to extract the real eigenvalues from the Hermitian matrix equation given by Eq. (4.75).

The resulting Hermitian eigenvalue problem can be represented as

$$[\bar{K}]\{\bar{Z}\} = [\omega^2][\bar{M}][\bar{Z}] \qquad (4.76)$$

The steps involved are similar to the procedure outlined in Section 3.8.2.

1. Trial eigenvectors on the right-hand side are assumed as $\{\bar{Z}_0\}$ and are orthonormalized with respect to mass matrix $[\bar{M}]$.
2. Vector

$$\{\bar{G}\} = [\bar{M}]\{\bar{Z}_0\} \qquad (4.77)$$

   is computed.
3. The reduced simultaneous equation of the form,

$$[\bar{K}]\{\bar{Z}_1\} = \{\bar{G}\} \qquad (4.78)$$

   is solved.
4. The new vector $[\bar{Z}_1]$ is orthonormalized with respect to $[\bar{M}]$.
5. The iteration matrix

$$[\Omega^2] = \{\bar{Z}_1\}^H [\bar{K}]\{\bar{Z}_1\} \qquad (4.79)$$

6. The orthonormalized new vector $[\bar{Z}_1]$ is used in step 2 for next iteration. This is repeated until the convergence of $[\lambda^2]$ is obtained.

The flow chart for the above scheme is given in Fig. 4.22.

In-core solution for eigenvalue problems using skyline approach is discussed in [4.16].

Number of multiplications

Let the size of each partition be equal to s and the number of eigenvectors be m. Here it is assumed that multiplication of a real

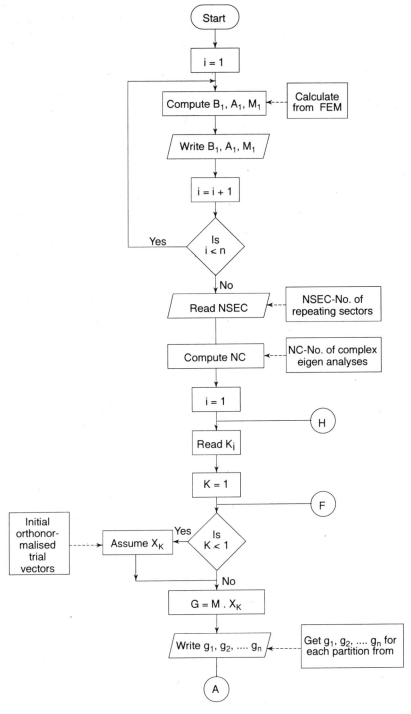

(Contd)

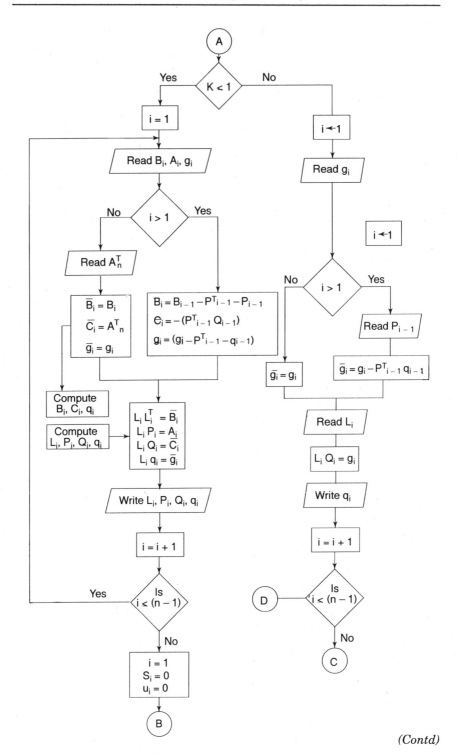

(Contd)

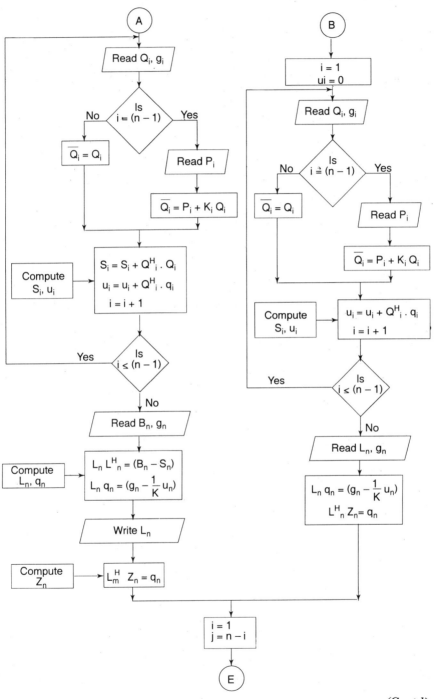

(Contd)

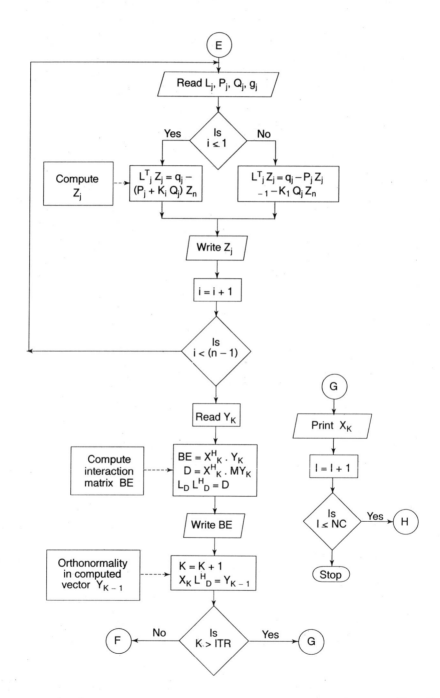

(Contd)

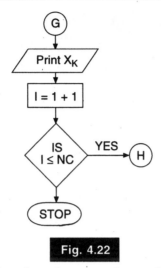

Fig. 4.22

Flow chart for eigenvalue problem

value by a complex value is equivalent to two real multiplications and multiplication of two complex values to four real operations. The labour involved is indicated in Table 4.7.

4.3.5 Physical Significance of the Real and Imaginary Parts

If we had considered the whole structure for analysis, we would have obtained an eigenvector that is real. The application of cyclic symmetry concept resulted in an eigenvector for the repeating sector that is complex. Now the obvious question that arises is "What is the meaning of the real and imaginary parts of the eigenvector?" Before attempting a discussion on it, it is worth noting that the complex eigenvalue problem of Eq. 4.69 can be considered as a real eigenvalue problem of double the size. NASTRAN implementation of the cyclic symmetric concept reportedly uses only real arithmetic throughout. The formulation into real eigenvalue problem (even of double size) is attractive owing to greater efficiency in solution than for complex eigenvalue problem. In the original formulation of the complex eigenproblem of cyclic symmetric structures as proposed by Thomas [4.1], all the matrices become complex. In the modified formulation proposed here, it is seen that only the corner submatrices [Eq. 4.74] become complex. This affords a large reduction in computer core and time. Hence, the complex notation is quite elegant and efficient since it also yields, as we shall presently see, a clear physical picture.

Let us first consider free wave propagation in infinitely long periodic structures such as a never ending continuous beam. It can be shown

Table 4.7

Total Number of Multiplications (Eigenvalue Problem)

	Modified submatrices elimination scheme
1. Decomposition	
1.1 Reduction Stage:	
(a) First partition	$[((7/6)s^3 + s^2m)]$
(b) Second to $(n - 1)$th partition	$[((8/3)s^3 + 3s^2m)(n - 2)]$
(c) nth partition	$[s^3(3n + 10)/6 + 2s^2$ $(1 + mn + m) + 32sm)$
1.2 Back Substitution:	
(a) nth partition	$[2s^2m]$
(b) $(n - 1)$th partition	$[5s^2m]$
(c) $(n - 2)$th partition to first partition	$[(5s^2m + 4sm)(n - 2)]$
Total	$[s^3(19n - 15)/6$ $+ s^2m(10n - 6)$ $+ 2sm(2n - 3) + 2s^2]$

Similarly, the number of major multiplications involved for subsequent iterations is as follows:

2. For Each Iteration	
2.1 Reduction Stage:	
(a) First partition	$[s^2m]$
(b) Second to $(n -1)$th partition	$[3s^2m(n - 2)]$
(c) nth partition	$[2s^2m(n + 1) + 2sm]$
2.2 Back Substitution:	
(a) nth partition	$[2s^2m]$
(b) $(n - 1)$th partition	$[5s^2m]$
(c) $(n - 2)$th partition to first partition	$[(5s^2m + 4sm)(n - 2)$
Total	$[s^2m(10n - 6)$ $+ sm(4n - 6)]$

3. Additional Operations for New Complex Constraint

$$[s^3(8/3) + 2s^2]$$

For example, if $s = 80$, $n = 8$ and $m = 10$, then

Unit operation	Modified submatrices elimination scheme
Decomposition	16.5×10^6
Each iteration	4.8×10^6
Complex constraint	1.4×10^6

that this wave propagation is characterised by a propagation constant which relates the displacement vector in one repeating sector (here the beam span $j, j = 1, 2, 3, ..., \not\subset$ to that before or after it. Mathematically, this is written as

$$\{u_j\} = \mu\{u_{j-1}\} \tag{4.80}$$

where μ, the propagation constant, in general is complex: $\mu = \pm (\mu_r + i\mu_i)$. The real part signifies the decay in the amplitude whereas the imaginary part relates to the phase of vibration. The plus or minus sign indicates the direction of propagation of the wave. It can further be shown that if μ is purely real, the wave decays and cannot propagate; for free (i.e. undecaying) propagation, μ has to be purely imaginary. So,

$$\{u_j\} = e^{\pm i\mu_i}\{u_{j+1}\} \tag{4.81}$$

However, for a finite periodic structure such as a bladed turbine disc or a gear wheel having N repeating sectors, repeated application of the above relation yields the condition that

$$e^{iN\mu_i} = 1 \text{ or } e^{i2\pi m} \qquad \text{where } m \text{ is an integer}$$

$$\therefore \qquad \mu_i = 2\pi m/N \qquad m = 0, 1, 2, ... \tag{4.82}$$

The ψ of our "$e^{i\psi}$ relationship" is merely this. For N repeating sectors, there are only $N/2$ (if N is even) or $(N + 1)/2$ (if N is odd) independent values of ψ, as other values merely indicate wave propagation in the opposite direction. Different values of m yield the complete set of eigenvectors.

An alternative interpretation

This is due to Thomas [4.6]. If there are N repeating sectors, the eigenvector of the entire structure would be of the form:

$$\{u\} = \{\{u^1\} \{u^2\} \{u^3\} \ldots \{u^1\} \ldots \{u^N\}\}^T \tag{4.83}$$

where the $\{u^j\}$ are all real. Let us now consider another vector,

$$\{u^1\} = \{\{u^N\} \{u^1\} \{u^2\} \ldots \{u^{N-1}\}\}^T \tag{4.84}$$

We notice that $\{u^1\}$ is $\{u\}$ merely rotated around one repeated structure. But in general $\{u^1\} \neq \{u\}$. And this process could be repeated $(N - 1)$ times to yield apparently different vectors, each of which can claim to be an eigenvector at that frequency. Let us remember, the eigenvectors are the least number of linearly independent vectors for the system. Hence, for the existence of such $\{u^1\}$, there must be a vector $\{\bar{u}\}$ orthogonal to $\{u\}$ such that all these $\{u^1\}$ can be expressed as a linear combination of $\{u\}$ and $\{\bar{u}\}$. The complex eigenvector we have taken is $\{z\} = \{u\} + i\{\bar{u}\}$. This is so because $\{u\}$ and $\{\bar{u}\}$ themselves are real orthogonal eigenvectors of the parent problem for the same eigenvalue. In fact, a real linear combination of $\{u\}$ and $\{\bar{u}\}$ also could be taken. But the complex vector $\{z\}$ yields a physical meaning—the actual deflected shape of the structure being $Re\{ze^{i\omega t}\}$ i.e. $[\{u\} \cos \omega t - \{\bar{u}\} \sin \omega t]$. At $t = 0$, this gives $\{u\}$ but at $\omega t = \psi$ this gives, for the instantaneous deflection shape, $\{u^1\}$! It is as if the complex eigenvectors describes modes of the structure that rotate' around the circumference. After every $t = \psi/\omega$, the deflection shape is rotated around yet another substructure.

4.3.6 Lanczos' Scheme of Reduced Labour

The procedure adopted here is an extension of 'Section 3.10. The mass matrix $[M]$ is expressed as

$$M = [B][B]^T \tag{4.85}$$

Let us consider the original cyclic symmetric problem of the type

$$[K]\{u\} = \omega^2[M]\{u\} \tag{4.86}$$

Here, K is Hermitian. Substituting

$$[B]^T\{u\} = \{v\} \tag{4.87}$$

we get

$$[B]^T[K]^{-1}[B]\{v\} = \lambda\{v\} \tag{4.88}$$

where

$$\lambda = 1/\omega^2 \tag{4.89}$$

This can be rewritten as

$$A\{u\} = \lambda\{v\} \tag{4.90}$$

where

$$A = [B]^T[K^{-1}][B] \tag{4.91}$$

To calculate α's and β's in Lanczos scheme, the following procedure is adopted [4.17].

If y_1 is the starting vector in complex form then

$$\alpha_1 = \{y_1\}^H A\{y_1\} \tag{4.92}$$

This is done in three stages. As a first step, $\{y_1\}$ is pre-multiplied by $[B]$

Let

$$\{x\} = [B]\{y_1\} \tag{4.93}$$

Let

$$[K]^{-1}\{x\} = \{w\} \tag{4.94}$$

This operation of computing $\{w\}$ is done using submatrices elimination scheme outlined in Sec. 4.2. As a third step, $\{q\}$ is computed from the equation

$$[B]^T\{w\} = \{q\} \tag{4.95}$$

Now

$$\alpha_1 = \{y_1\}^H\{q\} \tag{4.96}$$

As before

$$\{z_1\} = A\{y_1\} - \alpha_1\{y_1\} \tag{4.97}$$

Hence

$$\beta_1 = \{z_1\}^H\{z_1\}^{1/2} \tag{4.98}$$

and

$$y_2 = \{z_1\}\{\beta_1\} \tag{4.99}$$

Generalising for any i, we have

$$V_i = A\{y_i\} - \beta_{i-1}\{y_{i-1}\} \tag{4.100}$$

$$\alpha_i = \{y_i\}^H\{V_i\} \tag{4.101}$$

$$z_i = \{V_i\} - \alpha_i\{y_i\} \tag{4.102}$$

$$B_i = \{z_i\}^h\{z_i\}^{1/2} \tag{4.103}$$

$$\{y_{i+1}\} = \{z_i\}/\beta_i \tag{4.104}$$

Computation of labour involved

(i) To get the starting vector y_1 doing iteration t times

$$t[s^3(19n - 27)/6 + s^2(5n - 6)]$$

where s is the size of substructure, now the number of substructures in one repeated structure.

(ii) For solving

$$K^{-1}\{x\} = \{w\}$$

m times

$$m[s^3(19n - 27)/6 + s^2(5n - 6)]$$

(iii) Premultiplying by B and B^T, m times

$$2nm$$

(iv) Evaluating α_i and β_i

$$8nm$$

(v) For orthonormalisation of Lanczos vectors

$$nm^2$$

(vi) For simultaneous iteration of tridiagonal matrix

$$m\left(6m^2 + 2\frac{1}{2}m^3 + \frac{m^3}{6}\right)$$

(vii) **Transforming to original vector**

$$nm^2 + nb$$

Hence the total labour

$$(m + t)[s^3(19n - 27)/6 + s^2(5n - 6)] + 10nm + nm^2$$

$$+ m\left(6m^2 + 2\frac{1}{2}m^3 + \frac{m^3}{6}\right) + nm^2 + nb$$

$$= (m + t)[s^3(19n - 27)/6 + s^2(5n - 6)] + 10nm + 2nm^2$$

$$+ 6m^3 + 2.66m^4 + nb \qquad (4.105)$$

Example 4.4
Compute the first few natural frequencies corresponding to axisymmetric modes of a circular steel disc fixed at the inner radius and free at the outer radius as shown in Fig. 4.23 by using the Lanczos scheme and cyclic symmetric concept [4.17].

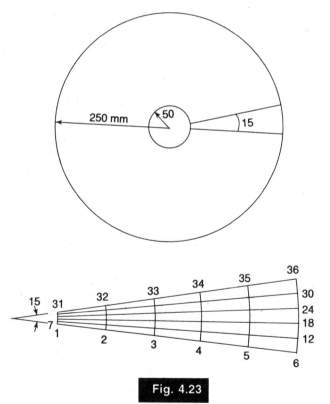

Fig. 4.23

Annular disc and finite element details

outer radius	250 mm
inner radius	50 mm
thickness	5 mm
sector angle	15°
number of nodes	36
number of Potters' partitions	6
size of each block	24
total number of nodes	120
number of iterations	15

The results obtained by this method are compared with those of simultaneous iteration in Table 4.8. Comparison of labour involved is shown in Fig. 4.24.

Table 4.8

Comparison of First Few Frequencies

Frequency	Potters' method with simultaneous iteration method (in rad/s)	Lanczos method with simultaneous iteration method (in rad/s)	Percentage difference
1	654.9926	654.9926	0
2	4133.1026	4133.1026	0
3	7737.2015	7737.2015	0
4	14015.2405	14015.2405	0
5	59649.4989	59651.5252	0.003
6	67956.7271	67947.1985	– 0.014

➤ 4.4 INDUSTRIAL APPLICATIONS

There are several reported applications of this approach to industrial problems [4.18 to 4.20]. Let us quote three specific cases in machine element design.

4.4.1 Analysis of a Fabricated Gear Wheel

The gear wheel taken up for analysis is shown in Fig. 4.25. This is subjected to concentrated load from the adjoining gear [4.21].

The gear wheel was divided into 8 sectors. The load 10 kN was expressed as the sum of 8 Fourier harmonics, each of magnitude 1.25 kN. The displacements and stresses were solved for the first sector using constant strain triangular element for in-plane stress

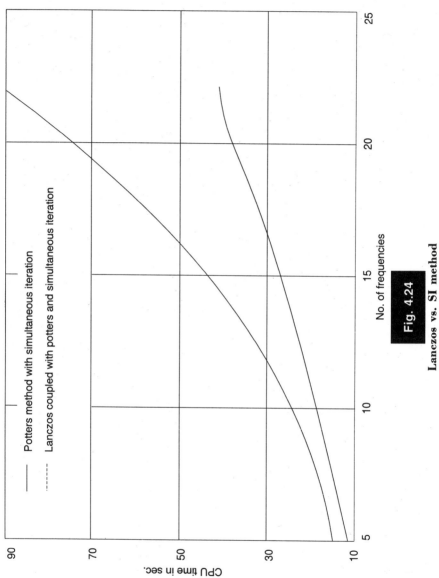

Fig. 4.24

Lanczos vs. SI method

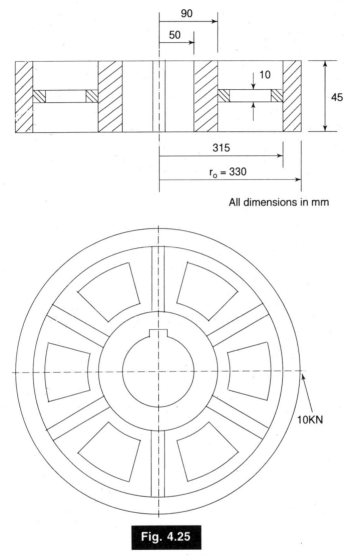

90

50

10

45

315

$r_o = 330$

All dimensions in mm

10KN

Fig. 4.25

Fabricated gear wheel

and a nine term polynomial for bending giving rise to six degrees of
freedom per node. The sector was divided into nine radial partitions
with 20 nodes per partition. The sector was thus discretized into 304
elements and 180 nodes. The size (m) of the problem was 912 and
the bandwidth (b) was 114. If one were to solve for the entire structure
using the same discretisation pattern, the size of the problem would
be 7296 and the bandwidth would be 396. The computer time required

to solve a problem of size m and bandwidth b is propotional to $m*b^2$ and the core needed is $m*b$. Thus, the cyclic symmetric approach reduces the computer time by a factor of about 88 and the core by a factor of about 26.

When cut-outs were introduced, the size of the problem became 732 with 150 nodes and 216 elements. The input data regarding the element connectivity, nodal coordinates, etc. were left unchanged and the presence of the cut-out was taken into account by simply ignoring the contributions of the elements in the zone of the cut-out while assembling the stiffness matrix. Ribs were introduced by changing the thickness of the elements along a strip of width equal to the rib thickness from its original thickness of the web. The gear with six ribs was analysed by dividing it into six sectors and the gear with four ribs had four sectors. In both these cases, the sector was discretised into 254 elements and 170 nodes. The number of partitions was increased to 10.

The input data is generated using a computer program. Any changes in the dimensions of the gear wheel, size of the cut-out, number of ribs, etc. can be easily accommodated. The computer program used for the present analysis can handle input data, so as to carry out a parametric study of heavy duty gear wheels with cut-outs and ribs.

To verify the results, the normal stress σ_{rr} has been plotted for the gear without cut-out (Fig. 4.26). The values are in good agreement with the results obtained by Ramamurti and Srinivasan [4.22]. The tangential stress values were found to be less than the normal stress values. The shear stress was maximum in the region close to the hub.

To have an idea of the stress distribution as the load moves along the circumferences, contour plots of σ_{rr} (Figs 4.27–4.30) have been presented. The regions close to the cut-outs are highly stressed. The effect of the movement of the load from one position to another in a sector is not felt by the sectors far from it. In those sectors, the values of σ_{rr} remain practically unchanged. Comparing Fig. 4.26 and Fig. 4.28, it can be seen that with the introduction of cut-outs, the region of high stress has shifted radially inward.

4.4.2   Eigenvalues of Impellers

The effectiveness of the method of analysis is demonstrated through an example. The unit analysed is a radial fan impeller used in a thermal power station. The sectional front and side views of the impeller are shown in Fig. 4.31. The impeller consists of 20 blades spaced 18° apart; 6 ribs placed 60° apart, a cover plate, a back plate

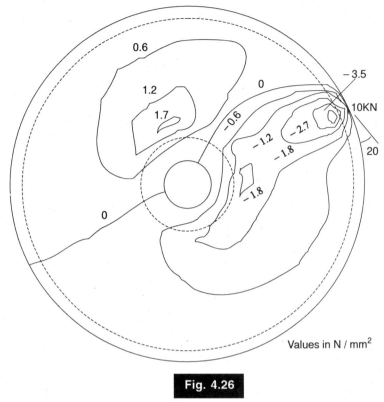

Values in N / mm^2

Fig. 4.26

Load at mid span of Ist sector σ_{rr}

and a ring. This rotates at a constant speed of 1480 rpm. The overall dimensions of various parts are given in Table 4.9.

The radial fan impeller constitutes a rotationally periodic structure and hence, it is enough to analyse one substructure to study the behaviour of complete structure. Hence, only one sector of 60° consisting of that portion of cover plate, back plate, ring, three blades and one rib is considered for investigation. The element used is a three noded triangular shell element having 6 degrees of freedom per node.

The assembly of element matrices for this substructure results in a stiffness matrix which is a real symmetric for static problem and complex Hermitian for the eigenvalue problem. The resulting equations are solved by the out-of-core submatrices elimination method. This algorithm is found to be very effective for the analysis of cyclic symmetric structures. The complete structure is taken for analysis when the NASTRAN and ASKA packages are used and the natural frequencies obtained are compared with the results by the present

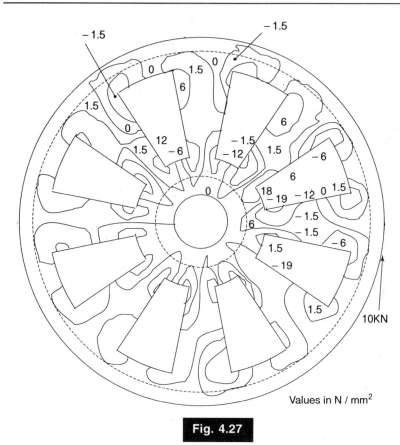

Fig. 4.27

Load at 1/4th span of 1st sector σ_{rr}

method [4.23]. Experiments have been conducted on a prototype of the impeller using Fast Fourier Transform analyser and the experimental results are obtained. The major principle stress contours for the various parts of the impeller are shown in Fig. 4.31. The calculated values and experimental results of the natural frequencies are listed in Table 4.10.

The values of the stresses, natural frequencies and mode shapes obtained with the use of triangular shell element and making use of cyclic symmetry agree well with the experimental results and hence, this program gives necessary data accurate enough for the design of the impeller. It may be pointed out that the effect of change in the thickness of the components of the impeller with reference to stress and natural frequencies can be very quickly computed once the program is perfected. This can be used to study the influence of erosion or corrosion on the thickness reduction of the impeller and as a consequence, the loss of strength can be quantified.

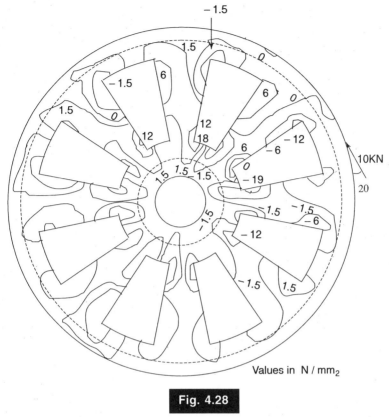

Values in N / mm$_2$

Fig. 4.28

Load at mid span of 1st sector σ_{rr}

4.4.3   Steady State Stress Analysis of Centrifugal Fan Impellers

The cyclic symmetry concept is used to analyse the centrifugal fan impeller. The impeller is shown in Plate 1: (Fig. 4.33) and one repeated sector in Fig. 4.34. It has eight backward curved blades and hence, only one-eighth of the impeller has been considered. The back plate, cover plate, blade and hub are the four components of the impeller. The back plate is assumed to be fixed to the hub. The number of elements and nodes taken for one sector are 140 and 88 respectively and it is shown in Fig. 4.35. Only centrifugal forces are considered as the force vector. The stresses are calculated at different speeds [4.24, 4.25].

To verify the above results, experiments were conducted using the strain gauge technique. Strain gauges along with slip rings and a six-channel carrier frequency amplifier are employed for the

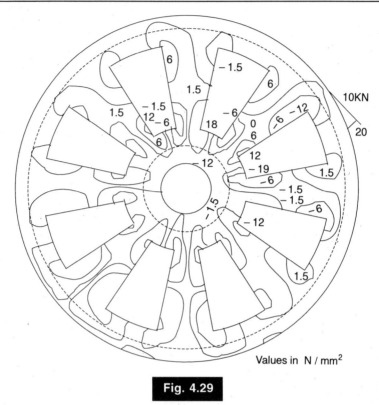

Values in N / mm²

Fig. 4.29

Load at 3/4th span of 1st sector σ_{rr}

experiment. The strains are measured at different points for varying speeds. These are plotted in Figs 4.36–4.39. All the figures except Fig. 4.39 show good agreement. The discrepancy in Fig. 4.39 is probably due to the very coarse finite element discretisation used in the shell portion. It may be pointed out, that if the cyclic symmetry concept would not have been employed, that problem would need at least eight times the core used for the present analysis.

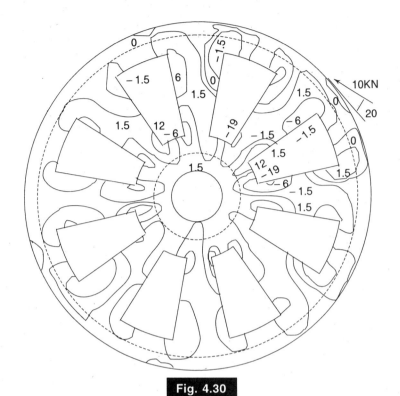

Fig. 4.30

Load at end of span 1st sector σ_{rr}

Table 4.9

Geometry of the Impeller

Part	Details		
1. Back plate	Inner radius	:	130 mm
	Outer radius	:	1120 mm
	Thickness	:	20 mm
2. Cover plate	Inner radius	:	730 mm
	Outer radius	:	1120 mm
	Thickness	:	12 mm
3. Blades	18 blades spaced 20° apart		
	Height: 206 mm at the outer end and 292 mm at the inner end		
	Thickness	:	16 mm
4. Rib	6 ribs placed 60°apart		
	Length	:	195 mm
	Height	:	90 mm
	Thickness	:	16 mm
5. Ring	Ring at the cover plate inner radius		
	Height	:	85 mm
	Thickness	:	65 mm

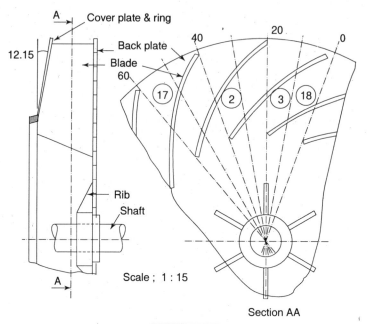

Fig. 4.31

Sectional view of the radial fan impeller

Table 4.10

Natural Frequencies in Hz

	ASKA	NASTRAN	Experimental	Present method
ID	18.01	18.45	19.5	22.02
IR	33.77	36.13	34.0	39.9
IIR	96.7	106.7	95.0	99.63
IID	183.16	169.2	135.0	148.74
IT	—	175.4		

D—diametral mode, R—ring mode, T—torsion mode

PLATE 1

Fig. 4.33 BSB and BCB impellers

Fig. 5.9 Circuit breaker

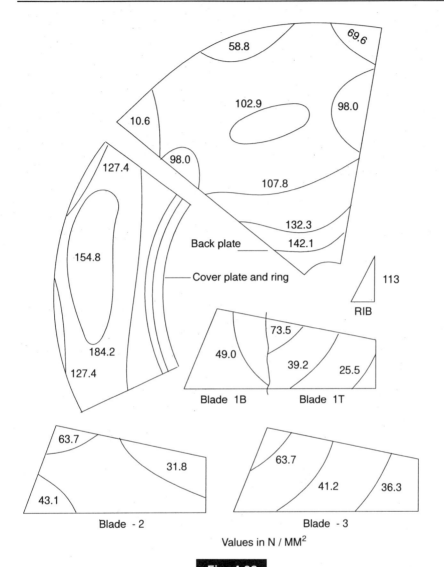

Back plate

Cover plate and ring

RIB

Blade 1B   Blade 1T

Blade - 2

Blade - 3

Values in N / MM^2

Fig. 4.32

Major principal stress contours

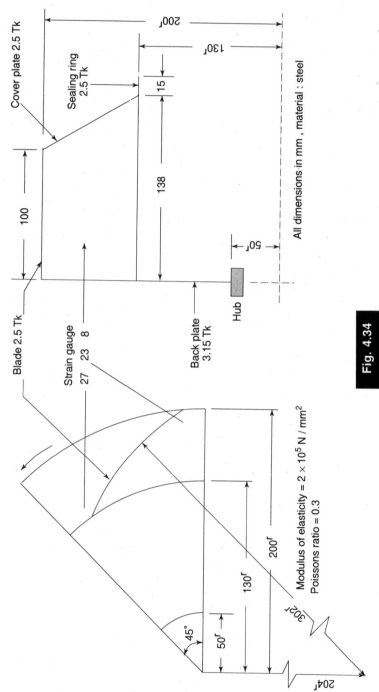

Fig. 4.34

One sector of the fan impeller

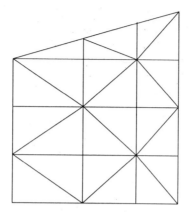

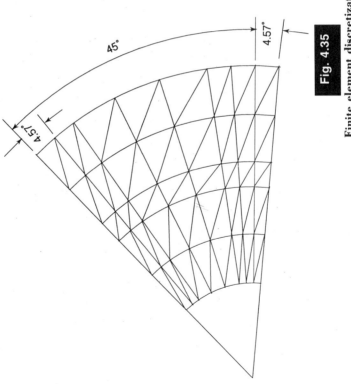

Fig. 4.35

Finite element discretization

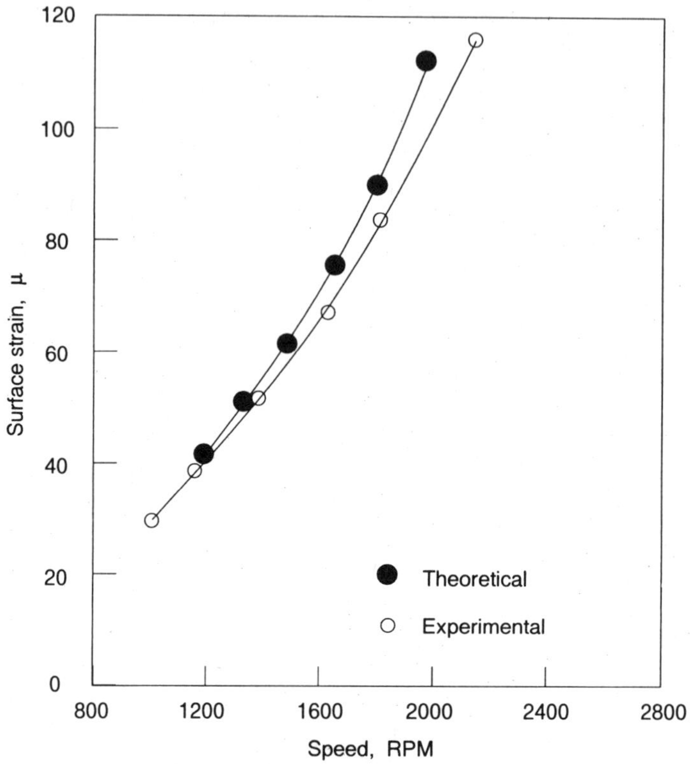

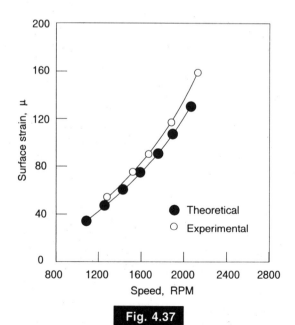

Fig. 4.36

Comparison of strains in cover plate

Fig. 4.37

Comparison of strain in back plate

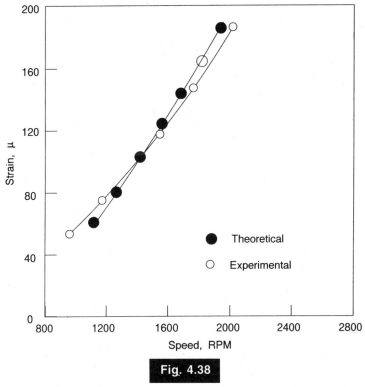

Fig. 4.38

Comparison of axial strains on convex surface of the blade

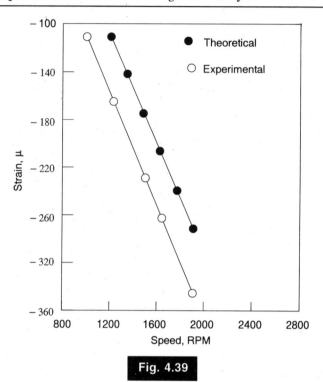

Fig. 4.39

Comparison of axial strains on the concave surface of the blade

REFERENCES

4.1 Thomas, D.L., Standing waves in rotationally periodic structures, *J. Sound and Vib*, vol. 37, pp 288–290, 1974.

4.2 Bullouin, L., *Wave Propagation in Periodic Structures*, 2nd edn, Dover Publications, New York, 1953.

4.3 Zienkiewicz, O.C. and F.C. Scott, On the principle of repeatability and its application in analysis of turbine and pump impellers, *Int. J. Num. Methods in Engg.*, vol. 4, pp 445–452, 1972.

4.4 Mead, D.J., Free wave propagation in periodically supported infinite beams, *J. Sound and Vib.*, vol. 11, no. 2, pp 181–197, 1970.

4.5 Heckl, M., Investigation on the vibration of grillages and other simple beam like structures, *J. Acous. Soc. America*, vol. 36, no. 7, pp 1335–1343, 1964.

4.6 Thomas, D.L., "Dynamics of rotationally periodic structures," *Int. J. Num. Methods in Engg.*, vol. 14, pp 81–102, 1979.

4.7 Mote, C.D., G.S. Shajer and W.Z. Wu, Band Saw and circular raw vibration and stability, *Shock and Vibration Digest*, vol. 14, no. 2, pp 19–25, 1982.

4.8 Ramamurti, V. and P. Balasubramanian, Static analysis of circumferentially periodic structures with Potters' scheme, *Computers and structures*, vol. 22, pp 427–431, 1986.

4.9 Oja, L.K., G.L. Kinzel and A.W. Leissa, Bending of an annually loaded square plate with a central circular hole, *ASME*, Paper No. 82-WA/ DE9.

4.10 Balasubramanian, P., H.K. Suhas and V. Ramamurti, Skyline solver for static analysis of cyclic symmetric structures, *Computers and Structures,* vol. 38, no. 3, pp 259–268, 1991.

4.11 Laxmiprasad, P., R. Nanda Kumar and V. Ramamurti, "Comparison of two different approaches for solution of large cyclic symmetric problems," Computers and Structures, communicated.

4.12 Ramamurti, V. and P. Seshu, On the principle of cyclic symmetry in machine dynamics, *Comm. in Applied Numerical Methods*, vol. 6, pp 259–268, 1990.

4.13 Mota Soares, C.A.M. Petyt and A.M. Salama, Finite element analysis of bladed disks, structural dynamic aspects of bladed disc assemblies, *ASME Winter Annual Meeting,* New York, 1976.

4.14 Balasubramanian, P. and V. Ramamurti, Frequency analysis of centrifugal fan impellers, *J. Sound and Vib.*, vol. 116, no. 1, pp 1–13, 1987.

4.15 Jennings, A., *Matrix Computation for Engineers and Scientists*, John Wiley and Sons, New York, 1977.

4.16 Balasubramanian, P., J.G. Jagadeesh, H.K. Suhas and V. Ramamurti, "Free vibration analysis of symmetric structures," *Comm. Applied Num. Methods*, vol. 7, pp 131–139, 1991.

4.17 Mangsuli, D.G. and V. Ramamurti, "Lanczos Method applied to solution of eigenvalue problems of cyclic symmetric structures," *Computers and Structures*, vol. 34, no. 2, pp 349–353, 1990.

4.18 Ramamurti, V. and V. Om Prakash, "Failure analysis of semi autogeneous grinding mills," *Zement Kalk Gips*, vol. 21, no. 2, pp 87–90, 1988.

4.19 Ramamurti, V., Y. Ramakrishna and K.S. Rama Prasad, "Design parameters of continuous sugar centrifugals," *J. Engg. for Industry*, ASME, vol. 111, no. 3, pp 291–294, 1989.

4.20 Henry, R. and M. Lalanne, Vibration analysis of rotating compressor blades, *J. Engg. for Industry*, ASME, vol. 96, no. 5, pp 1028–1035, 1974.

4.21 Ramamurti, V., S. Murugan and B. Shanker, Cyclic symmetric stress analysis of fabricated gear wheels, *Proc. 15th Design Automation Conf.*, ASME, Montreal 1989, vol. DE-19-3, pp 253–259.

4.22 Ramamurti, V. and V. Srinivasan, Stress analysis of web of solid wheels, *J. Strain Analysis*, vol. 16, no. 1, pp 1–8, 1981.

4.23 Ramamurti, V. and V. Om Prakash, Analytical and experimental investigations of practical radial fan impellers, *20th Mechanics Mid-Western Conf. Proc.*, Purdue Univ., pp 889–893, 1987.

4.24 Ramamurti, V. and P. Balasubramanian, Contribution to the design of centrifugal fan impellers, *Zement Kalk Gips*, vol. 39, pp 633–636, 1986.

4.25 Ramamurti, V. and P. Balasubramanian, Steady state stress analysis of centrifugal fan impellers, *Computers and Structures*, vol. 25, no. 1, pp 129–135, 1987.

CHAPTER 5

Dynamic Analysis

The problems discussed in Chapter 2 were connected with forces which did not change with time. But the majority of the forces encountered in mechanical engineering vary as a function of time. These forces can be periodic in time as in the case of unbalance in rotors or dynamic forces experienced by reciprocating machines like engines or compressors. These are the easiest to handle. These fall under the category of steady state vibration problems of the kind:

$$[M]\{\ddot{u}\} + [C]\{\dot{u}\} + [K]\{u\} = \{f \sin \omega t\} \qquad (5.1)$$

For pure torsional vibration problems encountered in internal combustion engines with or without damper, the three terms in Eq. (5.1) will be inertia torques, damping torques and restoring torques and the term on the R.H.S. exciting torques on the system. These forces exist as long as the machine runs, may be for hours, days or months. They give rise to dynamic deflections and stresses which vary with the same frequency as the exciting force. When the solution to this problem is sought in the form $u \sin \omega t$, the resulting equation takes the form expressed in Eq. (3.27) and hence, can be solved by the methods enunciated in Chapter 2. 'The square matrix on the L.H.S. will be similar to the one indicated in Example 3.12 when the damping term C is present. When C is absent, this matrix need not always be positive definite. It is better to declare it as complex. This avoids encountering the square root of a negative quantity. This will enable the computation of the resultant displacement $\{u\}$ as a complex quantity. Using the displacement, stresses can be computed.' The dynamic stresses computed by solving equations of the type as Eq. (5.1) are the ones which complete several millions of complete reversal. Hence, the endurance strength of the material will be the criterion for the design as opposed to yield used in Chapter 2.

In this chapter, we will discuss two other types of forces which vary with time. The first is the transient force which acts for a short period of time. Examples are: off-shore structures subjected to impact of drifting ice, shock loads on barrels of rocket launchers, transient behaviour of rotors while accelerating or decelerating and the transient dynamic forces due to punching, forging or shearing on the respective machines. It may be pointed out that these forces (which are mathematically predictable) act for a short period of time and die down before the occurrence of the next disturbance. Let us now qualify these periods of time. Ice may hit an off-shore structure for just a second but may cause extensive damage due to the impact. Punching may be done during 0.1 and 0.2 seconds, but these punches may be 30 times in one minute in continuous operation, leaving a time gap of 1.8 to 1.9 seconds between two punching operations, sufficient enough for the disturbance to die down. Similarly, in rocket launching, the explosion may take place within about 0.3 to 0.4 seconds. The time gap between two successive explosions may be 2 seconds, ensuring the full recovery of the launchers from the shock. The design of these units becomes critical. For example, being a mobile unit, the launcher has to be an optimum combination of light weight and high structural rigidity. Rigidity is essential for guaranteeing the desired level of firing accuracy and light weight to keep the power requirement to a minimum. Disturbances of this kind fall under the category of transient forces. Even when the forces are repetitive, (as in the case of continuous firing), the time gap between two excitations warrants the disturbance due to the former to diappear before the occurrence of the latter.

The methods of solution to these types will be grouped under two heads, the time integration scheme and modal superposition scheme. Detailed survey on direct integration methods is made by Dokainish and Subbaraj [5.1, 5.2]. The third type of dynamic force is the force which is random in nature. The disturbances experienced by a vehicle owing to undulations of the road are truly random. Even though the forces are not deterministic, these random inputs can be considered to have stationary and ergodic properties. Hence, the analysis procedure formulated under random vibration theory can be applied to this class of problems, combining them with finite element modelling. These will be discussed at the end of the chapter.

➤ 5.1 DIRECT INTEGRATION METHOD

But there is another class of problems of practical significance. When the forcing function is a transient function of time, the solution is little more involved. One of the approaches is to consider this equation to hold good at discrete intervals of time and treat them as static problems in every one of these intervals [5.3]. This seems to be the most reasonable shortcut when one realises that the general form of the differential equation is of second order with constant coefficients and the complementary solution is really involved. Three distinct methods will be discussed here. A summary of the concepts involved in transient analysis is presented in Ref. [5.4] in a concise form.

5.1.1   Central Difference Method

Let the governing differential equation be given by

$$[M]\{\ddot{u}_t\} + [C]\{\dot{u}_t\} + [K]\{u_t\} = \{f(t)\} \tag{5.2}$$

u_t, $\dot{u}_t$ and $\ddot{u}_t$ are the displacement, velocity and acceleration at any time t. Everyone of them is of size $n \times 1$ (for any specific time), n being usually large. $[M]$, $[C]$ and $[K]$ are the mass, damping and stiffness matrices, each of size $n \times n$. $[K]$ for a finite element problem will be symmetric, $[M]$ and $[C]$ may or may not be diagonal. It is very easy to visualise or define $[M]$, but $[C]$ is very difficult to comprehend for a complicated system; Let us postpone this to a later stage. In the central difference scheme the velocities and acceleration are expressed in terms of the displacements at discrete intervals of time in the following manner,

$$\{\ddot{u}_t\} = \frac{1}{(\Delta t)^2} (\{u_{t+\Delta t}\} - \{2u_t\} + \{u_{t-\Delta t}\}) \tag{5.3}$$

$$\{\dot{u}_t\} = \frac{1}{2\Delta t} (u_{t+\Delta t} - \{u_{t-\Delta t}\}) \tag{5.4}$$

The error introduced in both these approximation is of the order $(\Delta t)^2$ [Sec. (1.4)].

Substituting Eqs (5.3) and (5.4) in Eq. (5.2), we have

$$\left(\frac{1}{(\Delta t)^2}[M] + \frac{1}{(2\Delta t)}[C]\right)\{u_{t+\Delta t}\} = f(t) - \left([K] - \frac{2}{(\Delta t)^2}[M]\right)\{u_t\}$$

$$- \left(\frac{[M]}{(\Delta t)^2} - \frac{[C]}{2\Delta t}\right)\{u_{t-\Delta t}\} \tag{5.5}$$

From Eq. (5.5) it is clear that if $\{u_t\}$ and $\{u_{t-\Delta t}\}$ are known, one can compute $\{u_{t+\Delta t}\}$. But to start the scheme, we adopt the following approach. At $t = 0$, from Eq. (5.2), we have

$$[M]\{\ddot{u}_0\} + [C]\{\dot{u}_0\} + [K]\{u_0\} = \{f(0)\} \tag{5.6}$$

Since at $t = 0$, $\{u_0\}$ and $\{\dot{u}_0\}$ would have been specified $\{\ddot{u}_0\}$ can be computed. In the same way for $t = 0$, from Eqs (5.3) and (5.4), we have

$$\{\ddot{u}_0\} = \frac{1}{(\Delta t)^2}(\{u_{\Delta t}\} - 2\{u_0\} + \{u_{-\Delta t}\})$$

$$\{\dot{u}_0\} = \frac{1}{2\Delta t}(\{u_{\Delta t}\} - \{u_{-\Delta t}\})$$

Hence, we can write

$$\{u_{-\Delta t}\} = \{u_0\} - \Delta t\{\dot{u}_0\} + \frac{(\Delta t)^2}{2}\{\ddot{u}_0\} \tag{5.7}$$

There is no physical meaning for $\{u_{-\Delta t}\}$ (since $\{u\}$ does not exist before $t = 0$). However, making use of the general philosophy of central differences, we are able to compute $\{u_{-\Delta t}\}$.

Referring to Eq. (5.5) one can start the process of computation. When $t = 0$, since $\{u_0\}$, $\{u_{-\Delta t}\}$ are known $\{u_{\Delta t}\}$ can be computed. This can be used for calculating $\{u_{2\Delta t}\}$, $\{u_{3\Delta t}\}$ and so on. The following observations can be made of Eq. (5.5). If $[M]$ and $[C]$ are diagonal matrices, computing $\{u_{t+\Delta t}\}$ from $\{u_t\}$ and $\{u_{t-\Delta t}\}$ is a case of straightforward multiplications (and division). For this reason this method is known as the explicit integration scheme.

The major defect of this scheme is that this is only conditionally stable. There is a definite Δt above which the computed values become unstable [5.3]. The various steps involved can be summarised as follows.

At time $t = 0$
1. Form $[K]$, $[M]$ and $[C]$
2. With the values of $\{u_0\}$ and $\{\dot{u}_0\}$, compute $\{\ddot{u}_0\}$ (Eq. 5.6)
3. Choosing a time step Δt, compute $u_{-\Delta t}$ (Eq. 5.7)
4. If $[M]$ and $[C]$ are not diagonal, triangularise

$$\frac{[M]}{\Delta t^2} + \frac{[C]}{2\Delta t}$$

and express it as $[L][D][L]^T$ (Eq. 5.5)

For every time step
1. Compute R.H.S. of Eq. (5.5) viz.

$$\{\hat{f}(t)\} = \{f(t)\} - \left([K] - \frac{2}{(\Delta t)^2}[M]\right)\{u_t\} - \left(\frac{[M]}{(\Delta t)^2} - \frac{[C]}{2\Delta t}\right)\{u_{t-\Delta t}\}$$

2. Solve for $\{u_{t+\Delta t}\}$

$$[L][D][L]^T\{u_{t+\Delta t}\} = \{\hat{f}(t)\}$$

If $[M]$ and $[C]$ are diagonal matrices this operation is a straight-forward division.

3. Compute velocity and acceleration at t

$$\{\dot{u}_t\} = (\{u_{t+\Delta t}\} - \{u_{t-\Delta t}\})/2\Delta t$$

$$\{\ddot{u}_t\} = (\{u_{t+\Delta t}\} - 2\{u_t\} + \{u_{t+\Delta t}\})/(\Delta t)^2$$

Example 5.1

Compute the response of the two-degree-freedom system given below by the central difference scheme

$$\begin{bmatrix} 1 & 0 \\ 0 & 1 \end{bmatrix} \{\ddot{u}\} + \begin{bmatrix} 2 & 2 \\ 2 & 5 \end{bmatrix} \{u\} = f(t)$$

Initial conditions are

$$u_0 = \left\{ \begin{array}{c} 1 \\ 0 \end{array} \right\}, u_0 = \left\{ \begin{array}{c} 0 \\ 0 \end{array} \right\} \quad \text{and} \quad f(t) = 0$$

The natural frequencies of this system are given by

$$(2 - \omega^2)(5 - \omega^2) - 4 = 0$$

$$\omega^2 = 1 \quad \text{or} \quad 6$$

Hence the actual periods of oscillation are 2π and $2\pi/\sqrt{6}$ s.

Let us choose Δt of 0.25 s which is less than 1/10th of second period of oscillation.

$$\{\ddot{u}_0\} = -[K]\{u_0\} = -\begin{bmatrix} 2 & 2 \\ 2 & 5 \end{bmatrix} \left\{ \begin{array}{c} 1 \\ 0 \end{array} \right\} = \left\{ \begin{array}{c} -2 \\ -2 \end{array} \right\}$$

$$\frac{[M]}{(\Delta t)^2} = \frac{1}{(0.25)^2} \begin{bmatrix} 1 & 0 \\ 0 & 1 \end{bmatrix} = \begin{bmatrix} 16 & 0 \\ 0 & 16 \end{bmatrix}$$

For every time step:

$$\{\hat{f}'(t)\} = -\left([K] - \frac{2[M]}{(\Delta t)^2}\right)\{u_t\} - \left(\frac{[M]}{(\Delta t)^2}\right)\{u_{t-\Delta t}\}$$

$$= \left\{\begin{bmatrix} 2 & 2 \\ 2 & 5 \end{bmatrix} - \begin{bmatrix} 32 & \cdot 0 \\ 0 & 32 \end{bmatrix}\right\}\{u_t\} - \begin{bmatrix} 16 & 0 \\ 0 & 16 \end{bmatrix}\{u_{t-\Delta t}\}$$

$$= \begin{bmatrix} 30 & -2 \\ -2 & 27 \end{bmatrix}\{u_t\} - \begin{bmatrix} 16 & 0 \\ 0 & 16 \end{bmatrix}\{u_{t-\Delta t}\}$$

$$\{u_{-\Delta t}\} = \left\{ \begin{array}{c} 1 \\ 0 \end{array} \right\} - 0 + \frac{(0.25)^2}{2} \left\{ \begin{array}{c} -2 \\ -2 \end{array} \right\}$$

$$= \left\{ \begin{array}{c} 0.9375 \\ -0.0625 \end{array} \right\}$$

$$\therefore \quad \{u_{t+\Delta t}\} = \left[\begin{array}{cc} 0.0625 & 0 \\ 0 & 0.0625 \end{array} \right] \left[\begin{array}{cc} 30 & -2 \\ -2 & 27 \end{array} \right] \{u_t\} - 16\left[\{u_{t-\Delta t}\}\right]$$

Choosing the starting value $t = 0$, $\{u_{\Delta t}\}$, $\{u_{2\Delta t}\}$, etc. are given as below:

Table 5.1

$t =$	0.25	0.50	0.75	1.00	1.25	1.50	1.75	2.00	2.25
u_1	0.9375	0.7656	0.5258	0.2714	0.4997	−0.1127	−0.2162	−0.2672	−0.3168
u_2	−0.0625	−0.2226	−0.4088	−0.5329	−0.5244	−0.3582	−0.0659	0.2739	0.5616

The differential equation of the response of this system is given by

$$\frac{d^2x_1}{dx^2} + x_1 + 2x_2 = 0$$

$$2x_1 + 5x_2 + (d^2x_2/dt^2) = 0$$

where x_1 and x_2 are the dynamic displacements.
Complementary solution to these equations is given by

$$x_1 = -2[c_1]\cos t + \frac{c_2}{2}\cos\sqrt{6}\,t$$

$$x_2 = [c_1]\cos t + [c_2]\cos\sqrt{6}\,t.$$

since at $t = 0$, $x_1 = 1$, $x_2 = 0$ we have

$$[c_1] = -0.4, \ c_4 = 0.4$$

Exact solution is given by

$$x_1 = 0.8\cos t + 0.2\cos\sqrt{6}\,t$$

$$x_2 = -0.4\cos t + 0.4\cos\sqrt{6}\,t$$

Exact values for $0.25 < t < 2.25$ s are tabulated in Table 5.2.

Table 5.2

t	0	0.25	0.50	0.75	1.00
x_1	1.00	0.9388	0.798	0.5327	0.2786
x_2	0.00	-0.0602	-0.2144	-0.3978	-0.5247
t	1.25	1.50	1.75	2.00	2.25
x_1	0.0530	-0.1167	-0.2252	-0.3613	-0.5019
x_2	-0.5247	-0.3726	-0.0939	0.2391	0.2612

5.1.2 Wilson θ Method

This is basically a linear acceleration scheme. The acceleration is assumed to be linear from time t to time $t + \theta \Delta t$, where $\theta \geq 1.4$ [5.5] as shown in Fig. 5.1. For any time τ less than $\theta \Delta t$, we have

$$\{\ddot{u}_{t+\tau}\} = \{\ddot{u}_t\} + \frac{\tau}{\theta \Delta t}(\{\ddot{u}_{t+\theta \Delta t}\} - \{\ddot{u}_t\}) \qquad (5.8)$$

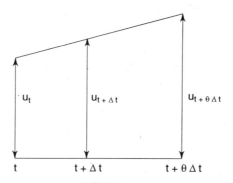

Fig. 5.1

Linear accelerations scheme

Integrating with respect to τ, we have

$$\{\dot{u}_{t+\tau}\} = \{\dot{u}_t\} + \{\ddot{u}_t\}\tau + \frac{\tau^2}{2\theta \Delta t}(\{\ddot{u}_{t+\theta \Delta t}\} - \{\ddot{u}_t\}) \qquad (5.9)$$

$$\{u_{t+\tau}\} = \{u_t\} + \{\dot{u}_t\}\tau + \left\{\frac{\ddot{u}_t \tau^2}{2}\right\} + \frac{1}{6\theta \Delta t}\tau^3(\{\ddot{u}_{t+\theta \Delta t}\} - \{\ddot{u}_t\}) \qquad (5.10)$$

When $\tau = \theta \Delta t$, we have

$$\{\dot{u}_{t+\theta \Delta t}\} = \{\dot{u}_t\} + \frac{\theta \Delta t}{2}(\{\ddot{u}_{t+\theta \Delta t}\} + \{\ddot{u}_t\}) \qquad (5.11)$$

$$\{u_{t+\theta \Delta t}\} = \{u_t\} + \theta \Delta t \cdot \{\dot{u}_t\} + \frac{\theta^2 (\Delta t)^2}{6}(\{\ddot{u}_{t+\theta \Delta t}\} + 2\{\ddot{u}_t\}) \qquad (5.12)$$

Equations (5.11) and (5.12) can be treated as two equations in two unknowns $\{\ddot{u}_{t+\theta\Delta t}\}$ and $\{\dot{u}_{t+\theta\Delta t}\}$ and the rest as known quantities. Solving for them, we get

$$\{\ddot{u}_{t+\theta\Delta t}\} = \frac{6}{\theta^2\Delta t^2}(\{u_{t+\theta\Delta t}\} - \{u_t\}) - \frac{6}{\theta\Delta t}\{\dot{u}_t\} - 2\{\ddot{u}_t\} \qquad (5.13)$$

$$\{\dot{u}_{t+\theta\Delta t}\} = \frac{3}{\theta\Delta t}(\{u_{t+\theta\Delta t}\} - \{u_t\}) - 2\{\dot{u}_t\} - \frac{\theta\Delta t}{6}\{\ddot{u}_t\} \qquad (5.14)$$

At a time $t + \theta\Delta t$, the equilibrium Eq. (5.2) gives

$$[M]\{\ddot{u}_{t+\theta\Delta t}\} + [C]\{\dot{u}_{t+\theta\Delta t}\} + [K]\{u_{t+\theta\Delta t}\} = \{f_{t+\theta\Delta t}\} \qquad (5.15)$$

Assuming linear variation for projected load, $\{f_{t+\theta\Delta t}\}$ can be written as

$$\{f_{t+\theta\Delta t}\} = \{f_t\} + \theta(\{f_{t+\Delta t} - f_t\}) \qquad (5.16)$$

Substituting Eq. (5.13) and (5.14) in Eq. (5.15) we can solve for $\{u_{t+\theta\Delta t}\}$.

$$[M]\left[\frac{6}{\theta^2\Delta t^2}\{u_{t+\theta\Delta t} - u_t\} - \frac{6}{\theta\Delta t}\{\dot{u}_t\} - 2\{\ddot{u}_t\}\right]$$

$$+ [C]\left[\frac{3}{\theta\Delta t}(\{u_{t+\theta\Delta t}\} - \{u_t\}) - 2\{\dot{u}_t\} - \frac{\theta\Delta t}{2}\{\ddot{u}_t\}\right]$$

$$+ [K]\{u_{t+\theta\Delta t}\} = \{f_t\} + \theta(\{f_{t+\Delta t} - f_t\})$$

Let $\quad c_0 = \dfrac{6}{(\theta\Delta t)^2} \qquad c_1 = \dfrac{3}{\theta\Delta t} \qquad c_2 = \dfrac{6}{\theta\Delta t} \qquad c_3 = \dfrac{\theta\Delta t}{2}$

Then

$$([K] + c_0[M] + c_1 C)\{u_{t+\theta\Delta t}\} = \{f_t\} + \theta(\{f_{t+\Delta t} - f_t\})$$
$$+ [M](c_0\{u_t\} + c_2\{\dot{u}_t\} + 2\{\ddot{u}_t\})$$
$$+ [C](c_1\{u_t\} + 2\{\dot{u}_t\} + c_3\{\ddot{u}_t\}) \qquad (5.17)$$

Having computed $u_{t+\theta\Delta t}$ from Eq. (5.17), one can compute $\{\ddot{u}_{t+\theta\Delta t}\}$ from Eq. (5.13) which can be used to calculate $\{\ddot{u}_{t+\Delta t}\}$, $\{\dot{u}_{t+\Delta t}\}$ and $\{u_{t+\Delta t}\}$ from Eqs (5.8), (5.9) and (5.10). It may be observed that the square matrix on the L.H.S. of Eq. (5.17), viz. $[K] + c_0[M] + c_1[C]$ is a matrix which is to be triangularised to compute $\{u_{t+\Delta t}\}$ and hence this scheme is an implicit integration scheme. It may also be observed that no special starting procedure is required since R.H.S. is a function of u_t, $\dot{u}_t$ and $\ddot{u}_t$ unlike in the case of central difference scheme where $u_{t-\Delta t}$ was also needed.

Summary

At time t = 0
1. Form $[K]$, $[M]$ and $[C]$
2. With the values of $\{u_0\}$ and $\{\dot{u}_0\}$, compute $\{\ddot{u}_0\}$ (Eq. 5.6)
3. Choosing a time step Δt, and θ (usually (1.4)) compute

$$c_0 = \frac{6}{(\theta \Delta t)^2} \qquad c_1 = \frac{3}{\theta \Delta t} \qquad c_2 = \frac{6}{\theta \Delta t} \qquad c_3 = \frac{\theta \Delta t}{2}$$

4. Form effective stiffness matrix of the L.H.S.

$$[\hat{K}] = [K] + c_0[M] + c_1[C]$$

5. Triangularise:

$$[\hat{K}] = [L][D][L]^T$$

For each step:
1. Calculate R.H.S.

$$\{\hat{f}_{t+\theta \Delta t}\} = \{f_t\} + \theta(\{f_{t+\Delta t} - f_t\}) + [M](c_0\{u_t\} + c_2\{\dot{u}_t\}$$
$$+ 2\{\ddot{u}_t\}) + [C](c_1\{u_t\} + 2\{\dot{u}_t\} + c_3\{\ddot{u}_t\})$$

2. Solve for $\{u_{t+\theta \Delta t}\}$ from

$$[L][D][L]^T\{u_{t+\theta \Delta t}\} = \{\hat{f}_{t+\theta \Delta t}\}$$

3. Calculate the displacements, velocities and accelerations at time $(t + \Delta t)$

$$\{\ddot{u}_{t+\Delta t}\} = \frac{c_0}{\theta}(\{u_{t+\theta \Delta t}\} - \{u_t\}) - \frac{c_2}{\theta}\{\dot{u}_t\} + \left(1 - \frac{3}{\theta}\right)\{\ddot{u}_t\}$$

$$\{\dot{u}_{t+\Delta t}\} = \{\dot{u}_t\} + \frac{\Delta t}{2}(\{\ddot{u}_{t+\Delta t}\} + \{\ddot{u}_t\})$$

$$\{u_{t+\Delta t}\} = \{u_t\} + \Delta t\{\dot{u}_t\} + \frac{\Delta t^2}{6}(\{\ddot{u}_{t+\Delta t}\} + 2\{\ddot{u}_t\})$$

Example 5.2
Solve the problem given in Example 5.1 by Wilson θ method.

$$\begin{bmatrix} 1 & 0 \\ 0 & 1 \end{bmatrix}\{\ddot{u}\} + \begin{bmatrix} 2 & 2 \\ 2 & 5 \end{bmatrix}\{u\} = \{f(t)\}$$

Initial conditions are

$$\{u_0\} = \begin{Bmatrix} 1 \\ 0 \end{Bmatrix} \qquad \{u_0\} = \begin{Bmatrix} 0 \\ 0 \end{Bmatrix} \qquad \text{and} \qquad \{f(t)\} = 0$$

Δt from the previous example = 0.25 s and $\{\ddot{u}_0\} = \begin{Bmatrix} -2 \\ -2 \end{Bmatrix}$

$\therefore$

$$c_0 = \frac{6}{(\theta \Delta t)^2} = \frac{6}{(0.35)^2} = 48.98$$

$$c_1 = \frac{3}{(\theta \Delta t)} = \frac{3}{(0.35)} = 8.57$$

$$c_2 = 2c_1 = 17.14$$

$$c_3 = \frac{\theta \Delta t}{2} = \frac{0.35}{2} = 0.175$$

$$[K] = [K] + c_0 [M] = \begin{bmatrix} 2 & 2 \\ 2 & 5 \end{bmatrix} + \frac{6}{(0.35)^2} \begin{bmatrix} 1 & 0 \\ 0 & 1 \end{bmatrix}$$

$$= \begin{bmatrix} 50.98 & 2 \\ 2 & 53.98 \end{bmatrix}$$

$$\{f_{t+\theta \Delta t}\} = [M][48.98\{u_t\} + 17.14\{\dot{u}_t\} + 2\{\ddot{u}_t\}]$$

$$\{\ddot{u}_{\Delta t}\} = 34.98(\{u_{t+\theta \Delta t}\} - \{u_t\}) - 12.26\{\dot{u}_t\} - 1.14\{\ddot{u}_t\}$$

$$\{\dot{u}_{\Delta t}\} = \{\dot{u}_t\} + 0.125(\{\ddot{u}_{t+\Delta t}\}) + \{\ddot{u}_t\})$$

$$\{u_{\Delta t}\} = \{u_t\} + \Delta t \cdot \{\dot{u}_t\} + 0.01041(\ddot{u}_{t+\Delta t}\} + 2\{\ddot{u}_t\})$$

Computed values of u are shown in Table 5.3.

Table 5.3

t	0.25	0.50	0.75	1.00	1.25
u_1	0.9397	0.771	0.526	0.268	0.052
u_2	− 0.0566	− 0.204	− 0.374	− 0.512	− 0.512
t	1.50	1.75	2.00	2.25	
u_1	− 0.092	− 0.235	− 0.352	− 0.492	
u_2	− 0.381	− 0.112	0.237	0.259	

5.1.3 Newmark Method

Newmark integration scheme is also an extension of linear acceleration scheme [5.6]. The velocity and displacement are expressed as

$$\{\dot{u}_{t+\Delta t}\} = \{\dot{u}_t\} + [(1 - \delta)\{\ddot{u}_t\} + \delta\ddot{u}_{t+\Delta t}]\,\Delta t \qquad (5.18)$$

$$\{u_{t+\Delta t}\} = \{u_t\} + \{\dot{u}_t\}\Delta t + \left[\left(\frac{1}{2} - \alpha\right)\{\ddot{u}_t\} + \alpha\{\ddot{u}_{t+\Delta t}\}\right] \qquad (5.19)$$

where α and δ are parameters to be used to obtain integration stability.

$$(\delta \geq 0.50, \quad \alpha \geq 0.25(0.5 + \delta)^2)$$

We have at any time $(t + \Delta t)$ the governing differential equation

$$[M]\{\ddot{u}_{t+\Delta t}\} + [C]\{\dot{u}_{t+\Delta t}\} + [K]\{u_{t+\Delta t}\} = \{f_{t+\Delta t}\} \qquad (5.20)$$

Following the procedure adopted in Wilson θ scheme, $\{\ddot{u}_{t+\Delta t}\}$ and $\{\dot{u}_{t+\Delta t}\}$ are obtained in terms of $\{u_t\}$, $\{\dot{u}_t\}$, $\{\ddot{u}_t\}$ and $\{u_{t+\Delta t}\}$ from Ëqs (5.18) and (5.19) and are substituted in Eq. (5.20). The steps can be summarised as follows:
At time $t = 0$

  1 For $[K]$, $[M]$ and $[C]$

  2. With the values of $\{u_0\}$ and $\{\dot{u}_0\}$, compute $\{\ddot{u}_0\}$ [Eq. (5.6)]

  3. Choosing Δt, α and δ ($\delta = 0.5$, $\alpha = 0.25(0.5 + \delta)^2$) compute the following constants

$$c_0 = \frac{1}{\alpha(\Delta t)^2} \qquad c_1 = \frac{\delta}{\alpha\,\Delta t} \qquad c_2 = \frac{1}{\alpha\,\Delta t} \qquad c_3 = \frac{1}{2\alpha} - 1$$

$$c_4 = \frac{\delta}{\alpha} - 1 \qquad c_5 = \frac{\Delta t}{2}\left(\frac{\delta}{\alpha} - 2\right) \qquad c_6 = \Delta t\,(1 - \delta)$$

$$c_7 = \delta\Delta t$$

  4. Form the effective stiffness matrix

$$[\hat{K}] = [K] + c_0[M] + c_1[C]$$

  5. Triangularise: $[\hat{K}] = [L][D][L]^T$.

For each time step:

  1. Calculate the effective load at time $t + \Delta t$

$$\{\hat{f}_{t+\Delta t}\} = \{f_{t+\Delta t}\} + [M]\,(c_0\{u_t\} + c_2\{\dot{u}_t\} + c_3\{\ddot{u}_t\}$$

$$+ [C]\,(c_1\{u_t\} + c_4\{\dot{u}_t\} + c_5\{\ddot{u}_t\}$$

2. Solve for $\{u_{t+\Delta t}\}$

$$[L][D][L]^T\{u_{t+\Delta t}\} = \{\hat{f}_{t+\Delta t}\}$$

3. Calculate accelerations and velocities at time $t + \Delta t$:

$$\{\ddot{u}_{t+\Delta t}\} = c_0(\{u_{t+\Delta t}\} - \{u_t\}) - c_2\{\dot{u}_t\} - c_3\{\ddot{u}_t\}$$

$$\{\dot{u}_{t+\Delta t}\} = \{\dot{u}_t\} + c_6\{\ddot{u}_t\} + c_7\{\ddot{u}_{t+\Delta t}\}$$

Example 5.3

Solve the problem given in Example 5.1 by Newmark method

$$\begin{bmatrix} 1 & 0 \\ 0 & 1 \end{bmatrix}\{\ddot{u}\} + \begin{bmatrix} 2 & 2 \\ 2 & 5 \end{bmatrix}\{u\} = \{f(t)\}$$

Initial conditions are

$$\{u_0\} = \begin{Bmatrix} 1 \\ 0 \end{Bmatrix} \qquad \dot{u}_0 = \begin{Bmatrix} 0 \\ 0 \end{Bmatrix} \qquad \text{and} \qquad f(t) = 0$$

t from the last example = 0.25 s

$$\{\ddot{u}_0\} = \begin{Bmatrix} -2 \\ -2 \end{Bmatrix}$$

Choosing $\alpha = 0.25$, $\delta = 0.5$, the constants are

$$c_0 = \frac{1}{\alpha \Delta t^2} = 64 \qquad c_1 = \frac{\delta}{\alpha \Delta t} = 8$$

$$c_2 = \frac{1}{\alpha \Delta t} = 16 \qquad c_3 = \frac{1}{2\alpha} - 1 = 1 \qquad c_4 = \frac{\delta}{\alpha} - 1 = 1$$

$$c_5 = \frac{\Delta t}{2}\left(\frac{\delta}{\alpha} - 2\right) = 0 \qquad c_6 = \Delta t(1 - \delta) = 0.125$$

$$c_7 = \delta \Delta t = 0.125$$

$$[K] = [K] + [C_0][M] = \begin{bmatrix} 2 & 2 \\ 2 & 5 \end{bmatrix} + 64\begin{bmatrix} 1 & 0 \\ 0 & 1 \end{bmatrix}$$

$$= \begin{bmatrix} 66 & 2 \\ 2 & 69 \end{bmatrix}$$

$$\{\hat{f}_{t+\Delta t}\} = [M]\,[64\{u_t\} + 16\{\dot{u}_t\} + 1\{\ddot{u}_t\}]$$

$$= 64\{u_t\} + 16\{\dot{u}_t\} + \{\ddot{u}_t\}$$

$$\{\ddot{u}_{t+\Delta t}\} = 64\,(\{u_{t+\Delta t}\} - \{u_t\}) - 16\{\dot{u}_t\} - \{\ddot{u}_t\}$$

$$\{\dot{u}_{t+\Delta t}\} = \{\dot{u}_t\} + 0.125\{\ddot{u}_t\} + 0.125\{\ddot{u}_{t+\Delta t}\}$$

The values of $\{u\}$ computed for various t are shown in Table 5.4.

Table 5.4

t	0.25	0.50	0.75	1.00	1.25	1.50
u_1	0.939	0.7699	0.529	0.264	0.0217	0.136
u_2	− 0.056	− 0.202	− 0.375	− 0.497	− 0.504	− 0.363
t	1.75	2.00	2.25			
u_1	− 0.243	− 0.373	− 0.499			
u_2	− 0.107	0.242	0.271			

Choice of time step Δt

The main point to be decided upon is the basis on which Δt is to be chosen. Any system we normally encounter in engineering is a continuous system with infinite degrees of freedom. But for purposes of computation we idealise the system as one having a finite number of degrees of freedom. More the number of degrees of freedom, more accurate will be results. The response consists of all the n modes of vibration. But it is the first m mode ($m \ll n$) which will mostly determine the overall response. If the natural period of the mth mode is T_m, choice of Δt equal to $T_m/10$ should give us a reasonable dynamic response up to the mth mode. This could be seen from Fig. 5.2.

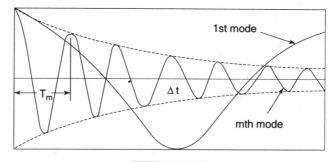

Fig. 5.2

Dynamic response

One wavelength corresponding to the mth response is approximated by ten intervals and hence the lower ones will be represented more accurately. But for the conditionally stable system (central difference scheme) Δt has to be smaller than the natural period of the practically all the n modes [5.1] to obtain reliable results. This makes the labour involved in computation enormous. But for unconditionally stable systems (Wilson θ and Newmark schemes), this choice gives stable results.

To quote a specific instance, let us assume that for turbomachinery up to 60,000 rpm is of interest to us. This gives rise to a time period of roughly 1/1000 s. If this has to be included in the response Δt to be chosen should be atleast 1/10th of this time period (roughly 10^{-4} s).

Multiplications Involved

For the Wilson θ scheme, let us assume the size of the problem to be n, bandwidth for $[M]$, $[C]$ and $[K]$ be b, carpet width c be $(2b - 1)$ and number of time steps m. Computing $[K]$ needs $nb^2/2$ multiplications computing $f_{t+\Delta t}$ needs $(6n + 2nc)$ multiplications, velocity $\dot{u}_{t+\Delta t}$ and acceleration $\ddot{u}_{t+\Delta t}$ another $6n$.

Hence the total $= \dfrac{nb^2}{2} + m2nc + 12n$

$$\approx \frac{nb^2}{2} + m4nb + 12n$$

Suppose we compute up to three waves of the third mode and that the third natural frequency is five times the first. If $\Delta t = T_3/10$, then the number of time steps m needed will be equal to 150.

For a typical case of $n = 500$, $b = 100$,

Total multiplications $= 500 \dfrac{(100)^2}{2} + 150 : 400.500 + 12.500$

$$= 35 \times 10^6 \text{ multiplications}$$

If time taken for each multiplication is 50×10^{-6} s, this needs around 1750 s ($\approx$ 29 min).

➤ 5.2 MODE SUPERPOSITION _____

From the foregoing discussions it is clear that the direct integration is really time-consuming since the bandwidth of the matrices involved is not very small and the number of time steps needed is large to get meaningful response. One way to improve the computational efficiency is to transform the governing differential equation into a convenient form in which the bandwidth gets considerably reduced. In the modal superposition technique it gets reduced to one, thereby making the $m(<< n)$ differential equations decoupled from each other. Then each of the differential equations is of the second order (linear) and hence it is easy to get their solutions in closed form. Before we discuss the method of approach it would be appropriate to recapitulate the response of single degree systems in transient phase.

5.2.1   Single Degree Freedom System Response

Excitation by a Force

Let us consider the system shown in Fig. 5.3. Its governing differential equation is given by

$$m\ddot{x} + kx + c\dot{x} = F_0 \qquad (5.21)$$

(F_0 —is constant force acting on the mass.)

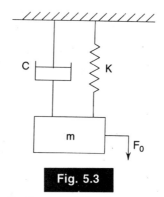

Fig. 5.3

Single degree of freedom system

Let the initial conditions be $x = x_0$ and $\dot{x} = v_0$. The solution of Eq. (5.21) is given by

$$x = \exp(-\xi\omega_n t)(c_1 \cos \omega_n t + c_2 \sin \omega_n t) + \frac{F_0}{k} \qquad (5.22)$$

where $\qquad \xi = $ damping ratio $= c/[2\sqrt{km}]$

and $\qquad \omega_n = \sqrt{k/m} \qquad (5.23)$

at $t = 0$, $x = x_0$, $\dot{x} = v_0$,

$$\therefore \qquad c_1 = \left(x_0 - \frac{F_0}{k} \right) \qquad c_2 = \frac{v_0 + \xi \omega_n \left(x_0 - \dfrac{F_0}{k} \right)}{\sqrt{(1 - \xi^2)}\, \omega_n} \qquad (5.24)$$

Velocity
$$\dot{x} = \omega_n \exp(-\xi \omega_n t)\, [\cos \omega_n t (c_2 - \xi c_1)$$

$$+ \sin \omega_n t (- c_1 - \xi c_2)] \qquad (5.25)$$

Equation (5.22) will be the response for an indefinite period of time. This response exponentially dies down the logarithmic decrement δ of the rate of decay being given by

$$\delta = \log \frac{X_n}{X_{n-1}} = \frac{2\pi \xi}{\sqrt{(1 - \xi^2)}} \qquad (5.26)$$

The value of ξ varies from 0 to 1, Beyond $\xi = 1$ this solution is not valid. Even for ξ as low as 0.1, δ will be equal to 0.628. Ratio of successive amplitudes of decay will be $e^{0.628} = 1.874$. In about 10 oscillations the original amplitude will get reduced by 534 times. Another point to be observed is the time period of oscillation which is $2\pi/\omega_n\sqrt{(1 - \xi^2)}$. Even for a structure having a natural frequency as low as 1 cycle/s, within 10 s (in 10 oscillations) the response will get considerably reduced. If the forcing function is not constant, but varies as shown in Fig. 5.4. The response can easily be inferred from Eqs (5.22) and (5.25).

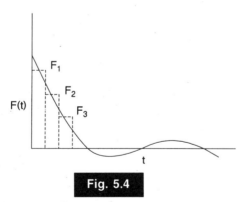

Fig. 5.4

Forcing functions

At $t = 0$, if $x = x_0$, $v = v_0$, and the force F_1 as shown in Fig. 5.4.
Treating the force as varying in steps, the response for $0 < t < \Delta t$ is given by

$$x = \exp(-\xi \omega_n t)\, (c_1 \cos \omega_n t + c_2 \sin \omega_n t) + \frac{F_1}{k}$$

$$x = \omega_n \exp(-\xi\omega_n t)(\cos \omega_n t (c_2 - c_1 \xi - \sin \omega_n t(c_1 + \xi c_2)) \quad (5.27)$$

where
$$c_1 = \left(x_0 - \frac{F_1}{k}\right)$$

$$c_2 = \frac{v_0 + \xi\omega_n(x_0 - F_1/k)}{\sqrt{(1 - \xi^2)}\,\omega_n} \quad (5.28)$$

At time $t = \Delta t$, let the displacement and velocity be x_1 and v_1. Then for $\Delta t < t < 2\Delta t$, we have

$$x = \exp(-\xi\omega_n t_1)(c_1 \cos \omega_n t_1 + c_2 \sin \omega_n t_1) + \frac{F_2}{k}$$

$$\dot{x} = \omega_n \exp(-\xi\omega_n t_1)(\cos \omega_n t_1(c_2 - c_1 \xi) \quad (5.29)$$
$$- \sin \omega_n t_1(c_1 + \xi c_2))$$

where
$$t_1 = (t - \Delta t)$$

$$c_1 = x_1 - \frac{F_2}{k}$$

$$c_2 = \frac{v_1 + \omega_n(x_1 - F_2/k)}{\sqrt{(1 - \xi^2)}\,\omega_n} \quad (5.30)$$

The argument can be extended to the other time periods.

Excitation by Ground

Let the system be excited by ground having a constant displacement X_g. Then the governing differential equation can be written as

$$m\ddot{x} + c\dot{x} + k(x - X_g) = 0$$

or
$$m\ddot{x} + c\dot{x} + kx = kX_g \quad (5.31)$$

Let the initial conditions be $x = 0$, $\dot{x} = 0$ when $t = 0$. Equation (5.31) is similar to Eq. (5.21). F_0 is replaced by kX_g. Hence the arguments advanced in Sec. 5.2.1 can also be applied to Sec. 5.2.1. Any system subjected to ground excitation which varies as a function of time can also be analysed. This is very useful in predicting the transient response of vehicles experiencing undulations from the ground.

5.2.2   Multidegree Freedom System Response

System with no Damping

Let the governing differential equation be given by

$$[M]\{\ddot{u}\} + [K]\{u\} = \{F(t)\} \quad (5.32)$$

$[M]$ and $[K]$ are matrices of size $n \times n$

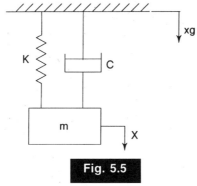

Fig. 5.5

Ground excitation

Let $\{u\}$ be expressed as

$$\{u\} = [X]\{p\} \qquad (5.33)$$

where $[X]$ is a matrix (size $n \times m$) of the first m eigenvectors $(m << n)$. $\{p\}$ is a generalised displacement vector of size $m \times 1$. The reason for this choice would become obvious when it is substituted in Eq. (5.32). The following explanation can be given for this choice. After all, when the complicated system is excited (excitation being transient), the system has to respond in one or more of its natural modes of vibration. It is well-known that the fundamental mode predominates, the other higher order ones gradually diminishing in the order of their contribution. By choosing m modes, we are only restricting the contribution to the first m modes only or the assumption that the rest do not contribute to the response. Obviously the elements of $\{p\}$, p_1, p_2, p_3 give an indication of the contribution from the first, second, third modes.

Substituting Eq. (5.33) in Eq. (5.32), we have

$$[M][X]\{\ddot{p}\} + [K][X]\{p\} = \{F(t)\} \qquad (5.34)$$

Premultiplying Eq. (5.34) by $[X]^T$, we have

$$[X]^T[M][X]\{\ddot{p}\} + [X]^T[K][X]\{p\} = [X]^T\{F(t)\} \qquad (5.35)$$

$[X]^T[M][X]$ is a diagonal matrix of size $(m \times m)$ with unity in the principal diagonal. $[X]^T[K][X]$ is also a diagonal matrix with $\omega_1^2, \omega_2^2 \ldots \omega_m^2$ (the square of the natural frequencies) in the diagonal. If $[X]^T[F(t)]$ (of size $m \times 1$) has its elements $\hat{f}_1(t), \hat{f}_2(t) \ldots \hat{f}_m(t)$ then we can write

$$\ddot{p}_1 + \omega_1^2 p_1 = \hat{f}_1(t)$$

$$\ddot{p}_2 + \omega_2^2 p_2 = \hat{f}_2(t)$$

$$\ddot{p}_m + \omega_m^2 p_m = \hat{f}_m(t) \qquad (5.36)$$

These m uncoupled equations in single degree can be solved by the method indicated in Sec. 5.2.1. After having obtained $\{p\}$, $\{u\}$ can be computed by using Eq. (5.33).

System with Damping

Let the governing differential equation be given by

$$[M]\{\ddot{u}\} + [C]\{\dot{u}\} + [K]\{u\} = \{F(t)\} \qquad (5.37)$$

If this equation has to be recast such that one gets m uncoupled equations in single degree, the equations which are to replace Eq (5.36) must have the form

$$\{\ddot{u}\} + 2\xi\omega_n\{\dot{u}\} + \omega_n^2\{u\} = \{\hat{f}(t)\} \qquad (5.38)$$

This is the form one obtains when Eq. (5.21) is divided throughout by m. But when Eq. (5.37) is recast substituting for $\{u\}$ from Eq. (5.33) and premultiplying by $[X]^T$, one gets

$$[X]^T[M][X]\{\ddot{p}\} + [X]^T[C][X]\{\dot{p}\} + [X]^T[K][X]\{p\} = [X]^T F(t) \quad (5.39)$$

Comparing Eq. (5.39) and Eq. (5.38) one can guess that $[X]^T[C][X]$ should also yield a diagonal matrix of size $m \times m$ whose diagonal elements being $2\xi_i\omega_i$ $(i = 1, 2, ..., m)$. Before we discuss as to how this is achieved, let us examine how the resulting uncoupled equations are solved.

The resulting uncoupled equations are given by

$$p_1 + 2\xi_1\omega_1 p_1 + \omega_1^2 p_1 = \hat{f}_1(t)$$

$$p_2 + 2\xi_2\omega_2 p_1 + \omega_2^2 p_2 = \hat{f}_2(t) \qquad (5.40)$$

$$\cdots \qquad \cdots \qquad \cdots$$

$$p_m + 2\xi_m\omega_m p_m + \omega_m^2 p_m = \hat{f}_m(t)$$

These m uncoupled equations can be solved as before by the method indicated in Sec. 5.2.1.

5.2.3 Rayleigh Damping

For the multidegree freedom system encountered the determination of the damping matrix $[C]$ has not been discussed so far. The material damping present for well-known materials has been reported by various investigators. This information for cast iron and mild steel is available in [5.7]. But the value of damping ratio reported is usually corresponding to the fundamental mode.

It is well-known that there are two ways of computing the damping present in a system [5.8] one by the free vibration record and another by the forced vibration method ($\sqrt{2}$ method). By the second method it is possible to obtain the damping ratio for the first few natural frequencies. But the elements of the damping matrix cannot be so explicitly expressed as the ones connected with mass or stiffness matrix. Since only $[M]$ and $[K]$ are available let us presuppose that $[C]$ can be expressed as linear combination of $[M]$ and $[K]$. We can then write

$$[C] = \alpha[M] + \beta[K] \tag{5.41}$$

α and β are constants to be determined. Let us postmultiply Eq. (5.41) by $[X]$ and premultiply the resulting equation by $[X]^T$. Then we have

$$[X]^T[C][X] = \alpha[X]^T[M][X] + \beta[X]^T[K][X] \tag{5.42}$$

If $[X]^T[C][X]$ is written as

$$[X]^T[C][X] = 2 \begin{bmatrix} \xi_1\omega_1 & & & 0 \\ & \xi_2\omega_2 & & \\ & & \ddots & \\ 0 & & & \xi_m\omega_m \end{bmatrix} \tag{5.43}$$

then we have

$$2\xi_1\omega_1 = \alpha + \omega_1^2\beta$$

$$2\xi_2\omega_2 = \alpha + \omega_2^2\beta$$

$$2\xi_m\omega_m = \alpha + \omega_m^2\beta \tag{5.44}$$

'Experiments were conducted [5.9] on a number of mild steel specimens (plates and beams of varying sizes and shapes with different boundary conditions) to study the relationship between the natural frequency ω and damping ratio ζ. The summary of observations for damping ratios ζ_1 and ζ_3 are shown in Figs 5.6 and 5.7. The best fit for these two damping ratios is given by the expressions $\zeta_1\omega_1 = 2.0$ and $\zeta_3\omega_3 = 3.2$, where ω_1 and ω_3 are the first and the third natural frequencies in rad/sec. Using these two expressions and the general expression for Rayleigh damping given by Eq. (5.44) fairly reasonable values of damping ratios for any other natural frequency can be obtained. This is explained through the Example 5.5'.

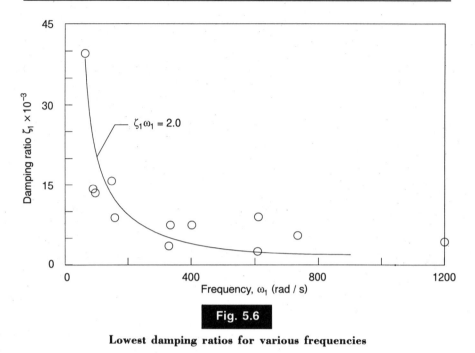

Fig. 5.6

Lowest damping ratios for various frequencies

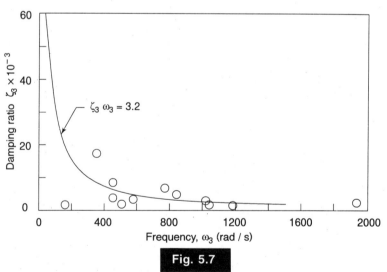

Fig. 5.7

Damping ratios for natural frequencies

Example 5.4
Compute the response of the system given below by modal superposition technique.

$$\ddot{x}_1 + 2x_1 + 2x_2 = 10$$

$$2x_2 + \ddot{x}_2 + 5x_2 = 0$$

x_1 and x_2 are the responses of the system. Initial conditions are $x_i = 0$, $\dot{x}_i = 0$ at $t = 0$.

Natural frequencies of this system with

$$[K] = \begin{bmatrix} 2 & 2 \\ 2 & 5 \end{bmatrix} \quad \text{and} \quad [M] = \begin{bmatrix} 1 & 0 \\ 0 & 1 \end{bmatrix}$$

are $\sqrt{1}$ and $\sqrt{6}$.

The eigenvectors are

$$\begin{Bmatrix} 2/\sqrt{5} \\ -1/\sqrt{5} \end{Bmatrix} \quad \text{and} \quad \begin{Bmatrix} 1/\sqrt{5} \\ 2/\sqrt{5} \end{Bmatrix}$$

$$f(t) = \begin{Bmatrix} 10 \\ 0 \end{Bmatrix}$$

Hence $\quad \hat{f}(t) = \begin{bmatrix} 2/\sqrt{5} & -1/\sqrt{5} \\ 1/\sqrt{5} & 2/\sqrt{5} \end{bmatrix} \begin{Bmatrix} 10 \\ 0 \end{Bmatrix} = \begin{Bmatrix} 20/\sqrt{5} \\ 10/\sqrt{5} \end{Bmatrix}$

The two decoupled equations are given by

$$\ddot{p}_1 + p_1 = 20/\sqrt{5}$$

$$\ddot{p}_2 + 6p_2 = 10/\sqrt{5}$$

Solution to the two equations are

$$p_1 = A \cos t + B \sin t + (20/\sqrt{5})$$

$$p_2 = C \cos \sqrt{6}t + D \sin \sqrt{6}t + (10/6\sqrt{5})$$

Since initial displacements and velocities are zero, we have

$$A = -\frac{20}{\sqrt{5}} \qquad B = 0 \qquad C = -\frac{10}{6\sqrt{5}} \qquad D = 0$$

$$\therefore \qquad \{p\} = \begin{Bmatrix} 20/\sqrt{5}(1 - \cos t) \\ -10/6\sqrt{5}(1 - \cos \sqrt{6}\, t) \end{Bmatrix}$$

Dynamic response $\{u\}$ is given by

$$\{u\} = [X]\{p\} = \begin{bmatrix} 2/\sqrt5 & 1/\sqrt5 \\ -1/\sqrt5 & 2/\sqrt5 \end{bmatrix} \begin{Bmatrix} 20/\sqrt6\,(1 - \cos t) \\ 10/\sqrt6\,(1 - \cos \sqrt6\,t) \end{Bmatrix}$$

$$= \begin{Bmatrix} 8(1 - \cos t) + \dfrac{2}{6}\,(1 - \cos \sqrt6\,t) \\ -4(1 - \cos t) + \dfrac{4}{6}\,(1 - \cos \sqrt6\,t) \end{Bmatrix}$$

It is very easy to verify that the complementary solution to the governing differential equation is also the same.

Example 5.5
Assume for a given system the damping ratios for the first and the third natural frequencies of 6 rad/sec and 18 rad/sec are 0.33 and 0.17. Estimate the constants α and β to be used for Rayleigh damping.

$$f_1 = 6.00 \text{ rad/sec}$$

$$f_3 = 18 \text{ rad/sec}$$

Using Eq. (5.44), we have

$$2\xi_1\omega_1 = \alpha + \beta\omega_1^2$$

$$2\xi_3\omega_3 = \alpha + \beta\omega_3^2$$

$$2(0.33)(6) = \alpha + \beta(6)^2$$

$$2(0.17)(18) = \alpha + \beta(18)^2$$

Hence $\alpha = 3.7/\text{sec}$ $\beta = 0.762 \times 10^{-2}$ sec

Hence the damping matrix $[C]$ is given by

$$[C] = \alpha[M] + \beta[K]$$

$$= 3.7[M] + 0.762 \times 10^{-2}[K]$$

The damping ratio for any other frequency is given by

$$\xi_i = \frac{(3.7) + (0.762 \times 10^{-2})\omega_i^2}{2\omega_i}$$

➤ 5.3 · CONDITION FOR STABILITY

In Sec. 5.1 the method of computing the dynamic response (displacement, velocity and acceleration) for a given initial value problem by three methods was discussed. The methods included were central difference, Wilson θ and Newmark. It was pointed out that the labour involved as a function of the chosen time interval Δt. Let us now discuss for what choice of Δt, do the results converge. These have been discussed extensively in [5.1].

Let us assume that the recurrence relationship for the response is given by

$$\{u_{t+\Delta t}\} = [A]\{u_t\} + [L]\{f_{t+\gamma}\} \tag{5.45}$$

where $\{u_t\}$, $\{u_{t+\Delta t}\}$ are the response (displacement, velocity and acceleration) at t and $t + \Delta t$ and $\{f_{t+\gamma}\}$ is the force vector $[A]$ and $[L]$ are appropriate square matrices whose values will be different for different approaches. When this operation is carried out n times, the response will be given by

$$\{u_{t+n\Delta t}\} = [A]^n\{u_t\} + [A]^{n-1}[L]\{f_{t+\gamma}\} + \dots [L]\{f_{t+(n-1)\Delta t+\gamma}\} \tag{5.46}$$

One can use Eq. (5.46) for studying the stability of the response. It can be easily seen that if the largest eigenvalue λ of $[A]$ is greater than 1, when $[A]$ is premultiplied n times (n being large) the results will be diverging. One has to make sure that the largest value λ is less than 1. Let us now derive Eq. (5.45) for the three different approaches listed.

5.3.1 Central Difference Method

Let us consider Eq. (5.40) which is the response of the uncoupled systems given by

$$\ddot{p}_1 + 2\xi_i \omega_i \dot{p}_i + \omega_i^2 p_i = \hat{f}_1(t) \tag{5.47}$$

We also have from Sec. 5.1.1,

$$\ddot{p} = \frac{1}{(\Delta t)^2}(p_{\Delta t} - 2p_0 - p_{-\Delta t}) \tag{5.48}$$

$$\dot{p} = \frac{1}{(2\Delta t)}(p_{\Delta t} - p_{-\Delta t}) \tag{5.49}$$

Substituting Eqs (5.48) and (5.49) in Eq. (5.47), we have

$$p_{t+\Delta t} = \frac{2 - \omega^2(\Delta t)^2}{1 + \xi\omega\Delta t} p_t - \frac{1 - \xi\Delta t}{1 + \xi\Delta t} p_{t-\Delta t} + \frac{(\Delta t)^2}{1 + \xi\omega\Delta t}\hat{f}_t \tag{5.50}$$

Equation (5.50) can now be written as

$$\begin{Bmatrix} p_{t-\Delta t} \\ p_t \end{Bmatrix} = [A] \begin{Bmatrix} p_t \\ p_{t-\Delta t} \end{Bmatrix} + [L]\{\hat{f}_t\} \tag{5.51}$$

where

$$[A] = \begin{bmatrix} \dfrac{(2 - \omega^2 \Delta t^2)}{(1 + \xi\omega\Delta t)} & -\dfrac{(1 - \xi\Delta t)}{(1 + \xi\Delta t)} \\ 1 & 0 \end{bmatrix} \tag{5.52}$$

and

$$[L] = \begin{bmatrix} \dfrac{\Delta t^2}{1 + \xi\omega\Delta t} \\ 0 \end{bmatrix} \tag{5.53}$$

5.3.2   Wilson θ Method

In the Wilson θ method acceleration is assumed to vary linearly over the time interval from t to $t + \theta\Delta t$, where $\theta \geq 1$. Hence the response can be written as

$$\{\ddot{u}_{t+\tau}\} = \{\ddot{u}_t\} + (\{u_{t+\Delta t}\} - \{\ddot{u}_t\})\frac{\tau}{\Delta t} \tag{5.54}$$

$$\{\dot{u}_{t+\tau}\} = \{\dot{u}_t\} + \{\ddot{u}_t\}\tau + (\{\ddot{u}_{t+\Delta t}\} - \{\ddot{u}_t\})\frac{\tau^2}{2\Delta t} \tag{5.55}$$

$$\{u_{t+\tau}\} = \{u_t\} + \{\dot{u}_t\}\,\tau + \frac{1}{2}\{\ddot{u}_t\}\tau^2 + (\{\ddot{u}_{t+\Delta t}\} - \{\ddot{u}_t\})\frac{\tau^3}{6\Delta t} \tag{5.56}$$

$$(0 < \tau < \theta\Delta t)$$

The response of the uncoupled systems at time $(t + \theta\Delta t)$ is given by

$$\{\ddot{u}_{t+\theta\Delta t}\} + 2\xi\omega\,\dot{u}_{t+\theta\Delta t} + \omega^2 u_{t+\theta\Delta t} = f_{t+\theta\Delta t} \tag{5.57}$$

Using Eqs (5.54), (5.55) and (5.56) at time $\tau = \theta\Delta t$ and substituting in Eq. (5.57) one can obtain $u_{i+\Delta t}$. Then the response can be recast in the following form

$$\begin{Bmatrix} \ddot{u}_{t+\Delta t} \\ \dot{u}_{t+\Delta t} \\ u_{t+\Delta t} \end{Bmatrix} = [A] \begin{Bmatrix} \ddot{u} \\ \dot{u}_t \\ u_t \end{Bmatrix} + [L]\{f_{t+\theta\Delta t}\} \tag{5.58}$$

where

$$[A] = \begin{bmatrix} 1 - \dfrac{\beta\theta^2}{3} - \dfrac{1}{\theta} - \eta\theta & \dfrac{1}{\Delta t}(\beta\theta - 2\eta) & \dfrac{1}{\Delta t^2}(-\beta) \\[3mm] \Delta t\left(1 - \dfrac{1}{2\theta} - \dfrac{\beta\theta^2}{6} - \eta\dfrac{\theta}{2}\right) & 1 - \beta\theta - \eta & \dfrac{1}{\Delta t}\left(-\dfrac{\beta}{2}\right) \\[3mm] \Delta t^2\left(\dfrac{1}{2} - \dfrac{1}{6\theta} - \dfrac{\beta\theta^2}{18} - \dfrac{\eta\theta}{6}\right) & \Delta t\left(1 - \dfrac{\beta\theta}{6} - \dfrac{\eta}{3}\right) & 1 - \dfrac{\beta}{6} \end{bmatrix}$$

(5.59)

$$[L] = \begin{bmatrix} \beta/\omega^2 \Delta t^2 \\[2mm] \beta/2\omega^2 \Delta t \\[2mm] \beta/6\omega^2 \end{bmatrix}$$

(5.60)

$$\beta = \left(\dfrac{\theta}{\omega^2 \Delta t^2} + \dfrac{\xi\theta^2}{\omega\Delta t} + \dfrac{\theta^3}{6}\right)^{-1}$$

$$\eta = \xi\beta/\omega\Delta t$$

(5.61)

5.3.3 Newmark Method

In Newmark's scheme velocity and displacement are expressed as

$$\{\dot{u}_{t+\Delta t}\} = \{\dot{u}_t\} + [(1 - \delta)\{\ddot{u}_t\} + \delta\{\ddot{u}_{t+\Delta t}\}]\,\Delta t \tag{5.62}$$

$$\{u_{t+\Delta t}\} = \{u_t\} + \{\dot{u}_t\}\Delta t + \left[\left(\dfrac{1}{2} - \alpha\right)\{\ddot{u}_t\} + \alpha\{\ddot{u}_{t+\Delta t}\}\right]\Delta t^2 \tag{5.63}$$

Following arguments similar to the earlier scheme, the response at $(t + \Delta t)$ can be written in the following form

$$\begin{Bmatrix} \ddot{u}_{t+\Delta t} \\ \dot{u}_{t+\Delta t} \\ u_{t+\Delta t} \end{Bmatrix} = [A] \begin{Bmatrix} \ddot{u}_t \\ \dot{u}_t \\ u_t \end{Bmatrix} + [L]\{f_{t+\Delta t}\} \tag{5.64}$$

where

$$[A] = \begin{bmatrix} -\left(\dfrac{1}{2} - \alpha\right)\beta - 2(1-\delta)\,\eta & \dfrac{1}{\Delta t}(-\beta - 2\eta) & \dfrac{1}{\Delta t^2}(-\beta) \\[3mm] \Delta t\left[1 - \delta - \left(\dfrac{1}{2} - \alpha\right)\delta\beta - 2(1-\delta)\delta\eta\right] & (1 - \beta\delta - 2\delta\eta) & \dfrac{1}{\Delta t}(-\beta\delta) \\[3mm] \Delta t^2\left[\dfrac{1}{2} - \alpha\left(\dfrac{1}{2} - \alpha\right)\alpha\beta - 2(1-\delta)\alpha\eta\right] & \Delta t(1 - \alpha\beta - 2\alpha\eta) & (1 - \alpha\beta) \end{bmatrix}$$

(5.65)

$$[L] = \begin{bmatrix} \beta/\omega^2 \Delta t^2 \\ \beta\delta/\omega^2 \Delta t \\ \alpha\beta/\omega^2 \end{bmatrix} \qquad (5.66)$$

and

$$\beta = \left(\frac{1}{\omega^2 \Delta t^2} + \frac{2\xi\delta}{\omega\Delta t} + \alpha \right)^{-1} \qquad \text{and} \qquad \eta = \xi\beta/\omega\Delta t \qquad (5.67)$$

5.3.4 Discussion on Stability

The largest eigenvalue of the square matrix [A] given by Eqs (5.52), (5.59) and (5.65) gives the indication for stability. It is observed from Eq. (5.52) that whenever $\omega\Delta t/2\pi \le 1/\pi$, the largest eigenvalue is less than 1. It means that the central difference scheme is stable only when $\Delta t/T \le 1/\pi$, where T is time period of nth the mode of vibration.

In the case of the Wilson θ method unconditional stability is obtained when $\theta = 1.37$. In the Newmark method the integration scheme is unconditionally stable if $\delta \ge 0.5$ and $\alpha \ge 0.25(\delta + 0.5)^2$.

➤ **5.4 RANDOM VIBRATION** _____

So far, we have considered disturbances which act for a short period of time and which die down subsequently owing to inherent damping present in the system. There is yet another class of disturbance which acts for a considerable period of time, but which is random in nature. A typical example is the case of disturbance experienced by a moving vehicle owing to the undulations of the road profile. These disturbances experienced by the vehicle moving at a uniform speed do not have any specific frequency of preference, but can be treated as a random process with ergodic and stationary properties [5.10]. The disturbances due to an earthquake are not, strictly speaking, random, since they act for a specific period of time, at the most for 30 seconds. To consider this as a random disturbance is in reality overemphasising its role. In case these are treated as random processes with ergodic and stationary properties, the power spectral density of input disturbance can always be obtained. Typical power spectral density for input disturbance are shown in Fig. 5.8. It may be observed that the power spectral density of the ground acceleration dies down rapidly beyond 8 cps.

For a typical sinusoidal input at a given frequency, it is easy to

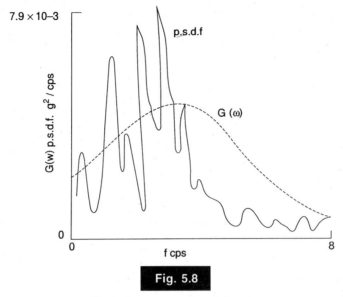

Fig. 5.8

Typical power spectral density

compute the response (displacement, velocity acceleration or stress) of a given structure at any given location by the methods indicated in Chapter 3. Let this factor $H(if)$ be computed for various frequencies of interest. This may be 0 to 100 cps for vehicles and in the range of 0 to 32 cps for earthquake disturbances. If input power spectral density is $S_x(f)$ and if the transfer function is $H(if)$, then using the random vibration theory, one can get the power spectral density of the output as

$$S_y(f) = |\,H(if)\,|^2\,S_x(f) \qquad\qquad (5.68)$$

The mean square value of the output is given by

$$\langle y^2 \rangle = \int_{f_1}^{f_2} |\,H(if)\,|^2 S_x(f)\,df \qquad\qquad (5.69)$$

Computation of this integral over a wide frequency range can be made very efficient by transputers. A several-fold reduction in computer time is possible by using multiple transputers [5.11]. The limits of integration have to be taken from the typical range for a practical system.

➤ 5.5   FIGURES OF SIGNIFICANCE

The frequencies of excitation due to unbalance in rotors range from 1 to 12,000 rpm (0.016 to 200 cps). Heavy machines like kilns run at 1 rpm (probably the lowest), induction motors normally run at 1440 rpm (50 cps supply frequency) or 1710 rpm (60 cps supply frequency). Step down units run at speeds from 1440 rpm (or 1710 rpm) downwards. Steam turbines run at speeds from 3000 to 8000 rpm. Reciprocating engines operate in the speed range of 1500 to 6000 rpm. If one is interested in dynamic response, it may be kept in mind that a vibration velocity of 15 mm/sec. (peak) is just the limit of tolerance. Depending upon the frequency of oscillation, one can compute the vibration amplitudes or acceleration values for velocities exceeding 15 mm/sec leading to situations of violent vibration.

In transient vibration, the following points have also to be kept in mind. If a system having a fundamental natural frequency of 100 cps is disturbed, it will complete 100 oscillations in one second. During this period, the disturbance introduced at the start will get reduced to nothing at the end of the 100th oscillation even when the damping ratio is as low as 0.01.

➤ 5.6   INDUSTRIAL APPLICATION

5.6.1   Earthquake Analysis of Circuit Breakers

The high voltage circuit breaker [shown in Plate 1: Fig. 5.9 and in Fig. 5.10)] was modelled as a space frame element allowing for six degrees of freedom at each node [5.12]. The discretised unit consisting of a porcelain superstructure with properties as shown in Table 5.5, with two branches at the top, was reduced to a multi-degree of freedom system having 216 degrees of freedom (Fig. 5.11). As boundary conditions, the deflections and slopes at the first node were assumed to be zero. The first six natural frequencies of this structure were computed using the simultaneous iteration scheme. The fundamental frequency at 2.48 Hz in the XY direction and the second natural frequency at 2.58 Hz in the XZ direction are within the expected range for such type of structures. The mode shapes have been shown in Fig. 5.12. The 50% mode coupling between the XY and XZ modes is probably due to the fact that the corresponding natural frequencies are so close to each other. The transfer function $H(if)$ for unit base acceleration input is obtained by repeatedly solving the set of simultaneous equations given by

$$[K - \omega^2 M + i\omega C]\{x_r\} = -[M]\{a\} \qquad (5.70)$$

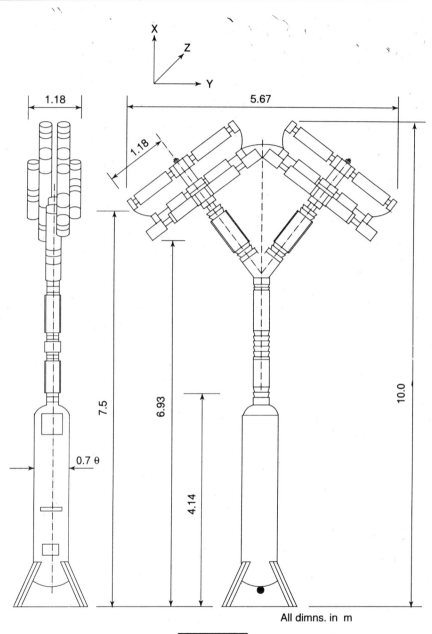

All dimns. in m

Fig. 5.10

Circuit breaker for analysis

Table 5.5

Properties of Electrical Porcelain

Formula	$K_2O\ Al_2O_3\ SiO_2$
Specific gravity	2.41
Compressive strength	700 N/mm^2
Tensile strength	40–60 N/mm^2
Modulus of rupture	100 N/mm^2
Modulus of elasticity	7×10^4 N/mm^2

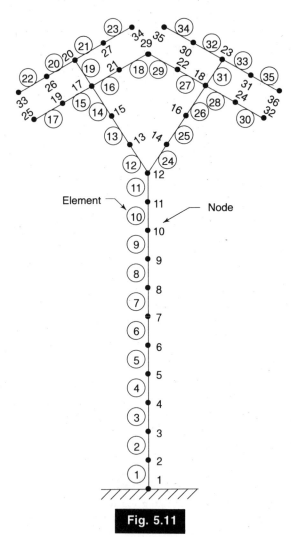

Fig. 5.11

Discretization and node numbering

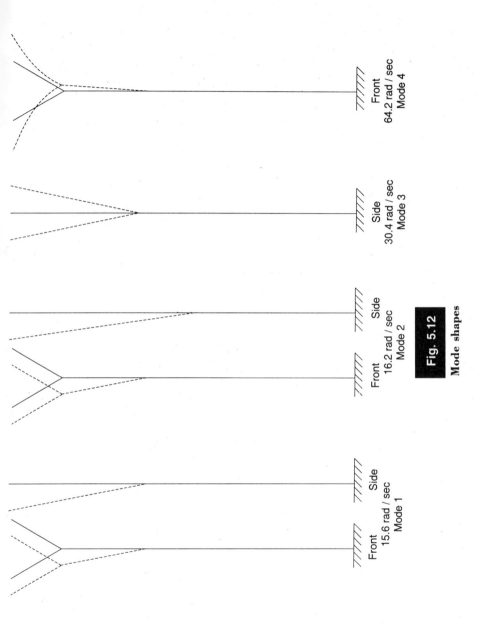

Fig. 5.12

Mode shapes

$[K]$: stiffness matrix of size (n, n)

$[M]$: mass matrix of size (n, n)

$[C]$: damping matrix of size (n, n)

$\{a\}$: column vector of ground acceleration of size (n, l). In each block of size degrees of freedom, the only one non-zero term will be the acceleration (unity) either in one of the two horizontal directions or in the vertical direction

ω : circular frequency in rad/sec.

i : complex operator

$\{x_r\}$: relative displacement vector of the structure with respect to the ground for various values of 'ω'—The damping matrix C is obtained using the concept of Rayleigh damping. The damping corresponding to the first three natural frequencies were estimated by conducting a forced vibration experiment on the structure and using the well-known $\sqrt{2}$ method.

The system of equations given by Eq. (5.70) are solved for unit acceleration in each of the three orthogonal directions independently. Stress is calculated from the response obtained. This stress in complex form is the transfer function sought. As earthquake excitation has components in the three orthogonal directions which are statistically independent, the spectral density of the stress owing to combined loading is obtained by adding the spectral density of stress due to individual components.

The El Centro accelerogram (acceleration in g as a function of time) given in Fig. 5.13 was discretised. Its power spectral density values

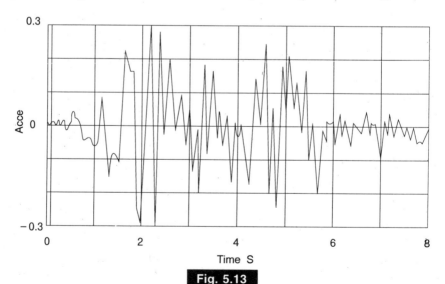

Time S

Fig. 5.13

Acceleration response

are as shown in Fig. 5.7. In order to compare the results obtained as above with results obtained by using previously recommended procedures, the widely quoted power spectral density formula can be used for typical earthquakes given by Tajima [5.14] and the stresses calculated. Figure 5.8 shows the smooth power spectral density curve given by Tajima's formula and the curve obtained by using the El Centro accelerogram.

Following the procedure outlined above, the R.M.S. stress at various nodes can be evaluated. Figure 5.14 shows a plot of stress at some typical nodes. The critical zone is the porcelain section where stresses in the order of 69.5 N/m^2 have been observed. At node 8, the material changes from steel to porcelain and this contributes to the observed discontinuity in stress.

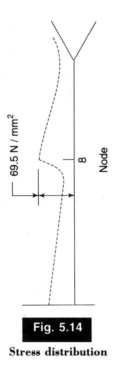

Fig. 5.14

Stress distribution

REFERENCES

5.1 Dokainish, M.A. and K. Subbaraj, A survey of direct integration method in computational structural dynamic—Part I—explicit methods, *Computers and Structures*, vol. 32, no. 6, pp 1371–1386, 1989.

5.2 Dokainish, M.A. and K. Subbaraj, A survey of direct integration method in computational structural dynamic—Part II—implicit methods, *Computers and Structures*, vol. 32, no. 6, pp 1387–1401, 1989.

5.3 Bathe, K.J. and E.L. Wilson, *Numerical Methods in Finite Element Analysis*, Prentice- Hall of India, New Delhi, 1978.

5.4 Hughes, T.J.R. and T. Belytschko, A precis of developments in computational methods for transient analysis, *J. App. Mechs.*, vol. 50, pp 1033–1040, 1983.

5.5 Wilson, E.L., I. Faihoomand and K.J. Bathe, Non-linear dynamic analysis of complex structures, *Int. J. Earthquake Engg. and Struct. Dyn.*, vol. 1, 1973, pp 241–252.

5.6 Newmark, N.M., A method of computation for structural dynamics, ASCE, *J. Engg. Mech. Div.*, vol. 85, 1959, pp 67–94.

5.7 Wilson, W. Ker, *Practical Solutions of Torsional Vibration Problems*, vol. 2, Chapman and Hall, London, 1963.

5.8 Thomson, W.T., *Theory of Vibration with Applications*, 2nd edn., Prentice-Hall of India, New Delhi, 1982.

5.9 Mohammad DRA, N.U. Khan and V. Ramamurti, on the role of Rayleigh damping, *J. Sound and Vibration*, 185(2), 207–218 (1995).

5.10 Robson, J.D., *Introduction to Random Vibrations*, Edinburgh, University Publications, Edinburgh, 1964.

5.11 Allik, H., E. O'Neil Moore and E. Tennenbaum, Finite element analysis on the B.B.N. butterfly multiprocessor, *Computers and Structures*, vol. 27, no. 1, pp 13–21, 1987.

5.12 Ramamurti, V., L. Balasubramanian and A. Thomas, Earthquake analysis of high voltage circuit breaker, *Proc. 7th Int. Modal Anal. Conf.*, Las Vegas, Jan. 1989, pp 208–212.

5.13 Tajima, H., A statistical method of determining the maximum response of a building structure during an earthquake, *Proc. 2nd World Cong. on Earthquake Engg.*, Tokyo, Kyoto, Japan, vol. 2, 1960.

EXERCISES

5.1 For the system given by the equation

$$[M]\{\ddot{u}\} + [K]\{u\} = f(t)$$

where [K] and [M] are as given in Exercise 3.1, predict the transient transpose for two full wave lengths corresponding to the fundamental period of oscillation when

$$u_0 = 0, \quad \dot{u}_0 = 0 \quad \text{and when}$$

$$f(t) = \begin{Bmatrix} 1 \\ 0 \\ 0 \\ 0 \\ 0 \end{Bmatrix}$$

by
  (i) Central difference scheme
  (ii) Wilson θ scheme
  (iii) Newmark scheme
Compare the results with the exact solution.
5.2 Compare the results of Exercise 5.1 with the modal superposition results by taking two modes from Exercise 3.4.
Note on Computer programs related to Chapter 5.
Programs 5.1 to 5.3 deal with the following:
5.1 Wilson θ method to obtain the transient response (Example 5.2).
5.2 Newmark method to obtain the transient response (Example 5.3).
For the above two programs mass and stiffness matrices are supplied in banded form.
5.3 Modal superposition method to obtain the transient response (Example 5.1).
Mass matrix is supplied in banded form; orthonormalized eigenvectors, natural frequencies and modal damping coefficients are supplied for truncated mode shapes.

6 Case Studies

As typical applications of the algorithms presented in the previous chapters, five case studies employing different methods are reported here. These cover illustrations in the areas of machine tool design, selection of torsional vibrations dampers, turbine bladed discs and vehicle dynamics. Experimental verifications, wherever available, have also been quoted.

➤ **6.1 DESIGN ASPECTS OF A 3-ROLL PLATE BENDING MACHINE**

A 3-roll bending machine (Fig. 6.1) belongs to the category of heavy

Fig. 6.1

3-Roll plate bending machine

duty machine tools which can roll plates into cylinders and cones. A 3-roll machine is the simplest member of the family of machines having 3 or more rolls. This machine can be used to roll I, T and angle sections. Its important applications are in the manufacture of cylindrical tanks, boiler equipment, fuel tanks for launch vehicles in space applications programs and special containers.

The capacity of the machine is limited by factors like the size of the roll and the horsepower of its motors. These machines are specified by the maximum thickness and width of the plate that can be rolled.

6.1.1 Machine Construction and Parts

The main working components of the machine are its forged steel wear resistant rolls. These rolls are supported on either side in the stands. The two bottom rolls receive power while the top roll idles. To bend plates into cylinders and cones, the bottom roll on either side must have independent vertical motion. This is obtained through a screw-nut drive. This drive receives power through a jaw clutch and a worm gear reduction. The worm gears are housed in an oil bath in the stands which are extensively ribbed and are made of thick plates. Figure 6.2 shows a sectional view of the left stand along with the other components (nondrive end), and Fig. 6.3 shows the sectional views of the fabricated right stand (drive end) alone. The bottom rolls must be inclined relative to the top roll when the cones are rolled. This is made possible by disengaging the jaw clutch. To compensate for the inclination of the roll and for transmitting power to it, the rolls are connected to the main drive gear box through universal joints.

The top roll is an idling roll. It does not receive power. As shown in Fig. 6.4, it carries more load than the two bottom rolls. It is, therefore, made of a larger diameter. It is supported on the right side in the right stand and on the left side in the hinge on the left stand. The hinge can be lowered to release the top roll. This is necessary to remove finished job from the machine. The bottom rolls held at each end in fixed bearings extend beyond the bearing at the drive end and are fitted with double helical gears which mesh with a common gear wheel which in turn is driven through a reduction gear box, generally by an electric motor. Reversible electric motor is used so that the rolls can be rotated in both directions. This is the main drive which causes the rolls to rotate.

The other drive, known as adjustment drive, causes the up and down motion of the rolls. The two drives are powered by different motors. The two bottom rolls can be raised or lowered independently.

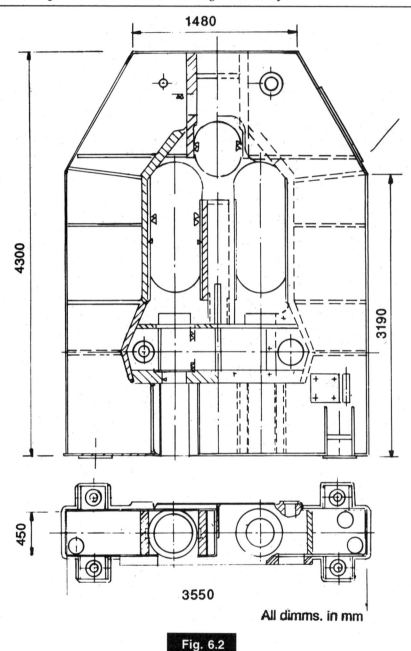

All dimms. in mm

Fig. 6.2

Sectional view of non-drive end

The two spindles which transmit power to the rolls for vertical adjustments are connected to the gear box through universal joints

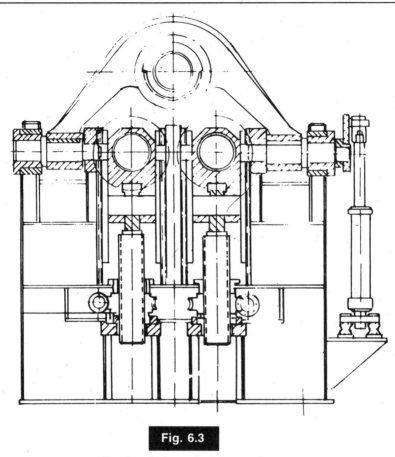

Fig. 6.3

Sectional view of right stand

so as to enable power transmission even while moving the rolls up or down. The vertical adjustments of the bottom roll make it possible to firmly clamp the plates between the rolls so that the plate can be pre-bent without taking it out of the machine and turned.

6.1.2 Design Considerations

The three major units connected with the three roll bending machines are the forged steel rolls, the drive for the machine and the stands housing the adjustable drive and supporting the rolls. The design of the rolls is a fairly straight-forward exercise. The rolls experience radial forces while rolling either for part of the length or for the entire length and tangential force due to friction. Since the rolls are simply supported at the two ends, they can be designed as beams for rigidity and strength.

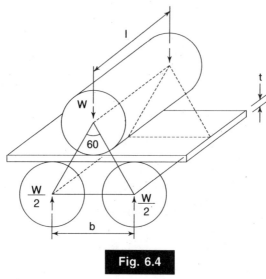

Fig. 6.4

Plate bending

The drive to the machine has two parts—main drive for rotating the two bottom rolls and two adjustment drives for lifting and lowering the rolls. Since the feed velocity of the plates and speed of lift of the rolls are specified, computing the horsepower of the prime mover and designing the machine elements like the shafts, worms and worm wheels is a straight-forward exercise. But designing the stands for rigidity and strength is really complicated. The stands are ribbed and fabricated, supported at the bottom and experience heavy loads having components in the X, Y and Z directions. This analysis is undertaken using finite element method.

6.1.3 Design Calculations

One of the important design parameters for the rolls is the folding pressure that comes into play during bending. Consider three rolls disposed with their axes at the corners of an isosceles triangle as shown in Fig. 6.4.

The bending moment to which the plate is subjected to is given by

$$M = \frac{W \cdot b}{4} \tag{6.1}$$

where W is the load exerted by the central top roll and b is the distance between the bottom rolls.

If the plate were to plastically deform at the point of contact, this bending moment should be equal to the plastic moment of resistance of the plate [6.2] (Fig. 6.5). Therefore

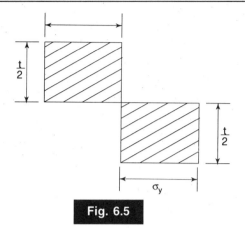

Fig. 6.5

Plastic moment

$$M = \frac{W \cdot b}{4} = \frac{\sigma l t^2}{4} \qquad (6.2)$$

where σ is the yield stress of the material

l is the width of the plate

and t is the thickness of the plate

Therefore,

$$W = \frac{\sigma l t^2}{b} \qquad (6.3)$$

If the radial load on the rolls is W, approximately $W \sin 30°$ will be its horizontal component (in X direction of the stand). When the machine is doing conical rolling, the component in the Z direction will be $W \sin \phi$ and ϕ is the maximum inclination of the roll from the horizontal. In general, the mill stand experiences loads in all the directions and the points of application of these loads vary depending on the size of the plates and position of the bottom rolls. Typical loading pattern for one of the machines analysed is shown in Fig. 6.6. Since the stand is subjected to such a complicated system of loading, it is being analysed by the finite element method. The stand is discretized by treating it as being made of triangular elements. Six degrees of freedom are allowed at each node [6.3]. The overall problem had 424 nodes. The skyline approach was used for solving the problem. The maximum band-width was 228 and the profile was round 500 K.

6.1.4   Numerical Results

The specification of the machine analysed is given in Table 6.1.

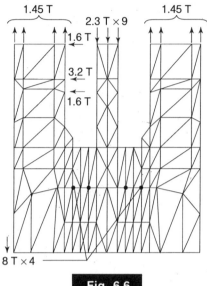

Fig. 6.6

Typical loading

Table 6.1

Specifications of the Two Size Analysed

Maximum pre-bending thickness	mm	10
Maximum width of plate (l)	mm	2500
Diameter of top roll	mm	295
Diameter of bottom rolls	mm	285
Distance between bottom rolls (b)	mm	330
Minimum bending diameter referred to top roll	mm	330
Bending speed (V_1)	mm/min	4000
Folding pressure (W)	tonnes	35
Angular adjustment of bottom roll (O)	degs	3.8
Vertical adjustment of bottom rolls	mm	175
Speed of adjustment of bottom rolls (V_2)	mm/min	125

The maximum stress for the left stand was 85.15 N/mm^2 on the inner plate. The nodal displacements are found to be more on the top half of the machine. This is understandable as the machine is fixed at the bottom. A maximum nodal displacement of 0.164 mm at the top of the stand was observed in the Z direction. The X and Y displacements are relatively small. The deformations and stresses experienced are shown in Figs 6.7 and 6.8 respectively. The displacements computed facilitate the tolerance to be specified in the machined guideways. The effect of location, the number and size of ribs on the rigidity and strength of fabricated stand can be elegantly studied by this approach.

➤ **6.2 TORSIONAL OSCILLATIONS OF A MULTICYLINDER ENGINE**

The response of multicylinder engines subjected to torsional oscillations and the determination of consequent stresses at the critical speeds can be obtained [6.4] by using the energy approach as described in References [6.5] and [6.6] but these methods are not applicable to speeds other than the critical ones, and are not convenient if hysteresis damping present in the crank shaft is to be taken into account.

In particular, it is not possible by the energy approach to study

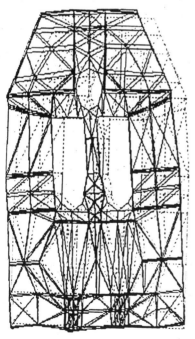
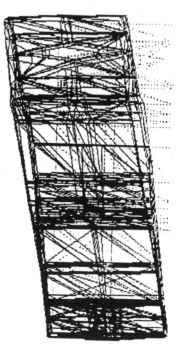

Fig. 6.7

Deformed shape

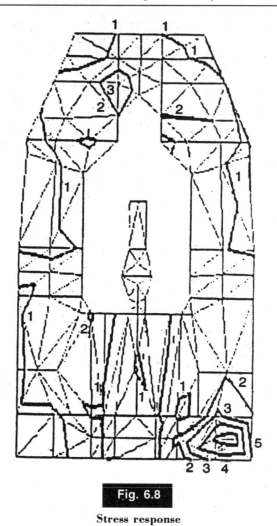

Fig. 6.8

Stress response

directly the effect of viscous or hysteristic damper installed on the crank shaft. In what follows here, a method of determining the engine dynamic response is described for a six-cylinder engine, in which the energies dissipated by the load, at the cylinder piston interfaces, and by a viscous or hysteretic crank shaft damper are all taken into account.

The schematic arrangement of the problem is shown in Fig. 6.9. The p-v diagram of the individual cylinders is shown in Fig. 6.10. The gas torque obtained from this diagram and the inertia torque computed for an engine running at a speed of 2400 rpm are plotted against θ in Fig. 6.11. Fourier analysis of these periodic functions is

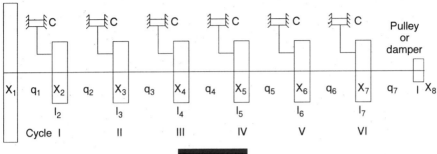

Fig. 6.9

Mathematical model of a six-cylinder engine

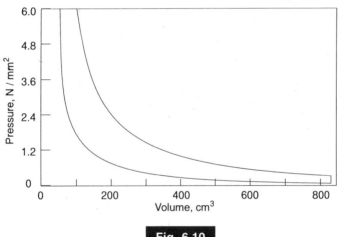

Fig. 6.10

p′–v diagram for a six-cylinder engine

Table 6.2

Mean Torque 7065 N mm, Resultant Fourier Coefficients (N mm) at 2400 rpm

Order	Fourier coefficient	Order	Fourier coefficient	Order	Fourier coefficient
0.5	156450	4.0	66840	7.5	12880
1.0	198280	4.5	56430	8.0	10550
1.5	176290	5.0	54100	8.5	8710
2.0	277520	5.5	35160	9.0	7290
2.5	132360	6.0	27520	9.5	6340
3.0	45650	6.5	21020	10.0	5780
3.5	89980	7.0	16180		

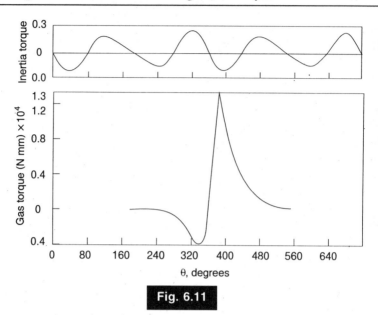

Fig. 6.11

T–θ diagram for a six-cylinder engine

carried out, and the resultant Fourier coefficients up to order 10 are presented in Table 6.2.

The mathematical formulation of the dynamic response is as follows:

$$
\begin{bmatrix}
-I_1\omega^2 + q_1 & -q_1 & & & \\
-q_1 & \begin{array}{c} -I_2\omega^2 + q_1 \\ + q_2 + ic\omega \end{array} & & & \\
 & & \cdots & & \\
 & & & \cdots & \\
 & & & & \cdots \\
 & & & & \cdots \\
 & & & \begin{array}{c} -I_7\omega^2 + q_6 \\ + ic + q_7 \end{array} & -q_7 \\
 & & & -q_7 & -I_8\omega^2 + q_7
\end{bmatrix}
$$

$$
\times \begin{Bmatrix} X_1 \\ X_2 \\ \vdots \\ X_7 \\ X_8 \end{Bmatrix} = \begin{Bmatrix} 0 \\ T_2 \lfloor \phi_2 \\ \vdots \\ T_7 \lfloor \phi_7 \\ 0 \end{Bmatrix} \tag{6.4}
$$

Here the term ic, where $i = \sqrt{-1}$, takes care of cylinder damping.

If the hysteretic damping of the crank shaft is also considered, then the stiffness matrix $\{q\}$ must be treated as a complex quantity being replaced by $\{q\}(1 + i\mu)$ where $1/\mu$ is the dynamic magnifier of the material of the crank shaft. This is around 280 for structural steels.

The phase angles ϕ_2, ϕ_3 ... ϕ_7 are determined according to the firing order of the multicylinder engine and the order of the Fourier coefficient under consideration. For instance, if the firing order is 1–5–3–6–2–4 then $\phi_2 = 0$; $\phi_6 = 4\pi m/n$, $\phi_4 = 2\phi_6$, $\phi_7 = 3\phi_6$, $\phi_3 = 4\phi_6$, and $\phi_5 = 5\phi_6$, where m is the order of the Fourier coefficient under consideration and n is the number of cylinders (six in this case); T_2 ... T_7 are the magnitudes of the torques (which are equal) for each Fourier coefficient.

The complex simultaneous equations given by Eqs (6.4) are solved by the Gaussian elimination method. Equations (6.4) are tridiagonal and hence the computations are not involved. Earlier, to solve the eigenvalue problem in order to determine the critical speeds, damping is ignored. The resultant real symmetric eigenvalue problem is easily solved by the Sturm sequence. Even in this case, the matrix is tridiagonal.

The dynamic stresses in the shaft sections are calculated from the computed dynamic deflections. The cumulative stresses in the shaft sections are the summation of stresses obtained for various harmonics.

For purposes of obtaining the numerical results, engine specifications were taken as follows: operating range of speed 600–2600 rpm, brake horsepower (max) 83 at 2400 rpm, stroke 127 mm, piston weight 11.42 N, connecting rod weight at piston end 4.66 N, connecting rod weight at crank pin 12.15 N, L/R ratio 4, journal length 30.5 mm, crank pin length 39.00 mm, web thickness 16.00 mm, web width (mean value) 90 mm, web height 120 mm, journal diameter 69 mm, crank pin diameter 56 mm, density of crank shaft 7.85×10^{-5} N/mm^3, bore 88.9 mm, flywheel moment of inertia 577 N/mm s^2, firing order 1–5–3–6–2–4, compression ratio 16.5.

Computed inertia and stiffness values (Fig. 6.8) are as follows: $I_1 = 577$ N mm s^2/rad; $I_2 = I_3 = I_4 = I_5 = I_6 = I_7 = 20$ N mm s^2/rad, $I_8 = 2.278$ N mm s^2/rad (pulley); $q_7 = 9.33 \times 10^8$ N mm/rad; $q_2 = q_3 = q_5 = q_6 = 6.644 \times 10^8$ N mm/rad; $q_4 = 6.278 \times 10^8$ N mm/rad; $q_7 = 1.372 \times 10^8$ N mm/rad. The computed first and second undamped natural frequencies are 1537.07 rad/s and 4196.15 rad/s.

When the pulley is replaced by a rubber damper, it has the following characteristics: inertia of the inner ring 230 N mm s^2 inertia of the outer ring 14 N mm s^2; stiffness of the rubber element in between 0.117×10^8 $(1 + i0.07)$ N mm/rad.

The computed dynamic stresses on the crank shaft, (with and without the damper) for the few orders of interest are plotted in Fig. 6.12 and the dynamic response at a speed of 2400 rpm is shown in Fig. 6.13. The cumulative stresses at a speed of 2400 rpm are shown in Table 6.3. The entire computation is performed in less than 15 seconds in IBM 370/155.

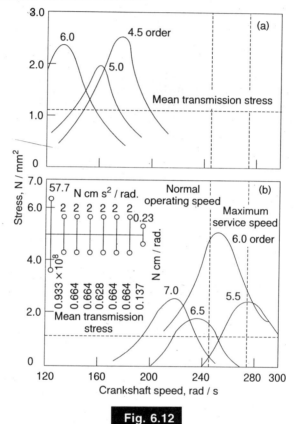

Fig. 6.12

Stress vs. crank shaft speed

➤ 6.3 DESIGN CONSIDERATIONS OF A GUILLOTINE SHEAR

Over-crank (Plate 2: Fig. 6.14) and under-crank (Plate 2: Fig. 6.15) Guillotine shears are used to cut plates in sizes varying from 3 mm–15 mm for widths of upto 3000 mm. They are essentially driven by an induction motor varying from 5–50 horse power, running at a speed of 1440 rpm. Number of strokes of the machine during continuous operations is likely to be in the range of 40–55 rpm. This necessitates

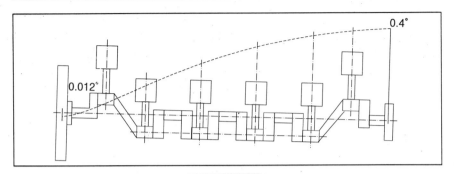

Fig. 6.13

Dynamic response of six-cylinder engine

Table 6.3

Stress Values (N/mm^2) at 2400 rpm without Damper

Section	Stress
1	11.4
2	20.0
3	18.7
4	14.5
5	8.8
6	6.0
7	0.4

a reduction of the order of 30–35. This is usually achieved through one vee belt reduction and two spur gear reductions. The machines are usually supplied with a massive flywheel to conserve energy during the non-cutting period. The blades are usually inclined at an angle of 1.5–3 degrees to enable gradual cutting of the plate from one end to another. This also, incidentally, enables reduction in the horse power requirement of the prime mover. The static part of the machine consists of the two sidestands, connecting bottom and top plates along with the hold down to withstand the forces during cutting. The deformations suffered and stresses experienced by the members must be within limits. In the analysis that follows, all these aspects have been critically examined.

The sidestands of the machine are solid steel plates to provide for maximum rigidity. The bed (bottom connecting member) is of closed box type construction with additional ribs to provide for rigidity.

6.3.1   Design Considerations

6.3.1.1   Existing Design Methodolgy

Since the force needed for shearing is known and since the shearing process is gradual, work done during cutting is easily determined. Since the duration of cutting is a fraction of the time taken for one revolution of the eccentric shaft, the fluctuation in speed is easily determined. From the maximum energy consumed during cutting, the Horse Power requirement of the prime mover is estimated. Once the speed of the intermediate shafts are assumed, design of the shafts and the associated antifriction bearings is a straightforward exercise. The only uncovered complex parts of the machine for which detailed analysis is called for are the stationary housing consisting of the two sidestands, hold down, the bottom plate and the moving blade carrier. It is the analysis of these two units which is attempted here.

6.3.1.2   Cutting Force

The blade of the Guillotine shear is inclined at an angle to the horizontal in order to enable gradual cutting of the plate. This helps in reducing the shear area at any instant of time as shown in Fig. 6.16. For a plate of thickness t to be cut, the width of cut at any instant of time is t/tan. Since this will be small in comparison with the maximum width to be cut (w), the cut takes place during a finite period of time. If the mean velocity of the blade is v, then the total time needed for cutting the maximum width of the plate will be:

$$T = w \tan \alpha/v \qquad (6.5)$$

The cutting force experienced by the blade will be given by

$$F = t^2 \tau/\tan \alpha \qquad (6.6)$$

where τ is the ultimate shearing strength of the material.

6.3.1.3   Free Body Diagram

In the case of an over-crank shear (Fig. 6.17), the eccentric shaft is located in the top half of the shearing machine along with the connecting link and the blade carrier. The line of stroke of the blade passes through the centre-line of the eccentric shaft.

In the case of an under-crank shear (Fig. 6.18), the eccentric shaft is located in the bottom half with the blade carrier in the top portion. Hence the height of the over-crank shear is always more than that of the under-crank shear. Moreover, in the case of under-crank shears,

PLATE 2

Fig. 6.14 Over crank guillotine shear

Fig. 6.15 Under crank guillotine shear

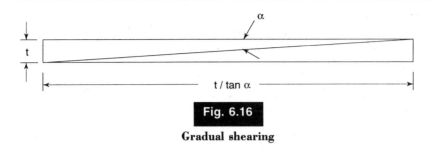

Fig. 6.16

Gradual shearing

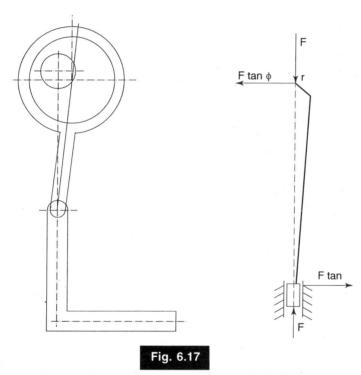

Fig. 6.17

Single slider crank mechanism-over crank shear

there is an offset (*e*) between the line of stroke of the blade carrier
and the centre-line of the eccentric shaft. The offset is due to the fact
that the space below the bottom dead centre position of the moving
top blade is occupied by the stationary bottom blade.

The free body diagrams of the stationary housing is shown in
Fig. 6.19. The stationary frames experience a downward vertical
force of *F* due to the plate being cut. This is uniformly distributed
in the zone of shearing (Fig. 6.19). The two side reactions, *F* tan α,

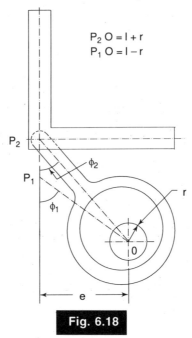

$$P_2 O = l + r$$
$$P_1 O = l - r$$

Fig. 6.18

Offset crank mechanism-under crank shear

in the zone of the blade carrier and along the eccentric shaft centre-line, result in a couple on the two side stands. In the case of the moving blade carrier, the axial force in the connecting link will be $F_i/\cos \alpha$, whereas the sliding faces of the blade carrier will experience a force of $F \tan \alpha$ and the blade, uniformly distributed upward load of F. The complex shape of the stationary and moving members necessitates the use of finite element method to determine the deformed shape and the stresses on the individual members.

6.3.2   FEM Model

The side plates, hold down, connecting plates and the bed are discretised with three-noded triangular plate elements. Each node is allowed six degrees of freedom—three translational and three rotational. Since the side plates are clamped at the ends at the bottom, the two end nodes of both the plates lose all the three translational degrees of freedom. The shape function for in-plane displacements is a linear polynomial in area co-ordinates, and for out-of-place displacements is a cubic polynomial [Sec. 1.2.3]

The problem is solved using the Cholesky method and the skyline

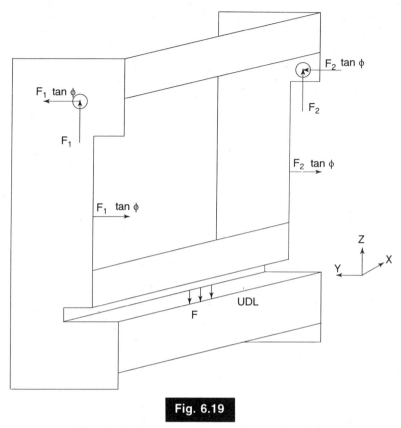

Fig. 6.19

Freebody diagram of the stationary housing

or variable bandwidth approach. In this method, the fact that the stiffness matrix $[K]$ is symmetric, helps in reducing the core and computational time required. Further, reduction in core and labour is achieved by stacking the elements of the stiffness matrix in a single dimensional array.

6.3.3 Details of Numerical Work

The details of the over-crank and under-crank shears analyzed, along with the computational details, are given in Table 6.4. The finite element discretization of the stationary and moving elements are shown in Fig. 6.20. The profile for the largest problem is around 3,06,000 [6.7]. The time taken for computing the deformations and stresses for each load case is negligibly small. The response is

Table 6.4

Details of the Machines

Specifications	Over-Crank	Under-Crank
Height of machine	3600 mm	1235 mm
Breadth of machine	1125 mm	900 mm
Maximum width cut	3000 mm	1250 mm
Maximum thickness cut	11 mm	5 mm
Cutting angle in degrees	2.5	1.65
Horse Power	40	5
Maximum cutting force	550 KN	191 KN
No. of strokes per minute	45	45
Connecting link length (l)	540 mm	320 mm
Eccentricity (r)	170 mm	40 mm
Offset (e)	—	225 mm

calculated for 12 different intervals of cutting, as the cutting progresses from the left end to the right end. In the case of the stationary housing, the nodes connected to the ground are assumed to have no translational degrees of freedom. In the case of the blade carrier, all the degrees of freedom are assumed to have been lost along the guideways. The load acts through the two connecting links at one end, and as a uniformly distributed load in the cutting zone. The cutting force is calculated using an ultimate shear strength of 200 MPa of the steel to be cut. The deformations and stresses obtained while the shear is half way through its travel are shown in Fig. 6.21 through Fig. 6.24.

6.3.4   Discussion of Results

The reaction on the stands $F_1 \tan \alpha$ and $F_2 \tan a$ are essentially in the Y direction. They cause lateral deflection of the two side stands, since the stands are fixed at the bottom. These values, for a given value of $F_i \tan \alpha$, will be less for the over-crank shear than for the under-crank shear. The two side stands essentially behave like plates subjected to in-plane forces in Z direction and bending forces in Y direction. This is evident from the deformation pattern as seen from Fig. [6.22]. Usually there is an increase in thickness in the zone in which the forces F_i and $F_i \tan \alpha$ act.

6.3.4.1   Deformation

The couple introduced by the two equal and opposite forces, $F_1 \tan a$

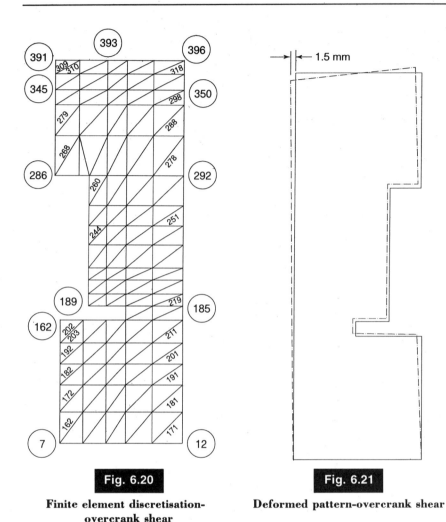

Fig. 6.20

Finite element discretisation-overcrank shear

Fig. 6.21

Deformed pattern-overcrank shear

on the left stand and $F_2 \tan \alpha$ on the right stand, are predominantly responsible for the lateral deflection in the Y direction. These couples have a smaller magnitude for the under-crank shear analysed here as compared to the over-crank shear, since the shearing force for cutting is small for the machine of smaller capacity. Besides, this couple acts at a higher level for the over-crank shear, since the eccentric shaft is located on the top half. As a consequence, the lateral deflection (deflection in the Y direction) experienced by the over-crank shear (Fig. 6.21), is higher than that for the under-crank shear (Fig. 6.22). When the blade is half way through its travel, the deflection in the X direction is expected to be negligibly small, whereas when the blade

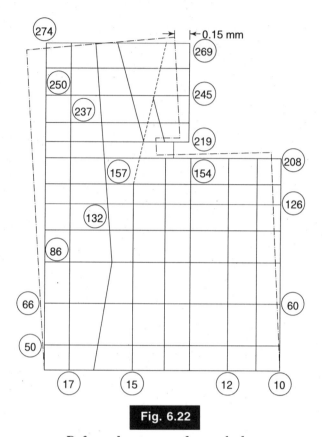

Fig. 6.22

Deformed pattern–undercrank shear

is at the two extremities of cutting, there is a possibility of the machine experiencing a deflection in the X direction. The maximum X deflection values experienced by the over-crank and under-crank shears as computed are 0.6 mm and 0.13 mm respectively. The machine is not expected to have any appreciable deflection in the Z direction. The computed values only strengthen this assumption.

6.3.6.2 Stresses

Von Mises [6.8] stresses have been computed for all the elements of the stand. This essentially controls the failure of ductile materials. Isostress lines corresponding to Von Mises stresses in MPa have been plotted when the blade is midway through its travel for both over-crank (Fig. 6.23) and under-crank (Fig. 6.24) shears. As may be seen from the figures, the maximum stress level experienced has

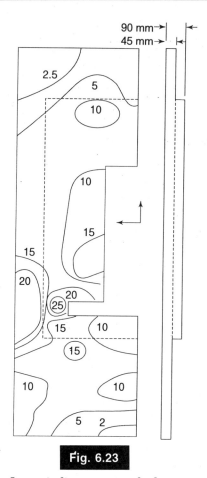

90 mm→| |←
45 mm→| |←

2.5

5

10

10

10

15

15

20

20

(25)

15   10

(15)

10   10

5   2

Fig. 6.23

Isostress lines-overcrank shear

not exceeded 25 MPa. Hence, from the strength point of view, both the machines are over designed. It may also be observed that the stress experienced by the stands in the zone above the eccentric shaft for over-crank shear and above the maximum travel position of the blade carrier for the under-crank shear is very low, since the bending moment in this section is zero. In the zone of the throat of the machine, as expected, a stress peak is observed. It may also be pointed out that in the zone of the throat, the thickness has been increased in order to reduce the maximum value of stress experienced.

It may be observed that the cutting force moves progressively from one end to the other along the length of the machine. This force can be modelled as a dynamic load and the consequent dynamic

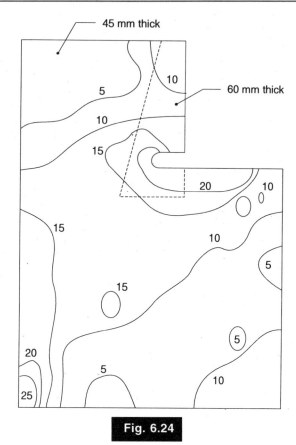

45 mm thick

60 mm thick

Fig. 6.24

Isostress line-undercrank shear

analysis carried out. The influence of the dynamic load on the response and the role of number of strokes per minute can be clearly determined.

➤ 6.4 TRANSIENT CHARACTERISTICS OF A BLADED DISK DURING RUN-UP

The transient response of a high-pressure stage turbo-machine bladed disk during run-up loaded by partial admission is studied by the finite element method exploiting the cyclic symmetric nature of the problem. The force acting on the entire structure is considered as a sum of finite number of spatial harmonics, each of which has the property that all substructures experience identical forces but with a fixed phase difference between neighbouring substructures. The modal superposition method is used to solve the equations of motion for transient vibration. The response is studied separately for each spatial Fourier harmonic on a single substructure. The results show the stress response of the bladed disk at the blade root under varying operating conditions [6.9].

The force acting on the kth d.o.f. of the jth substructure of the bladed disc at any time t can be written in the form of a finite Fourier series as

$$F_{jk}(t) = \sum_{p=1}^{N} A_{pk}(t) \, e^{i(j-1)(2\pi p/N)} \qquad (6.7)$$

We can show that

$$A_{pk}(t) = 1/N \sum_{j=1}^{N} F_{jk}(t) \, e^{-i(j-1)(2 \cdot p/N)} \qquad (6.8)$$

Given the forces acting on the kth d.o.f. of each substructure, we can use Eq. (6.8) to obtain the spatial Fourier harmonics of force for the kth d.o.f.

With this 'decoupling' of forces using Eq. (6.11), the equation of transient vibration for a given spatial Fourier harmonic is given by

$$M_p \ddot{u}_p^1(t) + C_p \dot{u}_p^1(t) + K_p u_p^1(t) = A_p(t) \qquad (6.9)$$

for
$$p = 0, 1, 2, \ldots, N \qquad (6.10)$$

where M, C and K are mass, damping and stiffness matrix of the substructure respectively, and superscript 1 denotes the substructure under consideration for analysis.

The deflection at any time t of the jth substructure is given by

$$d_j(t) = u_N^1(t) + 2 \sum_{p=1}^{N/2} Re(u_p^1(t) \cos{(j - 1)(2\pi p/N)} \qquad (6.11)$$
$$+ Im(u_p^1(t)) \sin{(j - 1)(2 \cdot p/N)}$$

The solution of Eq. (6.9) in time domain is obtained by the modal superposition method by which we get a set of uncoupled equations, each of which has the form

$$\ddot{q} + 2\xi_m \omega_m \dot{q} + \omega_m^2 q = g(t) \qquad (6.12)$$

where m takes the values 1, 2, ..., k, k being the number of modes taken for analysis. For a force F_0 acting during a period Δt, we have the system response in the form

$$q = e^{-\xi_m \omega_m t} (C_1 \cos{\omega_m \sqrt{(1 - \xi_m^2)} \Delta t}$$
$$+ C_2 \sin{\omega_m \sqrt{(1 - \xi_m^2)} \, t}) + (F_0/\omega_m^2) \qquad (6.13)$$

where

$$C_1 = [X_0 - (F_0/\omega_m^2)] \qquad (6.14)$$

$$C_2 = [V_0 + \xi_m(X_0 - (F_0/\omega_m^2)]/\omega_m \sqrt{(1 - \xi_m^2)} \qquad (6.15)$$

X_0 and V_0 are the modal displacement and velocity at the beginning of period Δt.

It is to be noted that for time dependent processes, the change in value of K_p in Eq. (6.9) due to rotation results in time-dependent values for natural frequencies and mode shapes and must be taken into consideration. However, for all practical purposes, the change in mode shapes is negligibly small and it is sufficient to take the variation of ω_m into consideration. The natural frequencies at different speeds can be obtained by taking geometric stiffness matrix into Eq. (6.9) corresponding to that speed. However, it is impractical to compute natural frequencies at each time step of modal analysis and hence, a quadratic interpolation of these values will suffice at the required time step.

6.4.1   Substructure Modelling

The choice of element is a question of engineering judgement based on the type of structure, number of degrees of freedom per node and number of nodes per element, and solution algorithms used. When plate elements represented by the mid-surface of the structure are used, for a non-zero stagger angle of blade, the mid-surface planes of disk and blade won't lie in the same plane (Fig. 6.14). But the inplane displacements of a node n_d on the disk at the blade-disk junction and a corresponding node on the blade n_b can be related by Love-Kirchhoff's hypothesis, according to which

$$w_b = w_d \qquad (6.16)$$

$$u_b = u_d + z(\partial_w/\partial_x) \qquad (6.17)$$

$$v_b = v_d + z(\partial_w/\partial_y)_d \qquad (6.18)$$

u, v are inplane displacements and w, the bending displacement.

6.4.2   Resonance Coincidence

It is well-known that the resonance between an exciting frequency and a natural frequency f_k corresponding to k nodal diameters occurs when the frequency of excitation equals the frequency of forward or backward travelling wave in the rotating bladed disk, i.e.

$$f_e = f_k \pm k f_n \qquad (6.19)$$

where f_n is the rotational frequency of the bladed disk.

It follows from Eq. (6.19) that for a static force with $f_e = 0$, the resonance takes place for the critical speeds of

$$f_n^{cri} = f_k/k \qquad (6.20)$$

6.4.3 Results and Discussion

With the theoretical formulation presented, all complicated geometric parameters like stagger, twist, taper and rotational effects can be easily incorporated. A twelve-bladed disk model of steel (with $E = 0.2168 \times 10^6$ N/mm^2, $s = 0.8228 \times 10^{-8}$ N– s^2/mm^4, $\gamma = 0.28$) with the following goemetry is considered for the transient vibration analysis during run-up loaded by partial admission.

Blade-disk junction radius	254 mm
Disk radii ratio	0.2
Blade aspect ratio	3
Blade stagger angle	30°
Blade twist at the tip	45°
Disk thickness	4 mm
Blade thickness	2 mm
Substructure angle	30°

The rotating blade and load due to two sectors of 90° of partial admission is shown in Fig. 6.25. The blade is subjected to load due to partial admission while passing through these two sectors and is unloaded during the other two sectors. The gas force of 200 N/mm of blade acts at 45° to the circumferential direction in the z—θ plane. Two operational accelerations are considered, viz. (i) attainment of a steady state speed of 10,000 rpm in 5 sec. (referred to as first case), (ii) attainment of a steady state speed of 20,000 rpm in 5 sec. (referred to as second case). A rectangular approximation to the force applied due to partial admission is used. With this, the force-time diagram for the first case is as shown in Fig. 6.26. The force has the character of a low-cycle step function in the beginning and with increasing time, the character of a rectangular excitation.

The lowest three frequencies corresponding to each value of n are taken for modal analysis which are proved to be sufficient. The natural frequencies for the stationary bladed disk are presented in Table 6.5. The time step corresponding to 1/10th of the lowest natural period comes to 0.2174×10^{-3} sec. as the basis for response calculation in time domain. As already explained, the natural frequencies increase with rotational speed due to the 'stiffening effect' and the variation of natural frequencies with speed is incorporated in Eq. (6.9) by

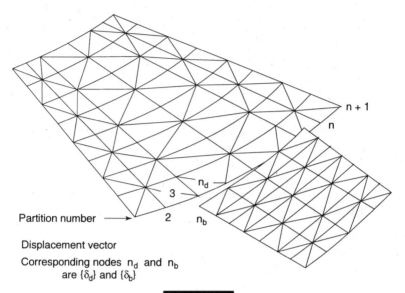

Fig. 6.25

Finite element model of substructure

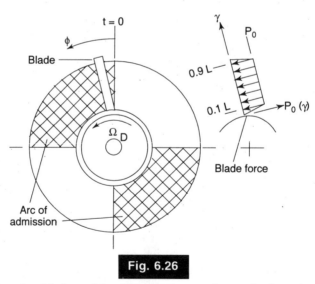

Fig. 6.26

Rotating blade and load due to areas of partial admission

computing these frequencies at each time step by a quadratic interpolation from the frequencies obtained by actual eigenvalue analysis at selected speeds (in this case 0, 5000, 10000 and 20000

Table 6.5

Natural Frequencies of the Model (Values in Hz)

Nodal Diameter	Frequency 1	Frequency 2	Frequency 3
0	30.385	80.010	207.77
1	29.553	78.712	211.29
2	32.865	93.791	221.33
3	37.607	138.720	247.91
4	39.418	172.853	312.41
5	40.097	185.124	400.17
6	40.281	188.054	458.36

rpm). The response is calculated separately for each spatial harmonic and recombined to get the resultant response. The stress response at the blade root for the blade on which the force is acting ($\theta = 0°$) is shown in Fig. 6.27.

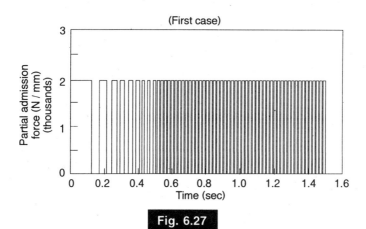

Fig. 6.27

Force torque diagram during run-up

➤ **6.5 RESPONSE OF VEHICLES OWING TO ROAD UNDULATIONS**

Let us consider the response of the bus to two random loadings $P(t)$ and $Q(t)$, acting simultaneously at the front and rear tyres as shown in Fig. 6.28. Let the resulting displacement of a given point in the vertical direction be $x(t)$. Let the corresponding receptances for harmonic excitation be α_{xP} and α_{xQ}. It can be shown that the spectral density of $x(t)$ can be expressed in terms of the spectral densities of $P(t)$ and $Q(t)$ as shown in Eq. (6.21).

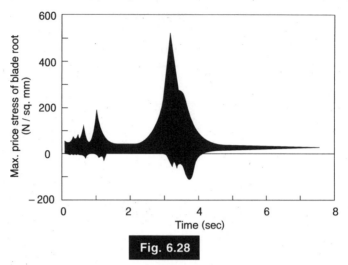

Fig. 6.28

Transient response during run-up

$$S_x(f) = (\alpha_{xP}^* \alpha_{xP} \, S_P(f) + \alpha_{xP}^* \alpha_{xQ} \, S_{PQ}(f)$$

$$+ \alpha_{xQ}^* \alpha_{xP} S_{QP}(f) + \alpha_{xQ}^* \alpha_{xQ} S_Q(f) \qquad (6.21)$$

where the * denotes complex conjugations and S_{PQ} and S_{QP} are the cross-spectral densities of forces $P(t)$ and $Q(t)$. For the specific case where $Q(t)$ reproduces $P(t)$ after a lag of τ_0 i.e. where $Q(t) = P(t + \tau_0)$ and where $S_P(f) = S_Q(f) = S(f)$ as is so in the case of the bus, we have

$$S_x(f) = (\alpha_{xP}^* \alpha_{xP} + e^{i2\pi f\tau_0} \, \alpha_{xP}^* \alpha_{xQ} \qquad (6.22)$$

$$+ e^{-i2\pi f\tau_0} \, \alpha_{xQ}^* \alpha_{xP} + \alpha_{xQ}^* \alpha_{xQ}) S_P(f)$$

The phase angle $\theta = 2\pi f\tau_0$ may be determined as discussed. In Fig. 6.28, let b represent the wheel base and V be the velocity of the bus. Let ω be the forcing frequency due to the road undulation input. Then one second corresponds to ω radians and hence, b/v seconds (i.e. the time taken to traverse distance b) corresponds to $\omega b/v$ radians or in other words, phase angle $\theta = 2\pi f\tau_0 = \omega b/v$ radians. Hence, $\tau_0 = b/v$ seconds in Eq. (6.22).

6.5.1   Vehicle Model

The bus has been idealised using a finite element model as shown in Fig. (6.29) The model takes into account the chassis members, the suspensions, the axles and the tyres. The chassis and axle members have been modelled as 3-dimensional beam elements with 6 d.o.f. at each node. For the chassis longitudinal members and the front and

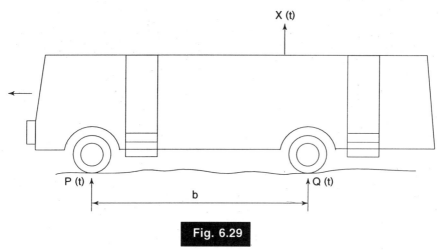

Fig. 6.29

Response of bus during random loading

rear axle elements, the degrees of freedom corresponding to twist were suppressed.

The suspensions and tyres have been modelled as linear springs with stiffnesses in the vertical direction alone. The leaf spring constants were determined experimentally in the laboratory. In Table 6.6 are indicated the spring stiffnesses and damping for front/rear suspensions and tyres. The mass of the engine has been lumped at the appropriate nodes. A constant mass of 0.980 N/cm/s^2 has been added to all nodes in the three translational degrees of freedom to account for the mass of the super structure (i.e. the body frame work, panelling, seats, glasses, etc).

6.5.2 Road Model

The input road power spectral densities used for the calculation were obtained using an Automatic Road Unevenness Recorder. An accelerometer was attached to an arm responding to road undulations and the output of this accelerometer was tape-recorded using an instrumentation tape-recorder. Some practical difficulties arose in the collection and interpretation of such random data. The selection of a representative stretch of road posed a problem. To ensure that the signal is stationary and ergodic, lengthy portions of data should be analysed but this is expensive in terms of the cost of time taken for the analysis. Hence, fairly long representative stretches, i.e. 1 km lengths of good and bad stretches of road were chosen. Figure

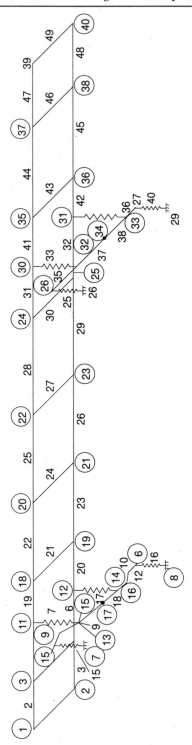

Fig. 6.30

FEM of chassis and axle

(6.30) shows the acceleration power spectral densities of road stretches used as input [6.10 to 6.12].

The vibration and stress responses were computed for 5 cases:

(1) Identical left and right tracks of good quality (0.495 g rms) and vehicle speed of 30 km/hr.
(2) Identical left and right tracks of good quality (1 98 g rms) and vehicle speed of 60 km/hr.
(3) Identical left and right tracks of bad quality (0.8 g rms) and vehicle speed of 30 km/hr.
(4) Identical left and right tracks of bad quality (3.2 g rms) and vehicle speed of 60 km/hr.
(5) Bad left track (3.2 g rms) and exceedingly good right track (zero g rms) and vehicle speed of 60 km/hr.

6.5.3 Computational Procedure

The road input acceleration power spectral density was sampled at 80 frequency steps in the range 0 to 50 Hz. The forces due to the road were fed to nodes 5, 6, 26 and 27 in Fig. 6.29. For example, an input force of $(1/2)K_{tf}X_g \sin \omega t$ was given to nodes 5 and 6 corresponding to the upper ends of the front tyres; here k_{tf} is the stiffness of the tyre as indicated in Table 6.6 and $\omega^2 X_g$ is the unit input acceleration at frequency ω. For this unit input, the output displacements at all

Table 6.6

Suspension and Tyre Characteristics

	Stiffness	Stiffness in N/cm	Damping	Damping in N/cm/sec
Front suspension	k_{sf}	9,300	c_{sf}	320
Rear suspension	k_{sr}	10,480	c_{sr}	360
Front tyre	k_{tf}	17,895	c_{tf}	22.4
Rear tyre	k_{tr}	35,790	c_{tr}	44.7

nodes and hence, the acceleration at all nodes and the stresses in all the elements were computed. These outputs correspond to α_{xP} in Eq. 6.21. If the unit input accelerations are given to nodes 26 and 27, corresponding to the upper ends of the rear tyres, the outputs correspond to α_{xQ} in Eq. (6.21). Using the values of α_{xP} and α_{xQ} thus obtained, the acceleration power spectral densities at all the nodes were computed using Eq. (6.22) for a vehicle speed of 30 km/hr. For a vehicle speed of 60 km/hr, the road acceleration input power spectral density values were modified in proportion to the square of the input speed, e.g. 4 times for 60 km/hr. For this speed, 160 frequency steps were taken in the range 0–50 Hz. These computations were made on a machine in which the time taken for one multiplication is 6 μs.

6.5.4   Experimental Verification

The accelerations and strains at a number of salient points were experimentally determined to validate the analytical findings. To determine the accelerations, inductive accelerometers were used. The conditioned outputs were recorded using an instrumentation tape recorder. The strains were found by pasting foil type linear gauges and connecting them in the arms of Wheatstone bridges. All instruments within the running bus were powered using an inverter which was fed by a 24 V battery. The tape-recorded data was later on analysed in the laboratory using a spectrum analyser. The rms values of acceleration and strain were determined using a computing voltmeter. Very good correlation between experimental and analytical results was obtained. Figure 6.31 shows a comparison of acceleration values obtained analytically and experimentally at 60 km/hr and

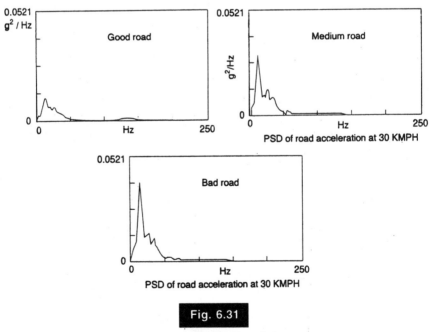

Fig. 6.31

Acceleration power spectral densities

72 km/hr on both stretches of roads. The points in the figure have been joined by straight lines for ease of visualisation. Figure 6.32

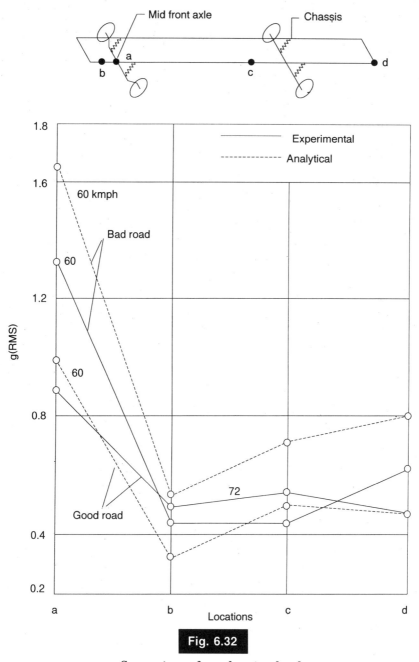

Fig. 6.32

Comparison of acceleration levels

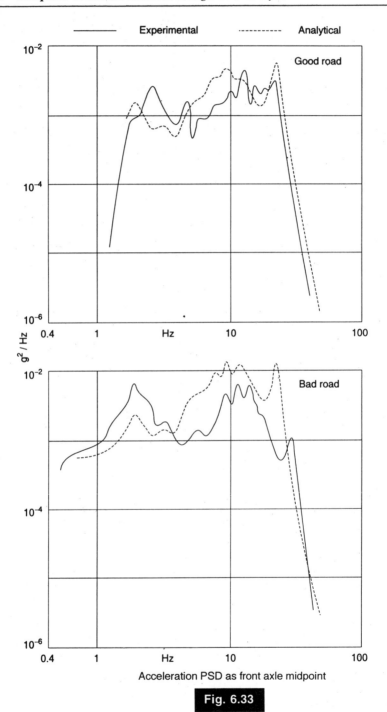

Acceleration PSD as front axle midpoint

Fig. 6.33

Comparison of power spectral densities

compares the analytically and experimentally obtained acceleration p.s.d. curves at the front axle mid-point at 30 km/hr on a good stretch of road and on a bad stretch of road. The stresses in the various members of the chassis longitudes are as shown in Fig. 6.34. Also shown in Fig. 6.34 are the stress variation along the front and rear axles.

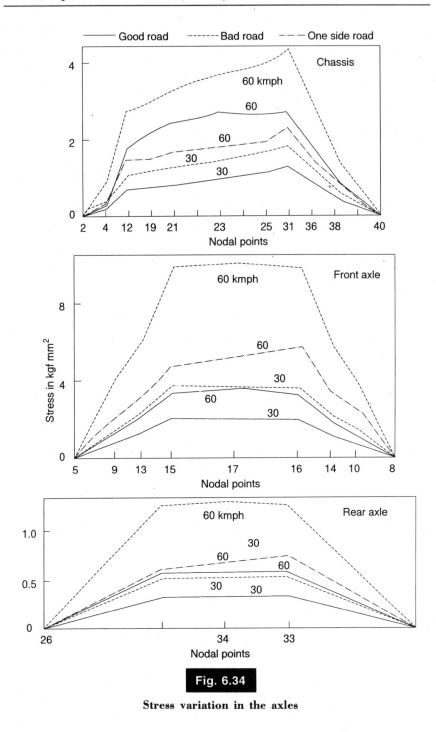

Fig. 6.34

Stress variation in the axles

REFERENCES

6.1 Ramamurti, V., V. Ravi Shankar Rao and N.S. Sriram, Design aspects and parametric study of heavy duty 3 role heavy duty plate bending machine, *J. Mat. Proc. Tech.* 32, pp 585–598 (1992).

6.2 Thomas, A.G., *High Productivity in Heavy Engineering*, Iliffe and Sons, London, 1960.

6.3 Zienkiewiez, O.C., *Finite Element Method in Engineering Science*, McGraw-Hill, London, 1971.

6.4 Kamath, M., R. Narasimhan, C. Nataraj and V. Ramamurti, Dynamic response of multicylinder engines with a viscous or hysteretic crank shaft damper, *J. Sound Vib.*, vol. 81, no. 3, pp 448–452, 1982.

6.5 Wilson, W. Ker, *Practical Solutions of Torsional Vibration Problems*, vol. 2, Chapman and Hall, London, 1963.

6.6 Macduff, J.N. and J.R. Curreri, *Vibration Control*, McGraw-Hill, New York, 1958.

6.7 Ramamurti, V. and S. Gowri, Design considerations of over-crank and under crank guillotine shears—a comparative study, *J. Mat. Proc. Tech*, vol. 57 no 3/4, pp 225–232, 1996.

6.8 Shigley J.E. and C.R. Mischke, *Mechanical Engineering Design*, McGraw-Hill, New York, 1989.

6.9 Om Prakash, V. Ramamurti, Transient characteristics of a plated disk during run up during partial admission, *Proc. 12th Biennial Vibr. Conf.*, V.DE18–3, pp 249–254, 1989.

6.10 Sujatha, C. and V. Ramamurti, Bus vibration study—experimental response to road undulation—*Int. J. Vehicle Design*, vol. 11, no. 4/5, pp 390-400, 1990.

6.11 Ramamurti, V. and C. Sujatha, Bus vibrations study—finite element modelling and determination of eigenpairs, *Int. J. Vehicle Design*, vol. 11, no. 4/5, pp 410–421, 1990.

6.12 Ramamurti, V. and C. Sujatha, Bus vibration study—theoretical response analysis and experimental verification, *Int. J. Vehicle Design*, vol. 11, no. 4/5, pp 401–409, 1990.

Summary

The concepts governing the topics covered in the preceding chapters will now be summarised. The introductory chapter indicates the general form of the governing equations for all problems associated with linear and non-linear systems subjected to static and dynamic loading. The five important finite element formulations, viz. 2-D beam element, 3-D beam element, six-degree-of-freedom, triangular shell element, axisymmetric element for semi-analytical approach and eight-noded isoparametric element are presented. Applications of these elements to problems in industry are also indicated. Pre- and post-processing of data are also discussed. The formulation of problems using finite difference scheme is also indicated.

The second chapter is on static linear problems. All these problems merely get reduced to solving simultaneous equations with many unknowns. Conventional methods like Cramer's rule or inversion are unsuitable from the point of view of core and labour. The Gaussian elimination scheme has the distinct advantage of reduced labour. For problems having a number of loading cases (not all appearing at the same time), this method is uneconomical since the labour is involved as often as there are instances of loading cases. Essentially, this is a blackboard and a scriber scheme since the intermediate steps are erased from the memory of the computer. The Cholesky scheme which does the same problem with a slightly different approach enables one to store the intermediate steps in the form of lower triangular matrix. This reduces the work involved in case of multiple loading. This can be treated as a paper and pencil scheme since the information is stored.

The sparseness of the matrices associated with engineering problems is the next to be exploited. The number of elements below the skyline of the lower triangular matrix is exactly the same as the number of elements below the skyline of the corresponding stiffness matrix. Hence, the core needed in solving the problem by Cholesky factorisation gets reduced further. Besides, the bandwidth variation from row to row enables substantial reduction in computing time.

When the core of the computer is not sufficient to accommodate the stiffness matrix (even when retaining the elements below the skyline), one has to resort to the out-of-core solution by substructuring the problems and transfering the information pertaining to inactive substructures to an auxiliary device like a tape or a disc. The details of this kind of approach (Potters' scheme) are discussed next. The mini problems associated with each substructure are solved by the Gaussian elimination technique. When the stiffness matrix is symmetric and when a number of problems have to be handled, the Gaussian elimination scheme used with each substructure can always be replaced by Cholesky factorisation. This, in effect, is exploiting the virtues of Potters' scheme of substructuring and Cholesky's scheme of recording the intermediate steps.

The next aspect considered is bandwidth optimisation. Overall core and labour involved in solving any problem directly depends on the sparseness of the matrix. One of the very efficient methods of achieving this is to use the reverse Cuthill McKee algorithm.

The following are the industrial applications presented:

(i) Stress analysis of 3 m dia. × 23.4 m long multilayer ammonia converter using finite element and finite difference methods.

(ii) Stress analysis of a 40 tonne C-frame hydraulic press using out-of-core solution.

(iii) *Gearbox casings:* Stress and deformation analysis of fabricated and cast heavy duty two stage gearbox casings.

(iv) Design considerations of heavy duty moulding boxes.

Chapter 3 is devoted to eigenvalue problems. The classical method of forming the characteristic polynomial and determining the roots of the associated equation is unsuitable for problems of the size encountered in complex machine elements. The use of Sturm sequence in conjunction with the method of bisection helps one to compute the first few eigenvalues of a large size problem with improved efficiency. The use of one of the transformation methods due to Jacobi, Givens and Householder facilitates the reduction of the eigenvalue problem of considerable bandwidth to a tridiagonal or diagonal form. But these, by themselves, are not useful in determining the eigenvectors.

Using iteration and computing Rayleigh's quotient is a very effective way of determining the largest or the smallest eigenvalue and associated eigenvector. To compute more than one vector, the Gram-Schmidt deflation technique is employed which prevents convergence to the vectors already computed.

Besides the fact that the iteration technique is very effective, computation of a set of vectors simultaneously combines in itself the advantage of the iteration scheme and saving in computer time. This calls for orthonormalising the chosen vectors at the end of each iteration. This is achieved by two approaches slightly different from each other, viz. simultaneous iteration and subspace iteration schemes.

In all iteration schemes, every stage of iteration involves the solution of many simultaneous equations. The most efficient method of achieving this is to use the concepts spelt out in Chapter 2 either for an in-core or out-of-core solution. Hence, a clear understanding of static problems of large size is a prerequisite for handling eigenvalue problems of a large size. Another efficient method of extracting the eigenpairs is to use the routine created by Lanczos. This method, in effect, transforms the original matrix into a tridiagonal form.

The following are the industrial applications presented:

(i) Determination of critical speeds of lateral oscillation of shafts of varying cross-section, including the gyroscopic effect.

(ii) Computing the natural frequencies and response of a plane frame supporting a boiler.

(iii) *Frequencies:* Computing the eigenvalues of a compressor disc consisting of two thin diaphrams with an outer rim. This uses the simultaneous iteration scheme for computing the first few natural frequencies.

(iv) Determining the first few critical speeds of an industrial sugar centrifugal and studying the influence of liquid inside.

Chapter 4 is devoted exclusively to cyclic symmetric structures. In the first part, static or steady state analysis of such structures is taken up. The necessity for out-of-core solution is highlighted. Static analysis for generalised loading involves Hermitian matrices. A method of solution involving minimum amount of complex numbers computation is presented. The study is later on extended to eigenvalue problems. All the solution schemes make use of the detailed programming techniques discussed in Chapters 2 and 3. A comparison of time and core needed by different schemes is also outlined. The Lanczos scheme of reducing the computational labour for eigenpair extraction is also presented. The following industrial applications are presented:

(i) Steady state stress analysis of centrifugal fan impellers;

(ii) Frequency analysis of industrial fan impellers; and

(iii) Stress analysis of heavy duty fabricated gear wheels. In the first two cases, analytical values are compared with experimental findings.

Chapter 5 is on transient vibration problems. Direct integration methods like central differences, Wilson θ and Newmark are discussed. Computational labour at every time step is optimised using the logic presented in Chapter 2 on simultaneous equations. The concept of model superposition as a viable alternative to the direct integration technique is presented. The advantages derived by using the mode of superposition for large-size problems from the point of view of time are discussed.

The behaviour of structures subjected to random vibrations is discussed next. As a case of practical application, earthquake analysis of a high voltage circuit breaker is presented. The dynamic stress experienced by the circuit breaker is determined.

Chapter 6 is on case studies. Five different problems of interest to industry have been reported, encompassing the different methods presented in the earlier chapters. The first one is the behaviour of a fabricated stand of a heavy duty plate-bending roll. The use of variable bandwidth algorithm for solving a large-size problem using triangular elements with 6 degrees of freedom at each node has been made. The second one is on torsional vibration problems. It employes the Sturm sequence for eigenvalue determination and Potters' scheme for solving complex simultaneous equations.

The third one concerns the design aspects of over-crank and under-crank guillotine shears. Stress and deformation analysis of the sturdy side stands for the operating conditions is presented.

The fourth one is connected with the transient stress response of bladed discs due to partial admission during run-up using the concept of cyclic symmetry. The maximum dynamic stress experienced by the blades is predicted.

The last exercise is to model commercial buses using shell and 3-D beam elements. The road undulations are taken as random inputs to the problem. These are obtained from records taken from typical roads. The dynamic response of the front axle, the rear axle and the chassis are determined. R.M.S. values of stresses on these critical elements are computed and are compared with the results experimentally arrived at.

Programs

```
C******* PROGRAM : 2.1
C****** SOLVING THE BEAM PROBLEM OF EXAMPLE 2.10
C       (METHOD OF INVERSION BY PARTITIONING).
        IMPLICIT REAL*8(A-H,O-Z)
        DIMENSION AK(22,22),F(22),AX(22)
        DATA AK,F/484*0.0D0,22*0.0D0/
        D = 30.0
        E = 2.1E07
        B = 10.0
        AL = 100.0
        Q = 100.0
        A = 1000.0
        DEL = AL/A
        AI = B*D**3/12.0
        AF = Q*A**3/(E*AI)
        AM = E*AI/A
        AK(1,1) = 1.0
        AK(2,2) = 1.0
        AK(21,21) = 1.0
        AK(22,22) = 1.0
        DO 20 I = 3,19,2
        AK(I,I-2) = 1.0/(DEL*DEL)
        AK(I,I) = -2.0/(DEL*DEL)
        AK(I,I+1) = 1.0
        AK(I,I+2) = AK(I,I-2)
        AK(I+1,I-1) = AK(I,I-2)
        AK(I+1,I+3) = AK(I,I-2)
20      AK(I+1,I+1) = AK(I,I)
        DO 30 I = 4,20,2
30      F(I) = AF
        CALL INVERT(AK,22)
        DO 40 I = 1,22
        AX(I) = 0.0
        DO 40 K = 1,22
40      AX(I) = AX(I) + AK(I,K)*F(K)
        WRITE(6,50)
50      FORMAT(/2X,'SECTION',2X,'DISPLACEMENT',5X,
     1  'MOMENT')
        J = 1
        DO 70 I = 1,21,2
        AX(I+1) = AX(I+1)*AM
        AX(I) = AX(I)*A
```

```
          WRITE(6,75)J,AX(I),AX(I+1)
          J = J+1
70        CONTINUE
75        FORMAT(2X,I2,8X,F7.4,5X,D11.4)
          STOP
          END


          SUBROUTINE INVERT(AE,NZ)
C******METHOD:-
C******    INVERSION IS DONE BY PARTITIONING.
C******DESCRIPTION OF PARAMETERS:-
C******AE   INPUT MATRIX.DESTROYED IN COMPUTATION
C******     AND REPLACED BY THE RESULTANT INVERSE
C******     IT IS OF SIZE NZ BY NZ.
C******NZ   NUMBER OF ROWS IN AE.
C******BX & CE ARE WORKING VECTORS OF SIZE NZ.
C******SUBROUTINE & FUNCTION SUBPROGRAM REQUIRED
C******NONE.
C******REMARKS:-
C******     MATRIX AE MUST BE GENERAL.
C******     THIS ROUTINE OPERATES IN DOUBLE
C******     PRECISION ARITHMETIC.
C******
          IMPLICIT REAL*8 (A-H,O-Z)
          DIMENSION AE(NZ,NZ),BX(22),CE(22)
          NN = NZ-1
          AE(1,1)=1./AE(1,1)
          DO 70 M = 1,NN
          K = M+1
          DO 10 I = 1,M
          BX(I) = 0.
          DO 10 J = 1,M
10        BX(I) = BX(I) + AE(I,J)*AE(J,K)
          D = 0.
          DO 20 I = 1,M
20        D = D + AE(K,I)*BX(I)
          D = -D + AE(K,K)
          AE(K,K) = 1./D
          DO 40 I = 1,M
40        AE(I,K) = -BX(I)*AE(K,K)
          DO 50 J = 1,M
          CE(J) = 0.
```

```
        DO 50 I = 1,M
50      CE(J) = CE(J) + AE(K,I)*AE(I,J)
        DO 60 J = 1,M
60      AE(K,J) = -CE(J)*AE(K,K)
        DO 70 I = 1,M
        DO 70 J = 1,M
70      AE(I,J) = AE(I,J) - BX(I)*AE(K,J)
        RETURN
        END
```

SECTION	DISPLACEMENT	MOMENT
1	0.0000	0.0000D 00
2	-0.8730	-0.4500D 07
3	-1.6508	-0.8000D 07
4	-2.2593	-0.1050D 08
5	-2.6455	-0.1200D 08
6	-2.7778	-0.1250D 08
7	-2.6455	-0.1200D 08
8	-2.2593	-0.1050D 08
9	-1.6508	-0.8000D 07
10	-0.8730	-0.4500D 07
11	0.0000	0.0000D 00

COMPILE TIME=0.35 SEC, EXECUTION TIME=1.91 SEC

```
C******* PROGRAM : 2.2
C******* SOLVING THE BEAM PROBLEM OF EXAMPLE
C******* 2.10 BY GAUSS-JORDAN METHOD.
        IMPLICIT REAL*8(A-H,O-Z)
        DIMENSION AK(22,22),F(22),AX(22)
        DATA AK,F/484*0.0D0,22*0.0D0/
        D = 30.0
        E = 2.1E07
        B = 10.0
        AL = 100.0
        Q = 100.0
        A = 1000.0
        DEL = AL/A
        AI = B*D**3/12.0
        AF = Q*A**3/(E*AI)
        AM = E*AI/A
        AK(1,1) = 1.0
        AK(2,2) = 1.0
        AK(21,21) = 1.0
        AK(22,22) = 1.0
        DO 20 I = 3,19,2
        AK(I,I-2) = 1.0/(DEL*DEL)
        AK(I,I) = -2.0/(DEL*DEL)
        AK(I,I+1) = 1.0
        AK(I,I+2) = AK(I,I-2)
        AK(I+1,I-1) = AK(I,I-2)
        AK(I+1,I+3) = AK(I,I-2)
20      AK(I+1,I+1) = AK(I,I)
        DO 30 I = 4,20,2
30      F(I) = AF
        CALL GAUSJR(AK,22,F,1,DETERM,ID)
        DO 40 I = 1,22
40      AX(I) = F(I)
        WRITE(6,50)
50      FORMAT(/2X,'SECTION',2X,'DISPLACEMENT',5X,
     1  'MOMENT')
        J = 1
        DO 70 I = 1,21,2
        AX(I+1) = AX(I+1)*AM
        AX(I) = AX(I)*A
        WRITE(6,75)J,AX(I),AX(I+1)
        J = J+1
```

```
70      CONTINUE
75      FORMAT(2X,I2,8X,F7.4,5X,D11.4)
        STOP
        END

        SUBROUTINE GAUSJR(A,N1,B,M1,DETERM,ID)
C*****  METHOD:-
C*****      MATRIX INVERSION AND SOLUTION OF
C*****      EQUATIONS BY GAUSS-JORDAN METHOD.
C*****  DESCRIPTION OF PARAMETERS:-
C*****A  INPUT MATRIX OF COEFFICIENTS.DESTROYED
C*****   IN COMPUTATION AND REPLACED BY RESULTANT
C*****   INVERSE. IT IS OF SIZE N1 BY N1.
C*****B  MATRIX OF CONSTANTS(FORCES) OF SIZE
C*****   N1 BY M1.DESTROYED AND REPLACED BY THE
C*****   RESULTANT DISPLACEMENTS.
C*****N1      NUMBER OF EQUATIONS.
C*****M1      NUMBER OF COLUMNS OF B.
C*****DETERM  VALUE OF THE DETERMINANT.
C*****ID      CONDITION CODE.
C*****   1  .MATRIX SUPPLIED IS NON-SINGULAR.
C*****   2  .MATRIX IS SINGULAR.SOLUTION IS
C*****         NOT POSSIBLE.
C*****  REMARKS:-
C*****         INDEX IS A WORKING VECTOR(N1,3).
C*****
C*****  SUBROUTINE & FUNCTION SUBPROGRAM REQUIRED
C*****  NONE.
C*****
        IMPLICIT REAL*8(A-H,O-Z)
        DIMENSION A(N1,N1),B(N1,M1),INDEX(22,3)
        EQUIVALENCE (IROW,JROW),(ICOLUM,JCOLUM),
       1(AMAX,T,SWAP)
        ABS(X)=DABS(X)
C*****INITIALIZATION
        M=M1
        N=N1
        DETERM=1.0
        DO 10 J=1,N
     10 INDEX(J,3)=0
        DO 180 I=1,N
C*****SEARCH FOR PIVOT ELEMENT
        AMAX=0.0
```

```
      DO 55 J=1,N
      IF(INDEX(J,3)-1)20,55,20
  20 DO 50 K=1,N
      IF(INDEX(K,3)-1) 30,50,240
  30 IF(AMAX-ABS(A(J,K)))40,50,50
  40 IROW=J
      ICOLUM=K
      AMAX=ABS(A(J,K))
  50 CONTINUE
  55 CONTINUE
      INDEX(ICOLUM,3)=INDEX(ICOLUM,3)+1
      INDEX(I,1)=IROW
      INDEX(I,2)=ICOLUM
C*****INTERCHANGE ROWS TO PUT PIVOT ELEMENT ON
C*****DIAGONAL
      IF(IROW-ICOLUM)60,100,60
  60 DETERM=-DETERM
      DO 70 L=1,N
      SWAP=A(IROW,L)
      A(IROW,L)=A(ICOLUM,L)
  70 A(ICOLUM,L)=SWAP
      IF(M) 100,100,80
  80 DO 90 L=1,M
      SWAP=B(IROW,L)
      B(IROW,L)=B(ICOLUM,L)
  90 B(ICOLUM,L)=SWAP
C*****DIVIDE PIVOT ROW BY PIVOT ELEMENT
 100 PIVOT=A(ICOLUM,ICOLUM)
      DETERM=DETERM*PIVOT
      A(ICOLUM,ICOLUM)=1.0
      DO 110 L=1,N
 110 A(ICOLUM,L)=A(ICOLUM,L)/PIVOT
      IF(M)140,140,120
 120 DO 130 L=1,M
 130 B(ICOLUM,L)=B(ICOLUM,L)/PIVOT
C*****REDUCE NON-PIVOT ROWS
 140 DO 180 L1=1,N
      IF(L1-ICOLUM) 150,180,150
 150 T=A(L1,ICOLUM)
      A(L1,ICOLUM)=0.0
      DO 160 L=1,N
 160 A(L1,L)=A(L1,L)-A(ICOLUM,L)*T
```

```
      IF(M) 180,180,170
170 DO 175 L=1,M
175 B(L1,L)=B(L1,L)-B(ICOLUM,L)*T
180 CONTINUE
C*****INTERCHANGE COLUMNS
      DO 205 I=1,N
      L=N+1-I
      IF (INDEX(L,1)-INDEX(L,2))190,205,190
190 JROW=INDEX(L,1)
      JCOLUM=INDEX(L,2)
      DO 200 K=1,N
      SWAP=A(K,JROW)
      A(K,JROW)=A(K,JCOLUM)
      A(K,JCOLUM)=SWAP
200 CONTINUE
205 CONTINUE
      DO 220 K=1,N
      IF(INDEX(K,3)-1) 210,220,210
210 ID = 2
      GO TO 240
220 CONTINUE
      ID=1
240 RETURN
      END
```

SECTION	DISPLACEMENT	MOMENT
1	0.0000	0.0000D 00
2	-0.8730	-0.4500D 07
3	-1.6508	-0.8000D 07
4	-2.2593	-0.1050D 08
5	-2.6455	-0.1200D 08
6	-2.7778	-0.1250D 08
7	-2.6455	-0.1200D 08
8	-2.2593	-0.1050D 08
9	-1.6508	-0.8000D 07
10	-0.8730	-0.4500D 07
11	0.0000	0.0000D 00

COMPILE TIME= 0.54 SEC, EXECUTION TIME= 2.88 SEC

```
C*******PROGRAM : 2.3
C*******SOLVING THE BEAM PROBLEM OF EXAMPLE 2.10
C*******BY  GAUSS ELIMINATION METHOD.
        IMPLICIT REAL*8(A-H,O-Z)
        DIMENSION AK(22,22),F(22),AX(22),STR(11)
        DATA AK,F/484*0.0D0,22*0.0D0/
        D = 30.0
        E = 2.1E07
        B = 10.0
        AL = 100.0
        Q = 100.0
        A = 1000.0
        DEL = AL/A
        AI = B*D**3/12.0
        AF = Q*A**3/(E*AI)
        AM = E*AI/A
        AS = D/(2.*AI)
        AK(1,1) = 1.0
        AK(2,2) = 1.0
        AK(21,21) = 1.0
        AK(22,22) = 1.0
        DO 20 I = 3,19,2
        AK(I,I-2) = 1.0/(DEL*DEL)
        AK(I,I) = -2.0/(DEL*DEL)
        AK(I,I+1) = 1.0
        AK(I,I+2) = AK(I,I-2)
        AK(I+1,I-1) = AK(I,I-2)
        AK(I+1,I+3) = AK(I,I-2)
20      AK(I+1,I+1) = AK(I,I)
        DO 30 I = 4,20,2
30      F(I) = AF
        CALL GAUSS(AK,22,F,1,AX)
        WRITE(6,50)
50      FORMAT(/2X,'SECTION',2X,'DISPLACEMENT',5X,
     1  'MOMENT')
        J = 1
        DO 70 I = 1,21,2
        AX(I+1) = AX(I+1)*AM
        AX(I) = AX(I)*A
        WRITE(6,75)J,AX(I),AX(I+1)
        J = J+1
70      CONTINUE
```

```
75      FORMAT(2X,I2,8X,F7.4,5X,D11.4)
        STOP
        END

        SUBROUTINE GAUSS(B,M,G,IC,QC)
C******* METHOD AND PURPOSE:-
C*******      TO SOLVE FOR A SYSTEM OF LINEAR
C*******      EQUATIONS OF THE FORM    B*QC=G
C*******      BY GAUSS ELIMINATION METHOD.
C******* DESCRIPTION OF PARAMETERS:-
C******* B    MATRIX OF COEFFICIENTS.THESE ARE
C*******      DESTROYED IN THE COMPUTATION.
C*******      THE SIZE OF MATRIX B IS M BY M.
C******* M    NUMBER OF EQUATIONS
C******* G    MATRIX OF CONSTANTS OF
C*******      SIZE M BY IC
C******* IC   NUMBER OF COLUMNS OF G
C******* REMARKS:-
C*******      MATRIX B MUST BE GENERAL.THE METHOD
C*******      FAILS IF IN THE REARRANGED MATRIX B,
C*******      ONE OF THE PIVOTAL ELEMENTS IS ZERO.
C*******      THIS CONDITION IS INDICATED BY THE
C*******      ERROR MESSAGE ,'METHOD FAILS'
C******* SUBROUTINES REQUIRED:
C*******           NONE
C******* WORKING VECTORS:
C*******      JCOL(M),Y(M)--TO BE DIMENSIONED
C*******      APPROPRIATELY
C*******
        IMPLICIT REAL*8(A-H,O-Z)
        DIMENSION B(M,M),G(M,IC),QC(M,IC)
     *  ,JCOL(200),Y(200)
        ABS(X)=DABS(X)
        M1=M-1
        DO 10 I=1,M
10      JCOL(I)=I
        DO 180 I=1,M1
        II=I+1
C*******FIND PIVOT ELEMENT
        IROW=I
        ICOL=I
        BIG=0.0
        DO 30 J=I,M
```

```
          DO 30 K=I,M
          AB=ABS(B(J,K))
          IF(BIG-AB)20,30,30
20        BIG=AB
          IROW=J
          ICOL=K
30        CONTINUE
          IF(IROW-I) 80,80,40
C*******EXCHANGE ROW IROW WITH ROW I
40        DO 50 J=I,M
          TEMP=B(I,J)
          B(I,J)=B(IROW,J)
50        B(IROW,J)=TEMP
60        DO 70 J=1,IC
          TEMP=G(I,J)
          G(I,J)=G(IROW,J)
70        G(IROW,J)=TEMP
80        IF(ICOL-I) 110,110,90
C*******EXCHANGE COLUMN ICOL WITH COLUMN I
90        DO 100 J=1,M
          TEMP=B(J,I)
          B(J,I)=B(J,ICOL)
100       B(J,ICOL) =TEMP
          J=JCOL(I)
          JCOL(I)=JCOL(ICOL)
          JCOL(ICOL)=J
110       IF(B(I,I))120,240,120
120       RATIO=1.0/B(I,I)
C*******ELIMINATE COLUMNWISE
          DO 140 K=II,M
C*******DIVIDE ROW I BY PIVOT
          B(I,K)=B(I,K)*RATIO
          DO 140 J=II,M
          TERM=B(J,I)*B(I,K)
          B(J,K)=B(J,K)-TERM
          DIF=1.0E12*ABS(B(J,K))-ABS(TERM)
          IF(DIF) 130,130,140
130       B(J,K)=0.0
140       CONTINUE
150       DO 170 K=1,IC
          G(I,K)=G(I,K)*RATIO
          DO 170 J=II,M
```

```
         TERM=B(J,I)*G(I,K)
         G(J,K)=G(J,K)-TERM
         DIF=1.0E12*ABS(G(J,K))-ABS(TERM)
         IF(DIF) 160,160,170
160      G(J,K)=0.0
170      CONTINUE
180      CONTINUE
         IF(B(M,M)) 190,240,190
C*******SOLVE BY BACKSUBSTITUTION
190      RATIO=1.0/B(M,M)
200      DO 230 K=1,IC
         Y(M)=G(M,K)*RATIO
         DO 210 LL=1,M1
         I=M-LL
         II=I+1
         Y(I)=G(I,K)
         DO 210 J=II,M
210      Y(I)=Y(I)-B(I,J)*Y(J)
         DO 220 I=1,M
         J=JCOL(I)
220      QC(J,K)=Y(I)
230      CONTINUE
         GO TO 260
240      WRITE(6,250)
250      FORMAT(//,5X,'METHOD FAILS',//)
260      CONTINUE
         RETURN
         END
```

SECTION	DISPLACEMENT	MOMENT
1	0.0000	0.0000D 00
2	-0.8730	-0.4500D 07
3	-1.6508	-0.8000D 07
4	-2.2593	-0.1050D 08
5	-2.6455	-0.1200D 08
6	-2.7778	-0.1250D 08
7	-2.6455	-0.1200D 08
8	-2.2593	-0.1050D 08
9	-1.6508	-0.8000D 07
10	-0.8730	-0.4500D 07
11	0.0000	0.0000D 00

COMPILE TIME=0.58 SEC,EXECUTION TIME=2.68 SEC

```
C******* PROGRAM : 2.4
C******* SOLVING  HE BEAM PROBLEM OF EXAMPLE C*******
2.10 BY CROUT'S PROCEDURE.
        IMPLICIT REAL*8(A-H,O-Z)
        DIMENSION AK(22,23),AX(22)
        DATA AK/506*0.0D0/
        D = 30.0
        E = 2.1E07
        B = 10.0
        AL = 100.0
        Q = 100.0
        A = 1000.0
        DEL = AL/A
        AI = B*D**3/12.0
        AF = Q*A**3/(E*AI)
        AM = E*AI/A
        AK(1,1) = 1.0
        AK(2,2) = 1.0
        AK(21,21) = 1.0
        AK(22,22) = 1.0
        DO 20 I = 3,19,2
        AK(I,I-2) = 1.0/(DEL*DEL)
        AK(I,I) = -2.0/(DEL*DEL)
        AK(I,I+1) = 1.0
        AK(I,I+2) = AK(I,I-2)
        AK(I+1,I-1) = AK(I,I-2)
        AK(I+1,I+3) = AK(I,I-2)
20      AK(I+1,I+1) = AK(I,I)
        DO 30 I = 4,20,2
30      AK(I,23) = AF
        CALL CROUT(AK,AX,22,23)
        WRITE(6,50)
50      FORMAT(/2X,'SECTION',2X,'DISPLACEMENT',5X,
     1  'MOMENT')
        J = 1
        DO 70 I = 1,21,2
        AX(I+1) = AX(I+1)*AM
        AX(I) = AX(I)*A
        WRITE(6,75)J,AX(I),AX(I+1)
        J = J+1
70      CONTINUE
75      FORMAT(2X,I2,8X,F7.4,5X,D11.4)
```

```
        STOP
        END

        SUBROUTINE CROUT(A,X,N,NA)
C******* METHOD:-
C*******    CROUT'S METHOD FOR SOLVING EQUATIONS
C******* DESCRIPTION OF PARAMETERS:-
C******* A    AUGMENTED MATRIX OF THE COEFFICIENT
C*******      MATRIX AND THE RIGHT SIDE VECTOR OF
C*******      FORCES. IT IS OF SIZE N BY NA.
C*******      THIS IS DESTROYED IN COMPUTATION.
C******* X    THE RESULTANT SOLUTION VECTOR
C*******      (DISPLACEMENTS) OF SIZE N.
C******* N    NUMBER OF EQUATIONS.
C******* NA   N + 1
C*******
C******* REMARKS:-
C*******      SPECIALLY SUITED FOR UNSYMMETRIC
C*******      MATRICES.
C*******SUBROUTINE &FUNCTION SUBPROGRAM REQUIRED
C******* NONE.
C*******
        IMPLICIT REAL*8(A-H,O-Z)
        DIMENSION A(N,NA),X(N)
        N1 = N- 1
        NN = N + 1
C*******CALCULATE ELEMENTS IN FIRST ROW
        RATIO = 1./A(1,1)
        DO 10 J = 2,NN
 10     A(1,J) = A(1,J)*RATIO
        DO 50 J = 2,N
C******CALCULATE ELEMENTS IN COLUMNS 2ND ONWARDS
        J1 = J - 1
        JJ = J + 1
        DO 20 I = J,N
        DO 20 K = 1,J1
 20     A(I,J) = A(I,J) - A(I,K)*A(K,J)
C*******CALCULATES ELEMENTS IN ROWS 2ND ONWARDS
        RATIO = 1./A(J,J)
        DO 40 I = JJ,NN
        DO 30 K = 1,J1
 30     A(J,I) = A(J,I) - A(J,K)*A(K,I)
 40     A(J,I) = A(J,I)*RATIO
```

```
50      CONTINUE
C*******X(I) BY BACK SUBSTITUTION
        X(N) = A(N,NN)
        DO 60 L=1,N1
        I = N-L
        II = I + 1
        X(I) = A(I,NN)
        DO 60 J = II,N
60      X(I) = X(I) - A(I,J)*X(J)
        RETURN
        END
```

SECTION	DISPLACEMENT	MOMENT
1	0.0000	0.0000D 00
2	-0.8730	-0.4500D 07
3	-1.6508	-0.8000D 07
4	-2.2593	-0.1050D 08
5	-2.6455	-0.1200D 08
6	-2.7778	-0.1250D 08
7	-2.6455	-0.1200D 08
8	-2.2593	-0.1050D 08
9	-1.6508	-0.8000D 07
10	-0.8730	-0.4500D 07
11	0.0000	0.0000D 00

COMPILE TIME= 0.36 SEC, EXECUTION TIME= 0.85 SEC

```
C****** PROGRAM :2.5
C****** SOLVING THE BEAM PROBLEM OF EX.2.10 BY
C****** POTTERS' METHOD.
        IMPLICIT REAL*8(A-H,O-Z)
        DIMENSION B(2,2),A(2,2),C(2,2),G(2)
        NPART=11
        IN2=2
        REWIND IN2
        M=2
        ISYM=0
        D = 30.0
        E = 2.1E07
        B1 = 10.0
        AL = 100.0
        Q = 100.0
        A1 = 1000.0
        DEL = AL/A1
        DEL2=1.0/(DEL*DEL)
        AI = B1*D**3/12.0
        AF = Q*A1**3/(E*AI)
        AM = E*AI/A1
        WRITE(6,110)
        WRITE(6,120)
        WRITE(6,130)
        WRITE(6,120)
        DO 70 K=1,NPART
        DO 10 I=1,M
        G(I)=0.0
        DO 10 J=1,M
        A(I,J)=0.0
        B(I,J)=0.0
10      C(I,J)=0.0
        IF(K.GT.1)GOTO20
        B(1,1)=1.0
        B(2,2)=1.0
        GOTO40
20      IF(K.EQ.NPART)GOTO30
        C(1,1)=1.0*DEL2
        B(1,1)=-2.0*DEL2
        B(1,2)=1.0
        A(1,1)=C(1,1)
        C(2,2)=C(1,1)
```

```
            B(2,2)=B(1,1)
            A(2,2)=C(2,2)
            G(2)=AF
            GOTO40
30          B(1,1)=1.0
            B(2,2)=1.0
40          IF(K.GT.1)WRITE(IN2)((C(I,J),J=1,M),I=1,M)
            WRITE(IN2)((B(I,J),J=1,M),I=1,M )
            IF(K.LT.NPART)WRITE(IN2)((A(I,J),J=1,M),
     1      I=1,M)
            WRITE(IN2)(G(I),I=1,M)
            DO 50 I=1,M
50          WRITE(6,60)(C(I,J),J=1,M),(B(I,J),J=1,M),
     1      (A(I,J),J=1,M),G(I)
60          FORMAT(13X,7F6.1)
70          CONTINUE
            WRITE(6,120)
            WRITE(6,80)
80          FORMAT(/2X,'SECTION',2X,'DISPLACEMENT',5X,
     1      'MOMENT')
            CALL POTTER(NPART,M,ISYM,IN2)
            REWIND IN2
            DO 90 K=1,NPART
            K1=NPART+1-K
            READ(IN2)(G(I),I=1,M)
            G(1)=G(1)*A1
            G(2)=G(2)*AM
            WRITE(6,100)K1,G(1),G(2)
90          CONTINUE
100         FORMAT(2X,I2,8X,F7.4,5X,D11.4)
110         FORMAT(/30X,'INPUT MATRICES')
120         FORMAT(9X,53('-'))
130         FORMAT(18X, 'C(I)',8X,'B(I)',9X, 'A(I)'
     1      5X,'G(I)')
            STOP
            END

            SUBROUTINE POTTER(NPART,M,ISYM,IN2)
C******* METHOD:-
C*******    POTTERS' SCHEME TO SOLVE SIMULTANEOUS
C*******    EQUATIONS OF TYPE K*(X)=F
C******* DESCRIPTION OF PARAMETERS:-
C******* NPART-NUMBER OF BLOCKS.(PARTITIONS)
```

```
C*******M  SIZE OF THE BLOCK. M*NPART=NO.OF EQNS
C*******ISYM- PARAMETER SPECIFYING THE SYMMETRY
C*******       CONDITION.
C*******0 -UNSYMMETRIC CASE. & 1 -SYMMETRIC CASE
C******* IN1-REF.FILE IN WHICH P(I),Q(I) ARE STORED
C******* IN2 REF.FILE IN WHICH A(I),B(I),C(I)
C******* ARE STORED
C*******  P,CP,CB ARE OF WORKING MATRICES OF
C*******  SIZE M BY M.
C*******  DC,D,Q,CQ,DIS,PZ ARE VECTORS OF SIZE M
C*******  ST IS OF SIZE (M,2*M)
C******* REMARKS:-
C*******  1. THE STIFFNESS MATRIX IS WRITTEN IN
C*******      TERMS OF BLOCKS  A(I),B(I),C(I) AS
C*******      EXPLAINED IN THE TEXT.
C*******  2. THE FORCE MATRIX 'F' IS ALSO STORED
C*******      IN TERMS OF BLOCKS AS V(I).
C*******  3. THE MATRICES A(I),B(I),C(I),V(I)
C*******      ARE TO BE STORED IN REF.FILE IN2
C*******      THROUGH THE MAIN PROGRAM.
C*******  4. THE PROGRAM WORKS ONLY FOR
C*******      PARTITIONS OF EQUAL SIZE
C*******  5. THE FILE STORAGE ORDER IS AS FOLLOWS:
C*******          SYMM.CASE        UNSYMM.CASE
C*******            B(1)              B(1)
C*******            A(1)              A(1)
C*******            G(1)              G(1)
C*******                              C(2)
C*******            B(2)              B(2)
C*******            A(2)              A(2)
C*******            G(2)              G(2)
C*******             .                 .
C*******             .                 .
C*******             .                 .
C*******             .                 .
C*******                              C(N-1)
C*******            B(N-1)            B(N-1)
C*******            A(N-1)            A(N-1)
C*******            G(N-1)            G(N-1)
C*******                              C(N)
C*******            B(N)              B(N)
C*******            G(N)              G(N)
```

```
C*******    6. THE BLOCKS A(I),B(I),C(I),G(I) ARE
C*******       DESTROYED DURING COMPUTATION.
C*******       THE FINAL RESULTS ARE STORED
C*******       BLOCKWISE IN  THE REF.FILE IN2
C*******    7. THE PROGRAM OPERATES IN DOUBLE
C*******       PRECISION ARITHMETIC
C*******SUBROUTINE &FUNCTION SUBPROGRAM REQUIRED
C*******       INVERT.
        IMPLICIT REAL*8 (A-H,O-Z)
        DIMENSION P(2,2),CP(2,2),CB(2,2),DC(2),
      1 D(2),Q(2),CQ(2),DIS(2),
      2 ST(2,4),PZ(2)
        IN1 = 1
        REWIND IN2
        N = M
        MO = M
        M1=M+1
        M2 = 2*M
        DO 310 K = 1,NPART
        READ(IN2)((ST(I,J),J=1,M),I=1,M)
        IF(K.LT.NPART)READ(IN2)((ST(I,J),
      1 J=M1,M2),I=1,M)
        READ(IN2)(D(I),I=1,M)
 110    IF(K-1) 160,120,160
C*******FINDS P1 & Q1
 120    DO 130  I = 1,M
        DO 130 J = 1,M
 130    CB(I,J) = ST(I,J)
        CALL INVERT(CB,M)
        DO 140 I = 1,M
        DO 140 J = 1,N
        KK=J+M
        P(I,J)=0.0
        DO 140 L = 1,M
 140    P(I,J) = P(I,J) + CB(I,L)*ST(L,KK)
        DO 150 I = 1,M
        Q(I)=0.0
        DO 150 L = 1,M
 150    Q(I) = Q(I) + CB(I,L)*D(L)
        GO TO 240
 160    IF(K-NPART)  170,310,170
C*******FINDS P2,Q2 UP TO P(NPART-1),Q(NPART-1)
```

```
170     DO 180 I = 1,M
        DO 180 J = 1,M
        CP(I,J) = 0.
        DO 180 L = 1,MO
180     CP(I,J) = CP(I,J) + CB(I,L)*P(L,J)
        DO 190 I = 1,M
        CQ(I) = 0.
        DO 190 L = 1,MO
190     CQ(I) = CQ(I) + CB(I,L)*Q(L)
        DO 200 I = 1,M
        DIS(I) = D(I)-CQ(I)
        DO 200 J = 1,M
200     CB(I,J) = ST(I,J)-CP(I,J)
        CALL INVERT(CB,M)
        DO 210 I = 1,M
        DO 210 J = 1,N
        P(I,J) = 0.0
        KK = J+M
        DO 210 L = 1,M
210     P(I,J) = P(I,J) + CB(I,L)*ST(L,KK)
        DO 220 I = 1,M
        Q(I) = 0.
        DO 220 L = 1,M
220     Q(I) = Q(I) + CB(I,L)*DIS(L)
230     CONTINUE
240     WRITE(IN1)M,N
        WRITE(IN1)((P(I,J),I=1,M),J=1,N),
      1 (Q(I),I=1,M)
        IF(ISYM-1) 290,270,270
270     DO 280 I = 1,N
        KK = I+M
        DO 280 J = 1,M
280     CB(I,J) = ST(J,KK)
        GO TO 310
290     READ(IN2)((CB(I,J),J=1,M),I=1,M)
310     CONTINUE
C*******SOLUTION PART
        REWIND IN2
        DO 420 K = 1,NPART
        IF(K-1) 320,320,370
320     DO 330 I = 1,M
        DO 330 J = 1,M
```

```
          CP(I,J) = 0.
          DO 330 L = 1,MO
 330      CP(I,J) = CP(I,J) + CB(I,L)*P(L,J)
          DO 340 I = 1,M
          DC(I) = 0.
          DO 340 L = 1,MO
 340      DC(I) = DC(I) + CB(I,L)*Q(L)
          DO 350 I = 1,M
          PZ(I) = D(I)-DC(I)
          DO 350 J = 1,M
 350      CB(I,J) = ST(I,J)-CP(I,J)
          CALL INVERT(CB,M)
          DO 360 I = 1,M
          DIS(I) = 0.
          DO 360 L = 1,M
 360      DIS(I) = DIS(I) + CB(I,L)*PZ(L)
          GO TO 400
 370      READ(IN1)M,N
          READ(IN1)((P(I,J),I=1,M),J=1,N),
        1 (Q(I),I=1,M)
          BACKSPACE IN1
          BACKSPACEIN1
C*******FINDS Z(NPART-1) TO Z1
          DO 380 I = 1,M
          CQ(I) = 0.
          DO 380 L = 1,N
 380      CQ(I) = CQ(I) - P(I,L)*DIS(L)
          DO 390 I = 1,M
 390      DIS(I) = Q(I) + CQ(I)
 400      CONTINUE
          WRITE(IN2)(DIS(I),I=1,M)
          BACKSPACE IN1
          BACKSPACE IN1
 420      CONTINUE
          RETURN
          END
```

INPUT MATRICES

C(I)		B(I)		A(I)		G(I)
0.0	0.0	1.0	0.0	0.0	0.0	0.0
0.0	0.0	0.0	1.0	0.0	0.0	0.0
100.0	0.0	-200.0	1.0	100.0	0.0	0.0
0.0	100.0	0.0	-200.0	0.0	100.0	0.2
100.0	0.0	-200.0	1.0	100.0	0.0	0.0
0.0	100.0	0.0	-200.0	0.0	100.0	0.2
100.0	0.0	-200.0	1.0	100.0	0.0	0.0
0.0	100.0	0.0	-200.0	0.0	100.0	0.2
100.0	0.0	-200.0	1.0	100.0	0.0	0.0
0.0	100.0	0.0	-200.0	0.0	100.0	0.2
100.0	0.0	-200.0	1.0	100.0	0.0	0.0
0.0	100.0	0.0	-200.0	0.0	100.0	0.2
100.0	0.0	-200.0	1.0	100.0	0.0	0.0
0.0	100.0	0.0	-200.0	0.0	100.0	0.2
100.0	0.0	-200.0	1.0	100.0	0.0	0.0
0.0	100.0	0.0	-200.0	0.0	100.0	0.2
100.0	0.0	-200.0	1.0	100.0	0.0	0.0
0.0	100.0	0.0	-200.0	0.0	100.0	0.2
100.0	0.0	-200.0	1.0	100.0	0.0	0.0
0.0	100.0	0.0	-200.0	0.0	100.0	0.2
0.0	0.0	1.0	0.0	0.0	0.0	0.0
0.0	0.0	0.0	1.0	0.0	0.0	0.0

SECTION	DISPLACEMENT	MOMENT
11	0.0000	0.0000D 00
10	-0.8730	-0.4500D 07
9	-1.6508	-0.8000D 07
8	-2.2593	-0.1050D 08
7	-2.6455	-0.1200D 08
6	-2.7778	-0.1250D 08
5	-2.6455	-0.1200D 08
4	-2.2593	-0.1050D 08
3	-1.6508	-0.8000D 07
2	-0.8730	-0.4500D 07
1	0.0000	0.0000D 00

COMPILE TIME=1.03 SEC,EXECUTION TIME=0.64 SEC

```
C***** PROGRAM:2.6
C***** SOLVING THE RING PROBLEM OF EXAMPLE 2.12
C      BY FRONTAL TECHNIQUE
       IMPLICIT REAL*8(A-H,O-Z)
       DIMENSION A(8,3),B(8)
       READ 100,N,KB
       READ 200,((A(I,J),J=1,KB),I=1,N)
       READ 200,(B(I),I=1,N)
       PRINT 300,N,KB
       PRINT 350
       PRINT 400,((I,J,A(I,J),J=1,KB),I=1,N)
       PRINT 450
       PRINT 500,(I,B(I),I=1,N)
   100 FORMAT(14I5)
   200 FORMAT(3F10.3)
   300 FORMAT(//,10X,'N=',I3,10X,'KB=',I3,//)
   350 FORMAT(//,10X,'STIFNESS MATRIX',//)
   400 FORMAT(2(2X,'A(',I2,',',I2,')=',D13.5))
   450 FORMAT(//,10X,'LOAD VECTOR',//)
   500 FORMAT(10X,'B(',I2,')=',D15.7)
       CALL UPTRI(A,N,KB)
       CALL SOLUT(A,N,KB,B)
       CALL SOLU(A,N,KB,B)
       PRINT 600
       PRINT 500,(I,B(I),I=1,N)
   600 FORMAT(//,10X,'DISPLACEMENT VECTOR',//)
       STOP
       END
       SUBROUTINE UPTRI(CB,N,KBAND)
C***** METHOD & PURPOSE:-
C***** UPPER TRIANGULARISATION USING
C***** CHOLESKY'S TRANSFORMATION(UT.U)
C***** DESCRIPTION OF PARAMETERS
C***** CB MATRIX TO BE DECOMPOSED IN COMPUTATION
C***** N         NUMBER OF EQUATIONS
C***** KB        BANDWIDTH
C***** REMARKS:-
C***** THE PROGRAM OPERATES IN DOUBLE
C***** PRECISION ARITHMETIC. APPLICABLE FOR
C*****SYMMETRIC POSITIVE DEFINITE MATRICES ONLY.
       IMPLICIT REAL*8(A-H,O-Z)
       DIMENSION CB(N,KBAND)
```

```
      DO 100 I=1,N
      NI=N-I+1
      IN=I-1
      IF(KBAND-NI)10,20,20
   10 NI=KBAND
   20 DO 100 J=1,NI
      KBJ=KBAND-J
      IF(IN-KBJ)30,40,40
   30 KBJ=IN
   40 SUM=CB(I,J)
      IF(KBJ-1)70,50,50
   50 DO 60 K=1,KBJ
      KK=K+1
      KI=I-K
      JK=J+K
   60 SUM=SUM-CB(KI,KK)*CB(KI,JK)
   70 IF(J-1)80,90,80
   80 CB(I,J)=SUM*TEMP
      GO TO 100
   90 CB(I,J)=DSQRT(SUM)
      TEMP=1./CB(I,J)
  100 CONTINUE
      RETURN
      END
      SUBROUTINE SOLU(CB,N,KBAND,Q)
C***** METHOD & PURPOSE:-
C***** SOLVING CB(X)=Q BY BACK SUBSTITUTION
C***** DESCRIPTION OF PARAMETERS:
C***** CB        UPPERTRIANGULAR MATRIX IN
C*****           BANDED FORM
C***** N         NUMBER OF EQUATIONS
C***** KBAND     BANDWIDTH
C***** Q         FORCE VECTOR OF SIZE N,DESTROYED
C*****           IN COMPUTATION AND REPLACED BY
C*****           THE RESULTANT DISPLACEMENTS
      IMPLICIT REAL*8(A-H,O-Z)
      DIMENSION CB(N,KBAND),Q(N)
      DO 60 L=1,N
      I=N-L+1
      J=I+KBAND-1
      IF(J-N)20,20,10
   10 J=N
```

```
 20    SUM=Q(I)
       II=I+1
       IF(II-J) 30,30,50
 30    DO 40 K=II,J
       KK=K-I+1
 40    SUM=SUM-CB(I,KK)*Q(K)
 50    Q(I)=SUM/CB(I,1)
 60    CONTINUE
       RETURN
       END
       SUBROUTINE SOLUT(CB,N,KBAND,Q)
C**** METHOD & PURPOSE:-
C**** SOLVING CB(X)=Q BY FORWARD SUBSTITUTION
C**** DESCRIPTION OF PARAMETERS:-
C**** CB        UPPER TRIANGULAR MATRIX
C****           IN BANDED FORM
C**** N         NUMBER OF EQUATIONS
C**** KBAND     BANDWIDTH
C**** Q         FORCE VECTOR OF SIZE N,DESTROYED
C****           IN COMPUTATION AND REPLACED BY
C****           THE RESULTANT DISPLACEMENTS
       IMPLICIT REAL*8(A-H,O-Z)
       DIMENSION CB(N,KBAND),Q(N)
       DO 60 I=1,N
       J=I-KBAND+1
       II=I+1
       IF(II-KBAND) 10,10,20
 10    J=1
 20    SUM=Q(I)
       IN=I-1
       IF(IN-J) 50,30,30
 30    DO 40 K=J,IN
       IK=I-K+1
 40    SUM=SUM-CB(K,IK)*Q(K)
 50    Q(I)=SUM/CB(I,1)
 60    CONTINUE
       RETURN
       END

       N= 8          KB= 3
```

STIFNESS MATRIX

```
A( 1, 1)=  0.10000D 01   A( 1, 2)=  0.10000D 00
A( 1, 3)=  0.10000D 00   A( 2, 1)=  0.20000D 01
A( 2, 2)=  0.00000D 00   A( 2, 3)=  0.20000D 00
A( 3, 1)=  0.30000D 01   A( 3, 2)=  0.00000D 00
A( 3, 3)=  0.30000D 00   A( 4, 1)=  0.40000D 01
A( 4, 2)=  0.00000D 00   A( 4, 3)=  0.40000D 00
A( 5, 1)=  0.30000D 01   A( 5, 2)=  0.00000D 00
A( 5, 3)=  0.30000D 00   A( 6, 1)=  0.20000D 01
A( 6, 2)=  0.00000D 00   A( 6, 3)=  0.20000D 00
A( 7, 1)=  0.20000D 01   A( 7, 2)=  0.20000D 00
A( 7, 3)=  0.00000D 00   A( 8, 1)=  0.10000D 01
A( 8, 2)=  0.00000D 00   A( 8, 3)=  0.00000D 00
```

LOAD VECTOR

```
B( 1)=  0.1400000D 01
B( 2)=  0.4700000D 01
B( 3)=  0.7000000D 01
B( 4)=  0.1400000D 02
B( 5)=  0.1080000D 02
B( 6)=  0.1020000D 02
B( 7)=  0.9900000D 01
B( 8)=  0.6600000D 01
```

DISPLACEMENT VECTOR

```
B( 1)=  0.1000000D 01
B( 2)=  0.2000000D 01
B( 3)=  0.2000000D 01
B( 4)=  0.3000000D 01
B( 5)=  0.3000000D 01
B( 6)=  0.4000000D 01
B( 7)=  0.4000000D 01
B( 8)=  0.5000000D 01
```

COMPILE TIME= 0.37 SEC,EXECUTION TIME= 0.17 SEC

```
C******PROGRAM: 2.7
C******SOLVING THE RING PROBLEM OF EXAMPLE 2.12
C******USING VARIABLE BANDWIDTH SCHEME
C******DESCRIPTION OF PARAMETERS
C******N      NUMBERS OF EQUATIONS
C******A(NN) ARRAY CONTAINING VARIABLE BANDWIDTH
C******      MATRIX
C******KDIAG(N) ARRAY OF ADDRESSES WITHIN A OF
C****          THE DIAGONAL ELEMENTS
C**** I,J      ROW AND COLUMN INDICES
C**** L        COLUMN NUMBER FIRST STORED
C****          ELEMENT IN ROW I
C****LBAR   COLUMN NUMBER TO START THE SUMMATION
C****KI,KJ   FICTITIOUS ADRESSES OF ELEMENTS
C****K        COLUMN NUMBER FOR SUMMATION
C****SUBROUTINES & FUNCTION SUBPROGRAMS REQUIRED
C****          NONE
C****
      IMPLICIT REAL*8(A-H,O-Z)
      DIMENSION A(100),B(10),KDIAG(10)
      READ 100,N,NN
      READ 200,(A(I),I=1,NN)
      READ 200,(B(I),I=1,N)
      READ 100,(KDIAG(I),I=1,N)
  100 FORMAT(14I5)
  200 FORMAT(7F10.2)
      PRINT 300,N,NN
      PRINT 350
      PRINT 400,(I,A(I),I=1,NN)
  300 FORMAT(//,10X,'N=',I3,10X,'NN=',I3,//)
  350 FORMAT(//,10X,'STIFFNESS MATRIX',//)
  400 FORMAT(2(2X,'A(',I2,')=',D15.7))
      PRINT 450
  450 FORMAT(//,10X,'LOAD VECTOR',//)
      PRINT 500,(I,B(I),I=1,N)
  500 FORMAT(10X,'B(',I2,')=',D15.7)
C**** VARIABLE BANDWIDTH CHOLESKY DECOMPOSITION
      A(1)=DSQRT(A(1))
      DO 510 I=2,N
      KI=KDIAG(I)-I
      L=KDIAG(I-1)-KI+1
      DO 520 J=L,I
```

```
       X=A(KI+J)
       KJ=KDIAG(J)-J
       IF(J.EQ.1) GO TO 520
       LBAR=KDIAG(J-1)-KJ+1
       LBAR=MAX0(L,LBAR)
       IF(LBAR.EQ.J) GO TO 520
       J1=J-1
       DO 530 K=LBAR,J1
 530   X=X-A(KI+K)*A(KJ+K)
 520   A(KI+J)=X/A(KJ+J)
 510   A(KI+I)=DSQRT(X)
C******* FORWARD SUBSTITUTION
       B(1)=B(1)/A(1)
       DO 540 I=2,N
       KI=KDIAG(I)-I
       L=KDIAG(I-1)-KI+1
       X=B(I)
       IF(L.EQ.I) GO TO 540
       I1=I-1
       DO 550 J=L,I1
 550   X=X-A(KI+J)*B(J)
 540   B(I)=X/A(KI+I)
C******* BACK SUBSTITUTION
       DO560  IT=2,N
       I=N+2-IT
       KI=KDIAG(I)-I
       X=B(I)/A(KI+I)
       B(I)=X
       L=KDIAG(I-1)-KI+1
       IF(L.EQ.I) GO TO 560
       I1=I-1
       DO 570 K=L,I1
 570   B(K)=B(K)-X*A(KI+K)
 560   CONTINUE
       B(1)=B(1)/A(1)
       PRINT 600
 600   FORMAT(//,10X,'DISPLACEMENT VECTOR',//)
       PRINT 500,(I,B(I),I=1,N)
       STOP
       END


       N= 8          NN= 21
```

STIFFNESS MATRIX

```
A( 1)=  0.1000000D 01   A( 2)=  0.1000000D 00
A( 3)=  0.2000000D 01   A( 4)=  0.2000000D 00
A( 5)=  0.4000000D 01   A( 6)=  0.4000000D 00
A( 7)=  0.2000000D 01   A( 8)=  0.2000000D 00
A( 9)=  0.1000000D 01   A(10)=  0.2000000D 00
A(11)=  0.2000000D 01   A(12)=  0.3000000D 00
A(13)=  0.3000000D 01   A(14)=  0.1000000D 00
A(15)=  0.0000000D 00   A(16)=  0.0000000D 00
A(17)=  0.0000000D 00   A(18)=  0.0000000D 00
A(19)=  0.0000000D 00   A(20)=  0.3000000D 00
A(21)=  0.3000000D 01
```

LOAD VECTOR

```
B( 1)=  0.1400000D 01
B( 2)=  0.4700000D 01
B( 3)=  0.1400000D 02
B( 4)=  0.1020000D 02
B( 5)=  0.6600000D 01
B( 6)=  0.9900000D 01
B( 7)=  0.1080000D 02
B( 8)=  0.7000000D 01
```

DISPLACEMENT VECTOR

```
B( 1)=  0.1000000D 01
B( 2)=  0.2000000D 01
B( 3)=  0.3000000D 01
B( 4)=  0.4000000D 01
B( 5)=  0.5000000D 01
B( 6)=  0.4000000D 01
B( 7)=  0.3000000D 01
B( 8)=  0.2000000D 01
```

COMPILE TIME= 0.31 SEC,EXECUTION TIME=0.17 SEC

```
C***PROGRAM : 2.8
C***SOLVING THE E.O.T CRANE PROBLEM BY
C***VARIABLE BANDWIDTH METHOD
C***A PORTAL FRAME, THE VERTICAL MEMBERS OF
C***HEIGHT 10M.(6 ELELMENTS) EACH, HAVING BOX
C***SECTION 1M X 1M AND THICKNESS 10MM AND A
C***HORIZONTAL MEMBER OF LENGTH 20M.(9 ELEMENTS)
C***HAVING BOX SECTION 1M X 2M & THICKNESS 10 MM
C***IS ANALYSED FOR DEFLECTION AS A PLANE FRAME.
C***A LOAD OF 40T IS APPLIED AT 2/3 RD SPAN.
C***NNOD      NUMBER OF NODES OF THE STRUCTURE
C***NELEM     NUMBER OF ELEMENTS OF THE STRUCTURE
C***NFREE     GLOBAL D.O.F/NODE
C***NNOEL     NUMBER OF NODES/ELEMENT
C***ID(I,J)   BOUNDARY CONDITION CODE
C***NEQ       FINAL NO. OF EQUATIONS TO BE SOLVED
C***NOD(I,J)  GLOBAL NODE NUMBER
C***                  I: ELEMENT NO.
C***                  J: LOCAL NODE NO.
C***A(NN)     ARRAY CONTAINING VARIABLE BANDWIDTH MATRIX
C***KDIAG(N)  ARRAY OF ADDRESSES WITHIN A OF THE
C***          DIAGONALELEMENTS
        IMPLICIT REAL*8(A-H,O-Z)
        DIMENSION NOD(21,2),X(22,2),ID(22,3),
     1  A(300),F(60),C(6,6), T(6,6), TCT(6,6),
     1  DIS(6,1) ,KDIAG(60),KCHK(60)
        DATA NNOD,NELEM,NNOEL,NFREE/22,21,2,3/
        DO 10 I=1,NNOD
        DO 10 J=1,NFREE
 10  ID(I,J)=0
        DO 11 I=1,NFREE
        ID(1,I)=1
 11  ID(22,I)=1
        ISUM=0
        DO 40 I=1,NNOD
        DO 40 J=1,NFREE
        IF(ID(I,J)) 30,20,30
 20  ISUM=ISUM+1
        ID(I,J)=ISUM
        GOTO 40
 30  ID(I,J)=0
 40  CONTINUE
```

```
      NEQ=ISUM
      WRITE(*,*) 'THE ID MATRIX IS: '
      WRITE(*,*) '------------------'
      WRITE(*,31) (I,(ID(I,J), J=1,NFREE),
    1 I =1,NNOD)
   31 FORMAT(I5,3I5)
      DO 50 I=1,NELEM
      DO 50 J=1,NNOEL
      I1=I+J-1
   50 NOD(I,J)=I1
      WRITE(*,*) 'THE ELEMENT AND NODE
    1 CONNECTIVITY MATRIX IS: '
      WRITE(*,*) '--------------------------'
      WRITE(*,51) (I,(NOD(I,J),J=1,NNOEL),
    1 I=1,NELEM)
   51    FORMAT(I5,2I5)
      DO 60 I=1,7
      I1=(NNOD+1)-I
      X(I,1)= 0.
      X(I1,1)=20000.
      Y=(I-1)*10000./6.
      X(I,2)=Y
   60 X(I1,2)=Y
      DO 70 I=8,15
      X(I,2)=10000.
      X1 = (I-7)*20000./9.
   70 X(I,1)=X1
      WRITE(*,*) 'THE COORDINATES ARE: '
      WRITE(*,*) '--------------------'
      WRITE(*,71) (I,(X(I,J),J=1,2),I=1,NNOD)
   71 FORMAT(I5,2F10.3)
      DO 80 I=1,NEQ
      F(I)=0.
      KDIAG(I)=0
   80    KCHK(I)=0
      F(41)=-40000.*9.81
      DO 601 I=1,300
  601 A(I)=0.
      LK=0
      KDIAG(1)=1
      DO 202 KL=1,NELEM
      LK = LK + 1
```

```
C       DETERMINE LEAST DOF IN PARTICULAR ELEMENT
        LEAST=500000
        DO 203 J = 1, NNOEL
        M = NOD(LK,J)
        DO 204 I = 1, NFREE
        IF(ID(M,I).EQ.0) GOTO 204
        LEAST1 = ID(M,I)
        IF(LEAST1.GE.LEAST) GOTO 204
        LEAST = LEAST1
   204 CONTINUE
   203 CONTINUE
C       DETERMINE COLUMN HEIGHTS
        DO 205 J = 1, NNOEL
        M = NOD(LK,J)
        DO 205 K = 1, NFREE
        IF(ID(M,K).EQ.0) GOTO 205
        KOUNT=ID(M,K)
        KCHK(KOUNT) = KOUNT-LEAST+1
        IF(KDIAG(KOUNT).GT.KCHK(KOUNT)) GOTO 205
        KDIAG(KOUNT)=KCHK(KOUNT)
   205 CONTINUE
   202 CONTINUE
C       DETERMINE LOCATION OF DIAGONAL TERMS
        DO 206 I=2,NEQ
        KDIAG(I) = KDIAG(I) + KDIAG(I-1)
   206 CONTINUE
        WRITE(*,*) 'KDIAG(1)= ',KDIAG(1)
        WRITE(*,*) 'NEQ= ',NEQ,'KDIAG(NEQ)=',KDIAG(NEQ)
        LK = 0
        DO 214 KL = 1,NELEM
        LK = LK + 1
        CALL FEM(T,C,Z,LK,TCT)
        DO 216 IT = 1,NNOEL
        II = NOD(LK,IT)
        IM = NFREE*(IT-1)
        DO 217 I = 1,NFREE
        MMI = ID(II,I)
        IF(MMI.EQ.0) GOTO 217
        IMI = IM + I
        DO 218 JT = 1,NNOEL
       JJ = NOD(LK,JT)
         JN = NFREE * (JT-1)
```

```
          DO 219 J = 1,NFREE
          NJJ = ID(JJ,J)
          IF(NJJ.EQ.0) GOTO 219
          JNJ = JN + J
          NNJ = NJJ - MMI
          IF(NNJ) 219,220,220
  220 MI = KDIAG(NJJ)
      KK = MI - NNJ
          A(KK) = A(KK) + TCT(IMI,JNJ)
  219 CONTINUE
  218 CONTINUE
  217 CONTINUE
  216 CONTINUE
  214 CONTINUE
          WRITE(*,32) NEQ
   32 FORMAT(12X,'NUMBER OF EQUATIONS= ',I5/)
          CALL VFACT(NEQ,A,KDIAG)
          CALL VFOR(NEQ,A,F,KDIAG)
          CALL VBACK(NEQ,A,F,KDIAG)
          WRITE(*,*) 'THE DISPLACEMENTS ARE:
          WRITE(*,801) (F(I),I=1,60)
  801 FORMAT(3E11.4)
      DO 603 I=1,6
  603 DIS(I,1)=0.
          DO 560 LK=1,NELEM
          DO 610 I=1,NNOEL
          DO 610 J=1,NFREE
          II=(I-1)*NFREE+J
          LKK=NOD(LK,I)
          IJ=ID(LKK,J)
          IF(IJ.EQ.0) DIS(II,1)=0.
          IF(IJ.EQ.0) GOTO 610
          DIS(II,1) = F(IJ)
  610     CONTINUE
          STOP
          END


          SUBROUTINE FEM(T,C,Z,LK,TCT)
C         *****************************************
          IMPLICIT REAL*8(A-H,O-Z)
          DIMENSION C(6,6),T(6,6),TC(6,6),TCT(6,6),TT(6,6)
          IF(LK.LE.6.OR.LK.GE.16) THEN
```

```
          AI=.647E+10
          A=39600.
          Y=500.
          AL=10000./6.
      ELSE
          AI=.327E+11
          A=59600.
          Y=1000.
          AL=20000./9.
      ENDIF
      E=.21E+6
      PI = 4*ATAN(1.0D0)
      IF(LK.LE.6) THEN
          THETA=90.*PI/180.
      ELSEIF(LK.GE.7.AND.LK.LT.16) THEN
          THETA=0.
      ELSE
          THETA=-90.*PI/180.
      ENDIF
      DO 10 I=1,6
      DO 10 J=1,6
      T(I,J)=0.
10 C(I,J)=0.
      T(1,1)=COS(THETA)
      T(1,2)=SIN(THETA)
      T(2,1)=-T(1,2)
      T(2,2)=T(1,1)
      T(3,3)=1.
      DO 20 I=4,6
      I1=I-3
      DO 20 J=4,6
      J1=J-3
20    T(I,J)=T(I1,J1)
      CALL TRANSPOSE(T,TT)
      CON=E*AI/AL
      C(1,1)=A*E/AL
      C(1,4)=-A*E/AL
      C(2,2)=12.*CON/AL**2
      C(2,3)=6.*CON/AL
      C(2,5)=-12.*CON/AL**2
      C(2,6)=6.*CON/AL
      C(3,3)=4.*CON
```

```
        C(3,5)=-6.*CON/AL
        C(3,6)=2.*CON
        C(4,4)=A*E/AL
        C(5,5)=12.*CON/AL**2
        C(5,6)=-6.*CON/AL
        C(6,6)=4.*CON
        DO 30 I=1,6
        DO 30 J=I,6
     30 C(J,I)=C(I,J)
        CALL MULTI(TT,C,6,6,6,TC)
        CALL MULTI(TC,T,6,6,6,TCT)
        Z=AI/Y
        RETURN
        END

        SUBROUTINE TRANSPOSE(A,B)
C       ***************************
        IMPLICIT REAL*8(A-H,O-Z)
        DIMENSION A(6,6),B(6,6)
        DO 10 I=1,6
        DO 10 J=1,6
     10 B(I,J)=A(J,I)
        RETURN
        END

        SUBROUTINE MULTI(A,B,L,M,N,D)
        IMPLICIT REAL*8(A-H,O-Z)
C       *******************************
        DIMENSION A(L,M),B(M,N),D(L,N)
        DO 20 I=1,L
        DO 20 J=1,N
        D(I,J)=0.
        DO 20 K=1,M
     20 D(I,J) = D(I,J)+A(I,K)*B(K,J)
        RETURN
        END
C***************************************************
        SUBROUTINE VFOR(N,A,B,KDIAG)
C***************************************************
C       SUBROUTINE FOR FORWARD SUBSTITUTION
        IMPLICIT REAL*8(A-H,O-Z)
        DIMENSION A(300),B(60),KDIAG(60)
```

```
        B(1)=B(1)/A(1)
        DO 540 I=2,N
        KI=KDIAG(I)-I
        L=KDIAG(I-1)-KI+1
        X=B(I)
        IF(L.EQ.I)GOTO 540
        I1=I-1
        DO 550 J=L,I1
  550   X=X-A(KI+J)*B(J)
  540   B(I)=X/A(KI+I)
        RETURN
        END
C************************************************
        SUBROUTINE VBACK(N,A,B,KDIAG)
C************************************************
C       SUBROUTINE FOR BACKWARD SUBSTITUTION
        IMPLICIT REAL*8(A-H,O-Z)
        DIMENSION A(300),B(60),KDIAG(60)
        DO 560 IT=2,N
        I=N+2-IT
        KI=KDIAG(I)-I
        X=B(I)/A(KI+I)
        B(I)=X
        L=KDIAG(I-1)-KI+1
        IF(L.EQ.I)GOTO 560
        I1=I-1
        DO 570 K=L,I1
  570   B(K)=B(K)-X*A(KI+K)
  560   CONTINUE
        B(1)=B(1)/A(1)
        RETURN
        END
C************************************************
        SUBROUTINE VFACT(N,A,KDIAG)
C************************************************
C       CHOLESKY DECOMPOSITION
C       VARIABLES USED
C       N : NO.OF EQUATIONS
C*****   A:ARRAY CONTAINING THE ELEMENTS OF
C*****   ASSEMBLED  [K]
C       KDIAG : ARRAY CONTAINING THE LOCATION OF
C*****   DIAGONAL TERMS
```

```
      IMPLICIT REAL*8(A-H,O-Z)
      DIMENSION A(300),KDIAG(60)
      A(1)=DSQRT(A(1))
      DO 510 I=2,N
      KI=KDIAG(I)-I
      L=KDIAG(I-1)-KI+1
      DO 520 J=L,I
      X=A(KI+J)
      KJ=KDIAG(J)-J
      IF(J.EQ.1)GOTO 520
      LBAR=KDIAG(J-1)-KJ+1
      LBAR=MAX0(L,LBAR)
      IF(LBAR.EQ.J)GOTO 520
      J1=J-1
      DO 530 K=LBAR,J1
530   X=X-A(KI+K)*A(KJ+K)
520   A(KI+J)=X/A(KJ+J)
510   A(KI+I)=DSQRT(X)
      RETURN
      END
```

THE DISPLACEMENTS ARE:

```
-.8838E-01 -.8412E-02  .9738E-04
-.2957E+00 -.1682E-01  .1427E-03
-.5351E+00 -.2524E-01  .1360E-03
-.7200E+00 -.3365E-01  .7720E-04
-.7636E+00 -.4206E-01 -.3363E-04
-.5790E+00 -.5047E-01 -.1965E-03
-.5835E+00 -.5314E+00 -.2313E-03
-.5880E+00 -.1056E+01 -.2358E-03
-.5926E+00 -.1557E+01 -.2102E-03
-.5971E+00 -.1968E+01 -.1544E-03
-.6016E+00 -.2221E+01 -.6842E-04
-.6061E+00 -.2250E+01  .4775E-04
-.6106E+00 -.1986E+01  .1941E-03
-.6152E+00 -.1365E+01  .3706E-03
-.6197E+00 -.4214E+00  .4363E-03
-.6885E-01 -.3512E+00  .2334E-03
 .1873E+00 -.2809E+00  .8262E-04
 .2355E+00 -.2107E+00 -.1611E-04
 .1625E+00 -.1405E+00 -.6279E-04
 .5508E-01 -.7023E-01 -.5742E-04
```

```
C****   PROGRAM :2.9
C****   SOLVING THE BOILER FRAME PROBLEM OF
C****   EXAMPLE 2.11 BY POTTERS' METHOD.
C****   VARIABLES USED:
C****
C****   NPART      NUMBER OF PARTITIONS
C****   NBLK       SIZE OF EACH PARTITION
C****   NFTNOD     FIRST NODE          ]
C****   NLTNOD     LAST NODE           ]  OF
C****   NFTELM     STARTING ELEMENT    ] EACH
C****   NLTELM     ENDING ELEMENT      ] PARTITION
C****   ID(I,J)    BOUNDARY CONDITION CODE:
C****                I--NODE NO.,
C****                J--DEGREES OF FREEDOM
C****                EQ.0.RELEASED
C****                EQ.1.SUPPRESSED
C**** NOD(I,J)  GLOBAL NODE NUMBER
C****                I--ELEMENT NO.,
C****                J--LOCAL NODE NO.
C**** SF        PARTITION FORCE VECTOR
C**** STF       WORKING VECTOR OF SIZE (NBLK,KB)
C**** KB        BANDWIDTH OF THE SYSTEM
C**** EST       ELEMENT STIFFNESS MATRIX
C**** B         MATRIX B(I)
C**** A         MATRIX A(I)
C**** NELEM     NUMBER OF ELEMENTS OF THE
C****             STRUCTURE
C**** NNOD      NUMBER OF NODES OF THE STRUCTURE
C**** NNOEL     NUMBER OF NODES/ELEMENT
C**** NFREE     GLOBAL D.O.F/NODE
C**** NS        MAX.SIZE OF EACH PARTITION
C**** ISYM      SYMMETRY CONDITION
C****             EQ.1 SYMMETRIC STIFF.MATRIX
C****             EQ.0 UNSYMMETRIC
C**** NEQ     FINAL NO.OF EQUATIONS TO BE SOLVED
C**** L2    FILE REFERENCE NO.TO STORE A,B,C ETC.
C****
C**** THE PARTITION-WISE NODAL FORCES
C**** FOR THIS EXAMPLE PROBLEM ARE AS FOLLOWS:
C**** FIRST PART:2000.,0.,0.,3000., -1500 , 0.0
C**** 2ND PART: 0.0,-2000.0,0.0,0.0,-3000.0,0.0
C**** 3RD PART: 0.0,-2500.0,0.0,0.0,-3000.0,0.0
```

```
C**** 4TH PART: 0.0,0.0,0.0,0.0,-1500.0,0.0
      IMPLICIT REAL*8(A-H,O-Z)
      DIMENSION NBLK(4),NFTNOD(4),NLTNOD(4),
    * NFTELM(4),NLTELM(4),ID(10,3),NOD(10,2),
    * SF(6), STF(6,9),EST(6,6),B(6,6),A(6,6)
      DATA NBLK/4*6/
      DATA NFTNOD/1,4,6,8/
      DATA NLTNOD/3,5,7,10/
      DATA NFTELM/1,3,5,7/
      DATA NLTELM/4,6,8,10/
      DATA ID/30*0/
      DATA NOD/1,2,3,2,5,4,7,6,8,10,2,3,5,
    * 4,7,6,9,8,9,8/
      DO 10 J=1,3
      ID(1,J)=1
10    ID(10,J)=1
      NELEM=10
      NNOD=10
      NNOEL=2
      NFREE=3
      NPART=4
      NS=6
      KB=9
      ISYM=1
      ISUM=0
      DO 50 I=1,NNOD
      DO 40 J=1,NFREE
      IF(ID(I,J))30,20,30
20    ISUM=ISUM+1
      ID(I,J)=ISUM
      GOTO 40
30    ID(I,J)=0
40    CONTINUE
50    CONTINUE
      NEQ=ISUM
      PRINT 60
60    FORMAT(//,5X,'ANALYSIS OF A BOILER FRAME')
      PRINT 70,NELEM,NNOD,NNOEL,NFREE,NPART,ISYM
70    FORMAT(//,5X,'NUMBER OF ELEMENTS=',I3,
    * /,5X,'NUMBER OF NODES=',I3,
    * /,5X,'NODES PER ELEMENT=',I3,
    * /,5X,'GLOBAL DEGREES OF FREEDOM PER
```

```
      * NODE=', I3,
      * /,5X,'NUMBER OF PARTITIONS=',I3,
      * /,5X,'ISYM=',I2,/)
        PRINT 80
80      FORMAT(//,5X,'PARTITION',1X,'FIRST',1X,
      * 'LAST',2X,'FIRST',3X,'LAST',/,8X,'NO',6X,
      *'NODE',1X,'NODE',1X,'ELEMENT',1X,
      *'ELEMENT')
        DO 100 I2=1,NPART
90      FORMAT(7X,I2,6X,I2,4X,I2,4X,I2,6X,I2)
100     PRINT 90,I2,NFTELM(I2),NLTELM(I2),
      * NFTNOD(I2), NLTNOD(I2)
        PRINT 110
110     FORMAT(//,5X,'NODAL CONNECTIVITY'//, //,
      * 5X,'ELEMENT NO.',5X,'NODE 1',5X,'NODE 2')
        DO 120 I2=1,NELEM
120     PRINT 130,I2,(NOD(I2,J),J=1,2)
130     FORMAT(9X,I2,10X,I2,9X,I2)
        PRINT 140
140     FORMAT(//,5X,'EQUATION NUMBERS',/,5X,'I',
      * 5X,'ID(I,1)',5X,'ID(I,2)',5X,'ID(I,3)')
        DO 150 I=1,NNOD
150     PRINT 160,I,(ID(I,J),J=1,NFREE)
160     FORMAT(4X,I2,7X,I2,10X,I2,10X,I2)
        CALL GENER
        IZ=0
        MSZ=0
        L2=2
        REWIND L2
        DO 320 II=1,NPART
        READ 170,(SF(I),I=1,NS)
170     FORMAT(6F10.1)
        DO 200 I=1,NS
        DO 180 J=1,KB
180     STF(I,J)=0.0
        DO 190 J=1,NS
        B(I,J)=0.0
190     A(I,J)=0.0
200     CONTINUE
        NST=NFTELM(II)
        NEN=NLTELM(II)
        N1=NFTNOD(II)
```

```
          N2=NLTNOD(II)
          DO 280 LK=NST,NEN
          I1N=LK
          READ(08'I1N)((EST(I,J),J=1,6),I=1,6)
          DO 270 LL=1,NNOEL
          IV=NOD(LK,LL)
          IVN=IV-N1
          IF(IVN)270,210,210
210       CONTINUE
          IVN=IV-N2
          IF(IVN)220,220,270
220       CONTINUE
          IM=NFREE*(LL-1)
          DO 260 KK=1,NNOEL
          JJ=NOD(LK,KK)
          JN=NFREE*(KK-1)
          DO 250 I=1,NFREE
          MMI=ID(IV,I)
          IF(MMI.EQ.0)GOTO250
          IMI=IM+I
          DO 240 J=1,NFREE
          NJJ=ID(JJ,J)
          IF(NJJ.EQ.0)GOTO240
          NNJ=NJJ-MMI+1
          IF(NNJ)240,240,230
230       JNJ=JN+J
          MR=MMI-MSZ
          STF(MR,NNJ)=STF(MR,NNJ)+EST(IMI,JNJ)
240       CONTINUE
250       CONTINUE
260       CONTINUE
270       CONTINUE
280       CONTINUE
          M=NBLK(II)
          MSZ=MSZ+M
          N=NBLK(II)
          DO 310 I=1,M
          J1=M-I+1
          DO 290 J=1,J1
          J2=I+J-1
          B(I,J2)=STF(I,J)
          B(J2,I)=B(I,J2)
```

```
290      CONTINUE
         IF(II.EQ.NPART) GOTO 310
         J3=J1+1
         DO 300 J=J3,KB
         J4=I+J-1-M
         IF(J4.GT.M) GOTO 300
         A(I,J4)=STF(I,J)
300      CONTINUE
310      CONTINUE
         WRITE(L2) ((B(I,J),J=1,M),I=1,M)
         IF(II.LT.NPART) WRITE(L2) ((A(I,J),
     1 J=1,M), I=1,M)
         WRITE(L2) (SF(I),I=1,M)
320      CONTINUE
         CALL POTTER(NPART,M,ISYM,L2)
         PRINT 330
330      FORMAT(//,5X,'RESULTS',//,5X,'ACTIVE
     * D.O.F', 7X,'DISPL')
         DO 360 K=1,NPART
         BACKSPACE L2
         READ(L2) (SF(I),I=1,M)
         DO 350 I=1,M
         L1=(K-1)*M+I
         PRINT 340,L1,SF(I)
340      FORMAT(13X,I2,F13.4)
350      CONTINUE
         BACKSPACE L2
360      CONTINUE
         STOP
         END


         SUBROUTINE GENER
C***** THIS SUBROUTINE GENERATES ELEM. STIFFNESS
C***** MATRICES FOR ALL ELEMENTS
         IMPLICIT REAL *8 (A-H,O-Z)
         DIMENSION EST(6,6),KEL(6,6)
         REAL *8 KEL
         DEFINE FILE 08(15,3000,L,I1N)
         DATA A1,A2,EI1,EI2,AL1,AL2,E / 60.0D0,
     * 45.0D0,0.2D11,0.1D11,100.0D0,200.0D0,
     * 2.0D6/
         DO 10 I=1,6
```

```
         DO 10 J=1,6
         EST(I,J)=0.0
10       KEL(I,J)=0.0
         PRINT 20
20       FORMAT(//1X,'MEMBER',2X,'AREA OF C/S',
    *    2X,'BENDING RIGIDITY', 2X, 'LENGTH' /)
         PRINT 30,A1,EI1,AL1
30       FORMAT(/1X,'HORIZ.',11X,F7.3,7X,E12.3,7X
    *    ,F5.1)
         PRINT 40,A2,EI2,AL2
40       FORMAT(/1X,'VERT.',11X,F7.3,7X,E12.3,7X
    *    ,F5.1)
         A11=A1*E/AL1
         A12=A2*E/AL2
         A21=EI1/(AL1*AL1)
         A22=EI2/(AL2*AL2)
         EST(1,1)=A11
         EST(1,4)=-A11
         EST(2,2)=12.0*A21/AL1
         EST(2,3)=+6.0*A21
         EST(2,5)=-EST(2,2)
         EST(2,6)=EST(2,3)
         EST(3,3)=4.0*AL1*A21
         EST(3,5)=-EST(2,3)
         EST(3,6)=0.5*EST(3,3)
         EST(4,4)=A11
         EST(5,5)=EST(2,2)
         EST(5,6)=EST(3,5)
         EST(6,6)=EST(3,3)
         KEL(1,1)=12.0*A22/AL2
         KEL(1,3)=6.0*A22
         KEL(1,4)=-KEL(1,1)
         KEL(1,6)=KEL(1,3)
         KEL(2,2)=A12
         KEL(2,5)=-A12
         KEL(3,3)=4.0*AL2*A22
         KEL(3,4)=-KEL(1,3)
         KEL(3,6)=0.5*KEL(3,3)
         KEL(4,4)=KEL(1,1)
         KEL(4,6)=KEL(3,4)
         KEL(5,5)=KEL(2,2)
         KEL(6,6)=KEL(3,3)
```

```
        DO 50 I=1,6
        DO 50 J=1,I
        EST(I,J)=EST(J,I)
50      KEL(I,J)=KEL(J,I)
        DO 100 I1=1,10
        I1N=I1
        IF(I1.LE.2.OR.I1.GE.9) GOTO 90
        WRITE(08'I1N) ((EST(I,J),J=1,6),I=1,6)
        GOTO 100
90      WRITE(08'I1N) ((KEL(I,J),J=1,6),I=1,6)
100     CONTINUE
        RETURN
        END
```

ANALYSIS OF A BOILER FRAME

```
NUMBER OF ELEMENTS= 10
NUMBER OF NODES= 10
NODES PER ELEMENT=  2
GLOBAL DEGREES OF FREEDOM PER NODE=  3
NUMBER OF PARTITIONS=  4
ISYM= 1
```

PARTITION NO	FIRST NODE	LAST NODE	FIRST ELEMENT	LAST ELEMENT
1	1	4	1	3
2	3	6	4	5
3	5	8	6	7
4	7	10	8	10

NODAL CONNECTIVITY

ELEMENT NO.	NODE 1	NODE 2
1	1	2
2	2	3
3	3	5
4	2	4
5	5	7
6	4	6
7	7	9
8	6	8
9	8	9
10	10	8

```
EQUATION NUMBERS
 I     ID(I,1)    ID(I,2)    ID(I,3)
 1        0          0          0
 2        1          2          3
 3        4          5          6
 4        7          8          9
 5       10         11         12
 6       13         14         15
 7       16         17         18
 8       19         20         21
 9       22         23         24
10        0          0          0
```

MEMBER	AREA OF C/S	BENDING RIGIDITY	LENGTH
HORIZ.	60.000	0.200D 11	100.0
VERT.	45.000	0.100D 11	200.0

```
RESULTS
ACTIVE D O F      DISPL
          1      0.2588
          2     -0.0223
          3      0.0007
          4      0.4962
          5     -0.0348
          6     -0.0001
          7      0.2573
          8     -0.0286
          9     -0.0005
         10      0.4958
         11     -0.0808
         12     -0.0005
         13      0.2558
         14     -0.0623
         15      0.0000
         16      0.4955
         17     -0.0932
         18      0.0004
         19      0.2544
         20     -0.0077
         21      0.0011
         22      0.4951
         23     -0.0152
         24      0.0010
COMPILE TIME=1.69 SEC,  EXECUTION TIME= 1.42 SEC
```

```
C**** PROGRAM : 2.10
C**** SOLVING THE RING PROBLEM OF 2.12 USING
C**** OUT-OF-CORE SOLUTION WITH CHOLESKY
C**** FACTORISATION WITHIN EACH SUBSTRUCTURE
      IMPLICIT REAL*8(A-H,O-Z)
      DIMENSION ST(2,4),Q(2,2),Z(2,2),G(2,2)
      DATA IN1,IN2,IN3/1,2,3/
      REWIND IN1
      REWIND IN2
      REWIND IN3
      READ 200,NPART,M
      PRINT 210,M,NPART
      M1=M+1
      M2=M*2
      DO 20 K=1,NPART
      IF (K.EQ.NPART) GO TO 10
      READ 220,((ST(I,J),J=1,M2),I=1,M)
      WRITE(IN1)((ST(I,J),J=1,M),I=1,M)
      WRITE(IN1)((ST(I,J),J=M1,M2),I=1,M)
      GO TO 20
10    READ 220,((ST(I,J),J=1,M),I=1,M)
      WRITE(IN1)((ST(I,J),J=1,M),I=1,M)
20    CONTINUE
30    READ(5,200,END=400)IP,IREPET
      PRINT 240,IP,IREPET
      DO 40 K=1,NPART
      READ 220,((G(I,J),I=1,M),J=1,IP)
      WRITE(IN2)((G(I,J),I=1,M),J=1,IP)
40    CONTINUE
      REWIND IN1
      REWIND IN2
      IF (IREPET.EQ.0)WRITE(6,260)
      IF (IREPET.EQ.1)WRITE(6,270)
      WRITE(6,280)
      IF (IREPET.EQ.0)WRITE(6,290)
      IF (IREPET.EQ.1)WRITE(6,300)
      WRITE(6,280)
      DO 80 K=1,NPART
      READ(IN2)((G(I,J),I=1,M),J=1,IP)
      READ(IN1)((ST(I,J),I=1,M),J=1,M)
      IF (K.EQ.NPART) GO TO 60
      READ(IN1)((ST(I,J),I=1,M),J=M1,M2)
```

```
         DO 50 I=1,2
         PRINT 310,(ST(I,J),J=1,M2),G(I,1),G(I,2)
50       CONTINUE
         GO TO 80
60       DO 70 I=1,2
         PRINT 320,(ST(I,J),J=1,M),(G(I,J),J=1,IP)
70       CONTINUE
80       CONTINUE
         WRITE(6,280)
         CALL POTSYM(NPART,M,IP,IREPET,IN1,IN2)
         REWIND IN3
         DO 90 K=1,NPART
         READ(IN3)((Z(I,J),I=1,M),J=1,IP)
         K1=NPART-K+1
         PRINT 330,K1,((Z(J,I),I=1,M),J=1,IP)
90       CONTINUE
         WRITE(6,280)
         GO TO 30
200      FORMAT(2I2)
210      FORMAT(/10X,'M = ',I2,20X,'NPART = ',I2)
220      FORMAT(4F4.2)
240      FORMAT(//10X,'IP=',I2,19X,'IREPET = ',I2/)
260      FORMAT(/20X,'INPUT MATRICES')
270      FORMAT(/13X,'OUTPUT MATRICES',6X,'INPUT
       * MATRICES')
280      FORMAT(2X,54('-'))
290      FORMAT (8X,'B(I)',10X,'A(I)',6X,'G(I,1)',
       * 1X,'G(I,2)')
300      FORMAT(8X,'L(I)',10X,'P(I)',6X,'G(I,1)',
       * 1X,'G(I,2)')
310      FORMAT(2X,6(2X,F5.2))
320      FORMAT(2X,2(2X,F5.2),14X,2(2X,F5.2))
330      FORMAT(/2X,'THE VALUES OF Z IN THE
       * SUBSTRUCTURE',' NUMBER '' ',I2,' '' ARE
       * ',/2(5X,F16.10,5X))
400      STOP
         END
         SUBROUTINE POTSYM(NPART,M,IP,IREPET,IN1,IN2)
C** METHOD & PURPOSE:-
C** AN OUT OF CORE SOLUTION TO SOLVE A SET OF
C** SIMULTANEOUS EQUATIONS OF THE TYPE K*(X)=(F)
C** USING POTTER'S SCHEME WITH CHOLESKY
```

```
C** FACTORISATION IN EVERY PARTITION.
C** DESCRIPTION OF PARAMETERS:-
C** NPART   NUMBER OF BLOCKS(PARTITIONS)
C** M       SIZE OF THE BLOCK.M*NPART=NO.OF EQNS
C** IP      NUMBER OF FORCE VECTORS TO BE SOLVED
C**         AT A TIME
C** IREPET  PARAMETER SPECIFYING WHETHER
C**         REPETITION WITH A NEW SET OF FORCE
C**         VECTORS NEEDS TO BE DONE
C**         0 - SOLVING IT FOR THE FIRST TIME
C**         1 - REPEATING
C** IN1     REF.FILE IN WHICH B(I) AND A(I) ARE
C**         STORED. LATER THEY WILL BE REPLACED BY
C**         L(I) AND P(I).
C** IN2     REF.FILE IN WHICH FORCE MATRIX
C**         G(I,IP) IS STORED. LATER IT IS
C**         REPLACED BY DISPLACEMENT MATRIX Z(I,J)
C**         A,B,P ARE WORKING MATRICES OF SIZE M x M
C***        G,Q,Z ARE MATRICES OF SIZE M BY IP
C***
C***   REMARKS:-
C***1 .THE STIFFNESS MATRIX IS WRITTEN IN TERMS
C***    OF B(I),A(I) AS EXPLAINED IN THE REMARK-6
C***2. THE FORCE MATRIX 'F' IS ALSO STORED IN
C***    TERMS OF BLOCKS OF G(I).
C***3. THIS PROGRAM CAN SOLVE FOR ANY NUMBER OF
C***    FORCE VECTORS BY CALLING POTSYM ONLY ONCE.
C***    IP DENOTES THE NUMBER OF FORCE VECTORS
C***    TO BE SOLVED FOR.
C***4. THE MATRICES B(I),A(I) ARE TO BE STORED
C***    IN REF.FILE IN1 THROUGH THE MAIN PROGRAM
C****   AND THE MATRICES G(I,1) THROUGH G(I,IP)
C****   ARE TO BE STORED AS A SINGLE RECORD IN
C****   REF.FILE IN2 THROUGH THE MAIN PROGRAM.
C****5.THE PROGRAM WORKS ONLY FOR PARTITIONS OF
C****   EQUAL SIZE.
C****6.THE FILE STORAGE ORDER IS AS FOLLOWS:
C****   REF.FILES:-
C****-------------------------------------------
C**** IN1   ]   IN2          ]   IN3
C****-------------------------------------------
C**** B(1)  ]G(1,1)...G(1,IP) ]Z(N,1)...Z(N,IP)
```

```
C**** A(1)    ]G(2,1)...G(2,IP) ]        ...
C**** B(2)    ]        ...       ]        ...
C**** A(2)    ]        ...       ]        ...
C**** ..      ]        ...       ]Z(2,1)...Z(2,IP)
C**** ..      ]G(N,1)...G(N,IP) ]Z(1,1)...Z(1,IP)
C**** ..      ]                  ]
C**** B(N-1) ]                   ]
C**** A(N-1) ]                   ]
C**** B(N)   ]                   ]
C****---------------------------------------------
C****7.THE BLOCKS B(I),A(I),G(I,IP) ARE
C****  DESTROYED DURING COMPUTATION. THE FINAL
C****  RESULTS ARE STORED BLOCKWISE IN
C****  REF.FILE IN3.
C****8.THE PROGRAM OPERATES IN DOUBLE PRECISION
C****  ARITHMETIC.
C**** SUBROUTINE & FUNCTION SUBPROGRAM REQUIRED:
C****      LOWTRI,FORSUB,BACSUB.
      IMPLICIT REAL*8(A-H,O-Z)
      DIMENSION B(2,2),A(2,2),P(2,2),G(2,2),
     1Q(2,2),Z(2,2)
      INTEGER T,U,V,W
      IN3 = 3
      REWIND IN1
      REWIND IN2
      REWIND IN3
      N=M*NPART
      MO=M
      M1=M+1
      M2=2*M
      IF (IREPET.EQ.0) GO TO 70
      DO 60 JK=1,NPART
      IF(JK-1)20,10,20
10    READ(IN1)((B(I,J),J=1,M),I=1,M)
      READ(IN2)((G(I,J),I=1,M),J=1,IP)
      CALL FORSUB(B,G,M,IP,Q)
      BACKSPACE IN2
      GO TO 50
20    READ(IN1)((P(I,J),J=1,M),I=1,M)
      READ(IN1)((B(I,J),J=1,M),I=1,M)
      READ(IN2)((G(I,J),I=1,M),J=1,IP)
      DO 40 II=1,IP
```

```
        DO 40 I=1,M
        SUM=0.0
        DO 30 J=1,M
30      SUM=SUM+P(J,I)*Q(J,II)
40      G(I,II)=G(I,II)-SUM
        CALL FORSUB(B,G,M,IP,Q)
        BACKSPACE IN2
50      WRITE(IN2)((Q(I,J),I=1,M),J=1,IP)
60      CONTINUE
        GO TO 160
70      DO 150 K= 1,NPART
        READ(IN1)((B(I,J),J=1,M),I=1,M)
        IF(K.LT.NPART)READ(IN1)((A(I,J),J=1,M),
       1 I=1,M)
        READ(IN2)((G(I,J),I=1,M),J=1,IP)
C****   CALCULATES L(1) ,P(1) ,Q(1) USING CHOLESKY
        IF(K-1) 90,80,90
80      CALL LOWTRI(B,M)
        CALL FORSUB(B,A,M,MO,P)
        CALL FORSUB(B,G,M,IP,Q)
        GO TO 140
C****   DETERMINES L(2) UPTO L(N)
90      DO 110 I=1,M
        DO 110 J=1,M
        SUM = 0.0
        DO 100 L=1,MO
100     SUM = SUM + P(L,I) * P(L,J)
110     B(I,J)=B(I,J)-SUM
        CALL LOWTRI(B,M)
C****   FINDS Q(2) UPTO Q(N)
        DO 130 II =1,IP
        DO 130 I=1,M
        TEMP=0.0
        DO 120 J=1,M
120     TEMP= TEMP+P(J,I)*Q(J,II)
130     G(I,II)=G(I,II)-TEMP
        CALL FORSUB(B,G,M,IP,Q)
C****   FINDS P(2)UPTO P(N)
        CALL FORSUB(B,A,M,MO,P)
140     BACKSPACE IN1
        IF (K.LT.NPART)  BACKSPACE IN1
        WRITE(IN1)((B(I,J),J=1,M),I=1,M)
```

```
       IF(K.LT.NPART)WRITE(IN1)((P(I,J),J=1,M),
     1 I=1,M)
       BACKSPACE IN2
       WRITE(IN2)((Q(I,J),I=1,M),J=1,IP)
150    CONTINUE
160    DO 210 K=1,NPART
       IF(K-1)170,170,180
C**** FINDS Z(N)
170    BACKSPACE IN1
       BACKSPACE IN2
       READ(IN1)((B(I,J),J=1,M),I=1,M)
       READ(IN2)((Q(I,J),I=1,M),J=1,IP)
       CALL BACSUB(B,Q,M,IP,Z)
       BACKSPACE IN1
       BACKSPACE IN2
       WRITE(IN3)((Z(I,J),I=1,M),J=1,IP)
       GO TO 210
C**** FINDS Z(NPART-1) UPTO Z(1)
180    BACKSPACE IN1
       BACKSPACE IN1
       BACKSPACE IN2
       READ(IN1)((B(I,J),J=1,M),I=1,M)
       READ(IN1)((P(I,J),J=1,M),I=1,M)
       READ(IN2)((Q(I,J),I=1,M),J=1,IP)
       DO 200 II = 1,IP
       DO 200 I=1,M
       TEMP=0.0
       DO 190 J=1,M
190    TEMP=TEMP+P(I,J)*Z(J,II)
       Q(I,II)=Q(I,II)-TEMP
200    CONTINUE
       CALL BACSUB(B,Q,M,IP,Z)
       BACKSPACE IN1
       BACKSPACE IN1
       BACKSPACE IN2
       WRITE(IN3)((Z(I,J),I=1,M),J=1,IP)
210    CONTINUE
       RETURN
       END
       SUBROUTINE LOWTRI(B,M)
C**** METHOD & PURPOSE:-
C**** LOWER TRIANGULARISATION USING
```

```
C**** CHOLESKY FACTORISATION L.LT=K
C**** DESCRIPTION OF PARAMETERS
C**** B    MATRIX TO BE DECOMPOSED IN COMPUTATION
C**** M    SIZE OF SUB-STRUCTURE
C**** REMARKS:-
C****    THE PROGRAM OPERATES IN DOUBLE PRECISION
C****    ARITHMETIC AND IS APPLICABLE ONLY TO
C****    SYMMETRIC,POSITIVE-DEFINITE MATRICES.
C****    IT WORKS ON NON-BANDED MATRICES.
C****   SUBROUTINE & FUNCTION SUBPROGRAM REQUIRED
C****      NONE.
C****
      IMPLICIT REAL*8(A-H,O-Z)
      DIMENSION B(M,M)
      B(1,1)=DSQRT(B(1,1))
      DO 10 I=2,M
      B(I,1)=B(1,I)/B(1,1)
10    CONTINUE
      DO 60 J=2,M
      J1=J-1
      DO 20 K=1,J1
20    B(J,J)=B(J,J)-B(J,K)**2
      B(J,J)=DSQRT(B(J,J))
      IF (J.EQ.M) GO TO 60
30    J2=J+1
      DO  50 I=J2,M
      DO 40 K=1,J1
      B(I,J)= B(I,J)-B(I,K)*B(J,K)
40    CONTINUE
      B(I,J)=(B(I,J))/B(J,J)
50    CONTINUE
60    CONTINUE
      RETURN
      END
      SUBROUTINE FORSUB(B,F,M,IP,DELTA)
C**** METHOD & PURPOSE:-
C**** SOLVING B*(DELTA)=(F) BY FORWARD
C**** SUBSTITUTION
C**** DESCRIPTION OF PARAMETERS:-
C**** B    LOWER TRIANGULAR MATRIX
C**** F    FORCE VECTOR OF SIZE M BY IP. IT IS
C****      NOT DESTROYED DURING COMPUTATION.
```

```
C**** M    NUMBER OF EQUATIONS
C**** IP   NUMBER OF FORCE VECTORS TO BE SOLVED
C****      AT A TIME
C**** DELTA  RESULTANT DISPLACEMENT VECTOR FOUND
C****          BY FORWARD SUBSTITUTION
C**** REMARKS:-
C****  THE PROGRAM OPERATES IN DOUBLE PRECISON
C****  ARITHEMETIC.UNLIKE SOLUT IT PERFORMS
C****  FORWARD SUBSTITUTION ON A MATRIX RATHER
C****  THAN A VECTOR.  FOR EG. IT SOLVES
C****  L(I)*P(I)=A(I) BY CALLING ONLY ONCE.
C**** SUBROUTINE & FUNCTION SUBPROGRAM REQUIRED:
C****          NONE.
C****
      IMPLICIT REAL*8(A-H,O-Z)
      DIMENSION DELTA(M,IP),F(M,IP),B(M,M)
      DO 30  K = 1,IP
      DELTA(1,K)=F(1,K)/B(1,1)
      DO 20 I=2,M
      SUM = F(I,K)
      I1=I-1
      DO 10 J=1,I1
10    SUM=SUM-B(I,J)*DELTA(J,K)
20    DELTA(I,K)=SUM/B(I,I)
30    CONTINUE
      RETURN
      END
      SUBROUTINE BACSUB(B,Q,M,IP,Z)
C**** METHOD & PURPOSE:-
C**** SOLVING B*(Z)=(Q) BY BACKWARD SUBSTITUTION
C**** DESCRIPTION OF PARAMETERS:-
C**** B    LOWER TRIANGULAR MATRIX
C**** Q    FORCE VECTOR OF SIZE N BY IP.IT IS
C****      NOT DESTROYED DURING COMPUTATION
C**** M    NUMBER OF EQUATIONS
C**** IP   NO.OF FORCE VECTORS TO BE
C****      SOLVED AT A TIME
C**** Z    DISPLACEMENT MATRIX COMPUTED BY
C****      BACKWARD SUBSTITUTION
C**** REMARKS:-
C**** THIS PROGRAM OPERATES IN DOUBLE PRECISION
C**** ARITHMETIC.UNLIKE SOLU IT PERFORMS
```

```
C****   BACKWARD SUBSTITUTION ON A MATRIX RATHER
C****   THAN A VECTOR. FOR EG. IT SOLVES
C***    LT(I)*Z(I,IP)=Q(I,IP) BY CALLING ONLY ONCE
C****   SUBROUTINE & FUNCTION SUBPROGRAM REQUIRED:
C****            NONE.
C****
        IMPLICIT REAL*8(A-H,P-Z)
        DIMENSION Z(M,IP),Q(M,IP),B(M,M)
        DO 30 K = 1,IP
        Z(M,K)=Q(M,K)/B(M,M)
        DO 20 I=2,M
        TEMP=0.
        IM=M-I+1
        IM1=IM+1
        DO 10 J=IM1,M
10      TEMP=TEMP+B(J,IM)*Z(J,K)
        Z(IM,K)=(Q(IM,K)-TEMP)/B(IM,IM)
20      CONTINUE
30      CONTINUE
        RETURN
        END
```

	M = 2		NPART = 4		
	IP = 2		IREPET = 0		

INPUT MATRICES

B(I)		A(I)		G(I,1)	G(I,2)

1.00	0.10	0.10	0.00	1.40	1.50
0.10	2.00	0.00	0.20	4.70	4.90
3.00	0.00	0.30	0.00	7.00	10.60
0.00	4.00	0.00	0.40	14.00	18.80
3.00	0.00	0.30	0.00	10.80	18.00
0.00	2.00	0.00	0.20	10.20	15.20
2.00	0.20			9.90	17.10
0.20	1.00			6.60	10.60

THE VALUES OF Z IN SUBSTRUCTURE NUMBER '4' ARE

4.0000000000		7.0000000000
5.0000000000		8.0000000000

```
THE VALUES  OF Z IN SUBSTRUCTURE NUMBER '3' ARE
            3.0000000000          5.0000000000
            4.0000000000          6.0000000000


THE VALUES OF Z IN SUBSTRUCTURE NUMBER '2 ' ARE
            2.0000000000          3.0000000000
            3.0000000000          4.0000000000


THE VALUES  OF Z IN SUBSTRUCTURE NUMBER '1' ARE
            1.0000000000          1.0000000000
            2.0000000000          2.0000000000
-------------------------------------------------
        IP =  2                    IREPET =   1


            OUTPUT MATRICES     INPUT MATRICES
-------------------------------------------------

        L(I)            P(I)      G(I,1) G(I,2)
-------------------------------------------------

     1.00    0.10    0.10  -0.01   1.50   1.40
     0.10    1.41    0.00   0.14   4.90   4.70
     1.73    0.00    0.17  -0.00  10.60   7.00
     0.00    1.99    0.00   0.20  18.80  14.00
     1.72    0.00    0.17  -0.00  18.00  10.80
     0.00    1.40    0.00   0.14  15.20  10.20
     1.40    0.14                 17.10   9.90
     0.20    0.98                 10.60   6.60
-------------------------------------------------


THE VALUES  OF Z IN SUBSTRUCTURE NUMBER '4 ' ARE
            7.0000000000          4.0000000000
            8.0000000000          5.0000000000

THE VALUES  OF Z IN SUBSTRUCTURE NUMBER '3 ' ARE
            5.0000000000          3.0000000000
            6.0000000000          4.0000000000

THE VALUES  OF Z IN SUBSTRUCTURE NUMBER '2 ' ARE
            3.0000000000          2.0000000000
            4.0000000000          3.0000000000

THE VALUES  OF Z IN SUBSTRUCTURE NUMBER '1' ARE
            1.0000000000          1.0000000000
            2.0000000000          2.0000000000

COMPILE TIME= 1.20 SEC,EXECUTION TIME=  0.88 SEC
```

```
C**** PROGRAM: 2.11
C**** STIFFNESS AND MASS MATRIX FORMULATION FOR
C**** TRIANGULAR PLATE ELEMENT WITH 6 DOF/NODE
C**** USING ZIENKIEWICZ SHAPE FUNCTIONS
C**** THIS PROGRAM HAS BEEN DEVELOPED BY
C****              S. SREENIVASAMURTHY
      SUBROUTINE FEMP(YOM,H,PR,XI,BT,DEN,C,LK,
     1 CF,OMEGA,Q,AMT)
C//////////////////////////////////////////////
C   THIS SUBROUTINE FORMS ELEMENT STIFFNESS,
C   CENTRIFUGAL FORCE AND MASS MATRIX FOR A
C   CONSTANT THICKNESS THREE NODED 6 D.O.F /
C   NODE TRIANGULAR PLATE ELEMENT IN CYLINDRICAL
C   CO-ORDINATES
C//////////////////////////////////////////////
        IMPLICIT REAL*8(A-H,O-Z)
      DIMENSION DCM(3,3),DSHF(6,9)
      DIMENSION C(18,18),TEMP(18,18),TRM(18,18),
     1 XI(3,3)
      DIMENSION CP(6,6), CB(9,9),B(3,9),D(3,3),
     *DB(3,9,160) ,BTDB(9,9), COF(3,6),Q(18),
     *ANT1(18), BN(3,3),XC(3),ANT(18,3), CS(9,9)
     *,D1(3,3),XA(3,3),AN(3,18),SHF(9),
     *AMA(18,18),ASHF(3,18),AMT(18,18),
     *TEMP1(18,18) ,AML(18,18)
      COMMON/FIVE/BP,TRS,DB
        COMMON/SHPE/ASHF,SHF

C    VARIABLE        EXPLANATION
C    CP       INPLANE STIFFNESS MATRIX
C    CB       BENDING STIFFNESS MATRIX
C    XI       GLOBAL COORDINATES OF THREE NODES
C    XA       LOCAL COORDINATES OF THREE NODES
C    C,Q,AM   ELEMENT ELASTIC STIFFNESS,
C             CENTRIFUGAL FORCE AND
C             LUMPED MASS MATRICES
C    SHF      SHAPE FUNCTIONS VECTOR
C    TRM      ELEMENT TRANSFORMATION MATRIX
C    B        MATRIX REPRESENTING STRAIN-NODAL
C             DISPLACEMENT RELN
C    D        MATRIX REPRESENTING STRESS-STRAIN
C             RELATIONSHIP
```

```
          X21=XI(2,1)-XI(1,1)
          X31=XI(3,1)-XI(1,1)
          X32=XI(3,1)-XI(2,1)
              Y21=XI(2,2)-XI(1,2)
              Y31=XI(3,2)-XI(1,2)
              Y32=XI(3,2)-XI(2,2)
          Z21=XI(2,3)-XI(1,3)
          Z31=XI(3,3)-XI(1,3)
          Z32=XI(3,3)-XI(2,3)
          AREA=0.5*(DSQRT(((Y21*Z31)-(Z21*Y31))**2+
      *                  ((Z21*X31)-(X21*Z31))**2+
      *                  ((X21*Y31)-(Y21*X31))**2))
          EL13=DSQRT(X31*X31+Y31*Y31+Z31*Z31)
          EL12=DSQRT(X21*X21+Y21*Y21+Z21*Z21)
          EL23=DSQRT(X32*X32+Y32*Y32+Z32*Z32)
          COST=(EL13*EL13+EL12*EL12-EL23*EL23) /
      1   (2.0*EL13*EL12)
      C    'EL13,EL12,EL23 ' ARE THE SIDES OF THE
      C     ELEMENT.'COST' IS
      C     THE INCLUDED ANGLE BETWEEN SIDES 12 & 13
      C     - CALCULATED USING THE
      C     COSINE FORMULA.
          CF=AREA*H*7.85/10.**9
          XA(1,1)=0.0
          XA(1,2)=0.0
          XA(2,2)=0.0
          XA(2,1)=EL12
          XA(3,1)=EL13*COST
          XA(3,2)=DSQRT(EL13**2-XA(3,1)**2)
          A1=XA(3,1)-XA(2,1)
          A2=XA(1,1)-XA(3,1)
          A3=XA(2,1)-XA(1,1)
              B1=XA(2,2)-XA(3,2)
              B2=XA(3,2)-XA(1,2)
              B3=XA(1,2)-XA(2,2)
           DT=(4.0*AREA*(1.0-PR*PR))/(YOM*H)
           DT=1.0/DT
          CP(1,1)=((B1*B1)+(A1*A1*BT))*DT
          CP(1,2)=((PR+BT)*A1*B1)*DT
          CP(1,3)=((B1*B2)+(A1*A2*BT))*DT
          CP(1,4)=((PR*A2*B1)+(BT*A1*B2))*DT
          CP(1,5)=((B1*B3)+(A1*A3*BT))*DT
```

```
         CP(1,6)=((PR*A3*B1)+(A1*B3*BT))*DT
              CP(2,2)=((A1*A1)+(B1*B1*BT))*DT
              CP(2,3)=((PR*A1*B2)+(B1*A2*BT))*DT
              CP(2,4)=((A1*A2)+(B1*B2*BT))*DT
              CP(2,5)=((PR*A1*B3)+(B1*A3*BT))*DT
              CP(2,6)=((A1*A3 )+(B1*B3*BT))*DT
       CP(3,3)=((B2*B2)+(A2*A2*BT))*DT
       CP(3,4)=((A2*B2)*(PR+BT))*DT
       CP(3,5)=((B2*B3)+(A2*A3*BT))*DT
       CP(3,6)=((PR*B2*A3)+(A2*B3*BT))*DT
              CP(4,4)=((A2*A2)+(B2*B2*BT))*DT
              CP(4,5)=((PR*A2*B3)+(B2*A3*BT))*DT
              CP(4,6)=((A2*A3)+(B2*B3*BT))*DT
       CP(5,5)=((B3*B3)+(A3*A3*BT))*DT
       CP(5,6)=((A3*B3)*(PR+BT))*DT
              CP(6,6)=((A3*A3)+(B3*B3*BT))*DT
       DO 10 IL=2,6
           IM=IL-1
         DO 10 JL=1,IM
               CP(IL,JL)=CP(JL,IL)
10     CONTINUE
       DO 11 I=1,3
       DO 11 J=1,3
11     D1(I,J)=0.0
       D1(1,1)=1.0
       D1(2,2)=1.0
       D1(1,2)=PR
           D1(2,1)=PR
       D1(3,3)=(1.0-PR)/2.0
       DO 20 I=1,3
           DO 20 J=1,3
20     D1(I,J)=D1(I,J)*YOM/(1.0-PR*PR)
       AR=1.0/(2.0*AREA)
C      INPLANE STIFFNES GENERATED
       CALL DMAT(H,YOM,PR,BT,D)
       CALL TRANM(XI,EL12,AREA,X21,X31,Y21,Y31,
      *Z21,Z31,TRM,DCM,LK)
852    AL=-4.0*AREA*AREA
       AL=1.0/AL
       COF(1,1)=B1*B1*AL
       COF(1,2)=B2*B2*AL
       COF(1,3)=B3*B3*AL
```

```
          COF(1,4)=2.0*B1*B2*AL
          COF(1,5)=2.0*B2*B3*AL
          COF(1,6)=2.0*B3*B1*AL
                   COF(2,1)=A1*A1*AL
                   COF(2,2)=A2*A2*AL
                   COF(2,3)=A3*A3*AL
                    COF(2,4)=2.0*A1*A2*AL
                    COF(2,5)=2.0*A2*A3*AL
                    COF(2,6)=2.0*A3*A1*AL
          COF(3,1)=2.0*A1*B1*AL
          COF(3,2)=2.0*A2*B2*AL
          COF(3,3)=2.0*A3*B3*AL
          COF(3,4)=2.0*(A1*B2+A2*B1)*AL
          COF(3,5)=2.0*(A2*B3+A3*B2)*AL
          COF(3,6)=2.0*(A1*B3+A3*B1)*AL
          DO 31 I=1,9
             DO 21  J=1,9
                  CB(I,J)=0.0
                  CS(I,J)=0.0
 21       CONTINUE
 31    CONTINUE
          DO 611 I = 1,18
          DO 611 J = 1,18
          AML(I,J) = 0.0
611       AMT(I,J) = 0.0
          DO 601 I = 1,18
601    ANT1(I) = 0.
C***   GAUSSIAN QUADRATURE FOR (CB&CS) BEGINS HERE
       W=1.0/3.0
       ALFA=0.5
       BETA=0.5
       GAMA=0.0
       DO 121 IDL=1,3
       GO TO (50,60,70,75)IDL
 50       AL1=ALFA
          AL2=BETA
          AL3=GAMA
       GO TO 80
 60          AL1=BETA
             AL2=GAMA
             AL3=ALFA
       GO TO 80
```

```
70        AL1=GAMA
          AL2=ALFA
          AL3=BETA
       GO TO80
75        AL1=1.0/3.0
          AL2=1.0/3.0
          AL3=1.0/3.0
80     CALL BMAT(AL1,AL2,AL3,A1,A2,A3,B1,B2,B3,COF,B)
          DO 100 I=1,3
             DO 100 J=1,9
                   DB(I,J,LK)=0.0
                DO 90 L=1,3
          DB(I,J,LK)=DB(I,J,LK)+D(I,L)*B(L,J)
90              CONTINUE
100    CONTINUE
          IF(IDL.EQ.4)GO TO170
          DO 110 IT=1,9
             DO 110 JT=1,9
                   BTDB(IT,JT)=0.0
                DO 110 L=1,3
          BTDB(IT,JT)=BTDB(IT,JT)+B(L,IT)*DB(L,JT,LK)
110    CONTINUE
          DO 120 IT = 1,9
          DO 120 JT = 1,9
          CB(IT,JT) = BTDB(IT,JT)*AREA*W + CB(IT,JT)
120    CONTINUE
170    CONTINUE
          CALL SHAPE(A1,A2,A3,B1,B2,B3,AL1,AL2 ,AL3,SHF)
          DO 111 IT=1,3
          DO 111 JT=1,18
  111 AN(IT,JT)=0.
          AN(1,1)=AL1
          AN(1,7)=AL2
          AN(1,13)=AL3
          AN(2,2)=AL1
          AN(2,8)=AL2
          AN(2,14)=AL3
          DO 12 IL=1,9
          IF(IL.GE.1.AND.IL.LE.3)I1=IL+2
          IF(IL.GE.4.AND.IL.LE.6)I1=IL+5
          IF(IL.GE.7.AND.IL.LE.9)I1=IL+8
   12 AN(3,I1)=SHF(IL)
```

```
            OMEGA1=DCM(1,3)*OMEGA
            OMEGA2=DCM(2,3)*OMEGA
            OMEGA3=DCM(3,3)*OMEGA
            BN(1,1)=OMEGA2**2+OMEGA3**2
            BN(1,2)=-OMEGA1*OMEGA2
            BN(1,3)=-OMEGA1*OMEGA3
            BN(2,1)=BN(1,2)
            BN(2,2)=OMEGA3**2+OMEGA1**2
            BN(2,3)=-OMEGA2*OMEGA3
            BN(3,1)=BN(1,3)
            BN(3,2)=BN(2,3)
            BN(3,3)=OMEGA1**2+OMEGA2**2
            XC(1)=DCM(1,1)*XI(1,1)+ DCM(1,2)*XI(1,2) +
     *  DCM(1,3)*XI(1,3)+AL1*XA(1,1)+AL2*XA(2,1)+
     *  AL3*XA(3,1)
            XC(3)=DCM(3,1)*XI(1,1)+DCM(3,2)*XI(1,2)+
     *  DCM(3,3)*XI(1,3)
            XC(2)=DCM(2,1)*XI(1,1)+DCM(2,2)*XI(1,2)+
     *  DCM(2,3)*XI(1,3)+ 2AL1*XA(1,2)+AL2*XA(2,2)
     *  +AL3*XA(3,2)
            DO 13 IT=1,18
               DO 13 JT=1,3
               ANT(IT,JT)=0.
               DO 13 KT=1,3
    13 ANT(IT,JT)=ANT(IT,JT)+AN(KT,IT)*BN(KT,JT)
            DO 14 IT=1,18
               DO 14 JT=1,3
    14 ANT1(IT)=ANT1(IT)+ANT(IT,JT)*XC(JT)
     1 *DEN*AREA*W*H

C       CALCULATION OF CONSISTANT MASS MATRIX AMT
C       ******************************************
            CALL ALSHP(AL1,AL2,AL3)
            DO 113 IK1 = 1,18
            DO 113 JK1 = 1,18
            AMA(IK1,JK1) =0.0
            DO 113 KK1 = 1,3
            AMA(IK1,JK1) = AMA(IK1,JK1) +
     1 ASHF(KK1,IK1)*ASHF(KK1,JK1)
   113      CONTINUE
            DO 112 IK1=1,18
            DO 112 JK1=1,18
```

```
           AMA(IK1,JK1)=AMA(IK1,JK1)*W*AREA*H*DEN
              AMT(IK1,JK1)=AMT(IK1,JK1)+AMA(IK1,JK1)
  112      CONTINUE
   121     CONTINUE
           DO 15 IT=1,18
              Q(IT)=0.
              DO 15 JT=1,18
   15      Q(IT)=Q(IT)+TRM(JT,IT)*ANT1(JT)
           CALL ARR(CB,CP,TEMP)
           DO 180 I=1,18
           DO 180 J=1,18
           C(I,J)=TEMP(I,J)
  180      CONTINUE
           CON=0.05
           CONST=CON*YOM*H**3/((1.-PR**2)*12)
           DO 190 I=6,18,6
  190      C(I,I)=C(I,I)+0.1*CONST
           DO 230 I=1,18
           DO 230 J=1,18
           TEMP(I,J)=0.0
           TEMP1(I,J) = 0.0
           DO 220 L=1,18
              TEMP1(I,J)=TEMP1(I,J)+AMT(I,L)*TRM(L,J)
  220      TEMP(I,J)=TEMP(I,J)+C(I,L)* TRM(L,J)
  230      CONTINUE
           DO 250 I=1,18
           DO 250 J=I,18
           C(I,J)=0.0
              AMT(I,J) =0.0
           DO 240 L=1,18
              AMT(I,J) =AMT(I,J) +TRM(L,I)*TEMP1(L,J)
  240      C(I,J)=C(I,J)+TRM(L,I)*TEMP(L,J)
              AMT(J,I) = AMT(I,J)
  250      C(J,I) = C(I,J)
           RETURN
           END
C**********************************************
      SUBROUTINE TRANM (XI,EL12,AREA,X21,X31,
     1 Y21,Y31,Z21,Z31,TRM,DCM,1LK)
C**********************************************
      IMPLICIT REAL*8(A-H,O-Z)
C/////////////////////////////////////////////
```

```
C     TRANSFORMATION FROM LOCAL CARTISIAN TO
C     GLOBAL POLAR IS
C     DONE IN TWO STAGES. (DCM) SPECIFIES THE
C     TRANSFORMATION FROM LOCAL
C     CARTISIAN TO GLOBAL CARTISIAN.(T) GIVES
C     THE TRANSFORMATION FROM
C     GLOBAL CARTISIAN TO GLOBAL POLAR.IF
C     (X,Y,Z) IS THE LOCAL CARTISIAN
C      SYSTEM,
C//////////////////////////////////////////////
C                    | LX MX NX |
C          (DCM) =   | LY MY NY |
C                    | LZ MZ NZ |
C//////////////////////////////////////////////
C     LOCAL X IS ALONG (1-2) EDGE :HENCE ITS
C     DIRECTION COSINES
C     LOCAL Z IS PERPENDICULAR TO THE PLANE OF
C     THE ELEMENT:HENCE ITS DIRECTION COSINES
C     ARE OBTAINED AS RATIOS OF THE PROJECTED
C     AREAS (OF THE TRIANGULAR ELEMENT ONTO THE
C     COORDINATE PLANES) TO THE ACTUAL AREA.
C     REFER O.C.ZIENKIEWICZ CH.13
C//////////////////////////////////////////////
      DIMENSION XI(3,3),DCM(3,3),TRM(18,18),
     1T(18,18)
      DCM(1,1)=X21/EL12
      DCM(1,2)=Y21/EL12
      DCM(1,3)=Z21/EL12
              A1=1.0/(2.*AREA)
              DCM(3,1)=(Y21*Z31-Z21*Y31)*A1
              DCM(3,2)=(Z21*X31-X21*Z31)*A1
              DCM(3,3)=(X21*Y31-Y21*X31)*A1
      DCM(2,1)=DCM(3,2)*DCM(1,3)-DCM(3,3)
     1 *DCM(1,2)
      DCM(2,2)=DCM(3,3)*DCM(1,1)-
     1 DCM(3,1)*DCM(1,3)
      DCM(2,3)=DCM(3,1)*DCM(1,2)-
     1 DCM(3,2)*DCM(1,1)
      DO 10 I=1,18
         DO 10 J=1,18
10    TRM(I,J)=0.0
      DO 20 L=1,6
```

```
            DO 20 I=1,3
            DO 20 J=1,3
                IL=(L-1)*3+I
                JL=(L-1)*3+J
                    TRM(IL,JL)=DCM(I,J)
20      CONTINUE
        RETURN
        END
C**************************************************
        SUBROUTINE SHAPE(A1,A2,A3,B1,B2,B3,AL1,
    *   AL2,AL3,SHF)
C**************************************************
C****   SHAPE FUNCTIONS FOR LATERAL DISPLACEMENTS
        IMPLICIT REAL*8(A-H,O-Z)
        DIMENSION SHF(9)
        ST1=AL1*AL1*AL2
        ST2=AL2*AL2*AL1
        ST3=AL2*AL2*AL3
        ST4=AL3*AL3*AL2
        ST5=AL1*AL1*AL3
        ST6=AL3*AL3*AL1
        ST7=0.5*AL1*AL2*AL3
        SHF(1)=AL1-ST2+ST1-ST6+ST5
        SHF(2)=B3*(ST1+ST7)-B2*(ST5+ST7)
        SHF(3)=A3*(ST1+ST7)-A2*(ST5+ST7)
        SHF(4)=AL2-ST4+ST3-ST1+ST2
        SHF(5)=B1*(ST3+ST7)-B3*(ST2+ST7)
        SHF(6)=A1*(ST3+ST7)-A3*(ST2+ST7)
        SHF(7)=AL3-ST5+ST6-ST3+ST4
        SHF(8)=B2*(ST6+ST7)-B1*(ST4+ST7)
        SHF(9)=A2*(ST6+ST7)-A1*(ST4+ST7)
        RETURN
        END
C******************* .*************************
        SUBROUTINE DMAT(H,YOM,PR,BT,D)
C**************************************************
C       FINDS D MATRIX FOR BENDING STIFFNESS
        IMPLICIT REAL*8(A-H,O-Z)
        DIMENSION D(3,3)
        FR=(YOM*(H**3))/(12.0*(1-PR*PR))
        DO 10 I=1,3
            DO 10 J=1,3
```

```
10      D(I,J)=0.0
        D(1,1)=FR
        D(1,2)=PR*FR
                D(2,1)=D(1,2)
                D(2,2)=D(1,1)
        D(3,3)=BT*FR
        RETURN
        END
C*************************************************
        SUBROUTINE BMAT(AL1,AL2,AL3,A1,A2,A3,B1,
      * B2,B3,COF,B)
C*************************************************
C    OBTAIN B MATRIX FOR BENDING STIFFNESS MATRIX
        IMPLICIT REAL*8(A-H,O-Z)
        DIMENSION B(3,9),DSHF(6,9),COF(3,6)
        CALL DERISH(AL1,AL2,AL3,A1,A2,A3,B1,
      * B2,B3,DSHF)
        DO 20 I=1,3
           DO 20 J=1,9
                 B(I,J)=0.0
              DO 10 K=1,6
10               B(I,J)=B(I,J)+COF(I,K) * DSHF(K,J)
20      CONTINUE
        RETURN
        END
C*************************************************
        SUBROUTINE DERISH(AL1,AL2,AL3,A1,A2,A3,B1,
      * B2,B3,DSHF)
C*************************************************
C    CALCULATES S2 MATRIX FOR BENDING STIFFNESS
C    B MATRIX
        IMPLICIT REAL*8(A-H,O-Z)
        DIMENSION DSHF(6,9)
        DSHF(1,1)=2.0*(AL3+AL2)
        DSHF(1,2)=2.0*(B2*AL3-B3*AL2)
        DSHF(1,3)=2.0*(A2*AL3-A3*AL2)
        DSHF(1,4)=-2.0*AL2
        DSHF(1,5)=0.0
        DSHF(1,6)=0.0
        DSHF(1,7)=-2.0*AL3
        DSHF(1,8)=0.0
        DSHF(1,9)=0.0
```

```
        DSHF(2,1)=-2.0*AL1
        DSHF(2,2)=0.0
        DSHF(2,3)=0.0
        DSHF(2,4)=2.0*(AL1+AL3)
        DSHF(2,5)=2.0*(B3*AL1-B1*AL3)
        DSHF(2,6)=2.0*(A3*AL1-A1*AL3)
        DSHF(2,7)=-2.0*AL3
        DSHF(2,8)=0.0
        DSHF(2,9)=0.0
  DSHF(3,1)=-2.0*AL1
  DSHF(3,2)=0.0
  DSHF(3,3)=0.0
  DSHF(3,4)=-2.0*AL2
  DSHF(3,5)=0.0
  DSHF(3,6)=0.0
  DSHF(3,7)=2.0*(AL1+AL2)
  DSHF(3,8)=2.0*(B1*AL2-B2*AL1)
  DSHF(3,9)=2.0*(A1*AL2-A2*AL1)
        DSHF(4,1)=2.0*(AL1-AL2)
        DSHF(4,2)=(0.5*AL3*(B2-B3))-
*              (2.0*B3*AL1)
        DSHF(4,3)=(0.5*AL3*(A2-A3))-
*              (2.0*A3*AL1)
        DSHF(4,4)=2.0*(AL2-AL1)
        DSHF(4,5)=(0.5*AL3*(B3-B1))+
*              (2.0*B3*AL2)
        DSHF(4,6)=(0.5*AL3*(A3-A1))+
*              (2.0*A3*AL2)
        DSHF(4,7)=0.0
        DSHF(4,8)=0.5*AL3*(B1-B2)
        DSHF(4,9)=0.5*AL3*(A1-A2)
  DSHF(5,1)=0.0
  DSHF(5,2)=0.5*AL1*(B2-B3)
  DSHF(5,3)=0.5*AL1*(A2-A3)
  DSHF(5,4)=2.*(AL2-AL3)
  DSHF(5,5)=(0.5*AL1*(B3-B1))-(2.0*B1*AL2)
  DSHF(5,6)=(0.5*AL1*(A3-A1))-(2.0*A1*AL2)
  DSHF(5,7)=2.0*(AL3-AL2)
  DSHF(5,8)=(0.5*AL1*(B1-B2))+(2.0*B1*AL3)
  DSHF(5,9)=(0.5*AL1*(A1-A2))+(2.0*A1*AL3)
        DSHF(6,1)=2.0*(AL1-AL3)
        DSHF(6,2)=(0.5*AL2*(B2-B3))+
```

```
*                        (2.0*B2*AL1)
          DSHF(6,3)=(0.5*AL2*(A2-A3))+
*                        (2.0*A2*AL1)
          DSHF(6,4)=0.0
          DSHF(6,5)=(0.5*AL2*(B3-B1))
          DSHF(6,6)=(0.5*AL2*(A3-A1))
          DSHF(6,7)=2.0*(AL3-AL1)
          DSHF(6,8)=(0.5*AL2*(B1-B2))-
*                   ( 2.0*B2*AL3)
          DSHF(6,9)=(0.5*AL2*(A1-A2))-
*                        (2.0*A2*AL3)
      RETURN
      END
      SUBROUTINE ARR(CB,CP,TEMP)
C*************************************************
C     COMBINES INPLANE MATRIX AND BENDING MATRIX
      IMPLICIT REAL*8(A-H,O-Z)
      DIMENSION CP(6,6),CB(9,9),TEMP(18,18),
     * C(18,18)
      DO 10 I=1,18
          DO 10 J=1,18
              TEMP(I,J)=0.0
10    CONTINUE
C     REARRANGE INPLANE STIFFNESS MATRIX
      DO 20 I=1,6
          I1=(I-1)/2*4+I
          DO 20 J=1,6
              J1=(J-1)/2*4+J
              TEMP(I1,J1)=CP(I,J)
20    CONTINUE
C     REARRANGE BENDING STIFFNESS MATRIX
      DO 30 I=1,9
          I1=(I-1)/3*3+I+2
          DO 30 J=1,9
              J1=(J-1)/3*3+J+2
              TEMP(I1,J1)=CB(I,J)
30    CONTINUE
      RETURN
      END
      SUBROUTINE ALSHP(AL1,AL2,AL3)
C**** FOR COMPUTING THE CONSISTENT MASS
      IMPLICIT REAL*8 (A-H,O-Z)
```

```
        DIMENSION SHF(9),ASHF(3,18)
        COMMON/SHPE/ASHF,SHF
        DO 10 I=1,3
        DO 10 J=1,18
10      ASHF(I,J) = 0.0
        ASHF(1,1) = AL1
        ASHF(1,7) = AL2
        ASHF(1,13) = AL3
        ASHF(2,2) = AL1
        ASHF(2,8) = AL2
        ASHF(2,14) = AL3
        DO 20 I= 1,3
        DO 20 J= 1,3
        J1 = 3*(I-1) + J
        J2 = 6*(I-1) + J + 2
        ASHF(3,J2) = SHF(J1)
20      CONTINUE
        RETURN
        END
C***** PROGRAM:2.12
C***** THIS PROGRAM USES R.C.M ALGORITHM TO REARRANGE
C***** THE GIVEN NUMBERING FOR OPTIMUM BANDWIDTH.
C***** PROBLEM SOLVED IS GIVEN IN FIG.6.10
C***** PROGRAM USES SINGLE PRECISION ARITHMETIC.
C***** PROGRAM HOLDS FOR SINGLE D.O.FREEDOM
C***** THE PROGRAM USES THE FOLLOWING CORE:
C*****    FOUR ARRAYS WITH SIZE=NO.OF.NODES
C*****    A 2-DIM ARRAY OF SIZE(3,NO.OF ELEM)
C*****2 ARRAYS OF SIZE=MAX.NO.OF NODES CONNECTED
C***** TO ANY NODE(HERE UPTO 100 CAN BE HANDLED)
C***** THE PRESENT PROG. CAN HANDLE A MAX. OF
C***** 600 NODES AND A MAX.OF 600 ELEMENTS.
        INTEGER ELEMNT, ORDER(100),FIRST,COUNT,
     *  STORED,ORDRCW(100)
        LOGICAL PART,ELEMS2
        DIMENSION NA(2),NEWNOD(600)
        COMMON /A/X(600),Y(600)/B/NODES(3,600)
     *  C/M /D/MDIFF(600)
C**** ELEMNT THE ELEMENT CONTAINING REFERENCE OR
C**** STARTING NODE; ITS VALUE IS GOT BY
C*****SUBROUTINE SEARCH.
C*****ORDER:AN ARRAY WHICH STORES THE NODES
```

```
C***** CONN.TO THE STARTING NODE IN A CCW
C***** DIRECTION.
C*****COUNT : THE NUMBER OF NODES CONN.TO
C*****STARTING
C*****          NODE IN A CCW DIRECTION.
C***** PROFIL : STORES THE VALUE OF PROFILE.
C***** ORDRCW : ARRAY WHICH STORES THE NODES
C       CONNECTED
C*****  TO THE REFERENCE NODE IN A CW DIRECTION
C***** NEWNOD : AN ARRAY WHICH STORES THE
C****NEWNODE NO.S CORRESPONDING TO THE OLD NODES
C****NODES :ARRAY TO STORE THE NODAL
C****CONNECTIVITIES.
C*****  X,Y ARRAYS TO STORE THE NODAL CO-ORDS.
C***** NA,FIRST: TEMPORARY QUANTITIES.
C***** N=NO.OF NODES   M=NO.OF ELEMENTS
C*****          INPUT SECTION
       READ(3,*)N,M
       DO 100 I=1,N
       X(I)=0.0
       Y(I)=0.0
100    CONTINUE
C***** READING NODAL COORDINATES
       DO 110 J=1,N
       READ(3,*) K,X(J),Y(J)
110    CONTINUE
C****READ : NODAL COORDINATES INTO ARRAY 'NODES'
       DO 120 I=1,600
       NODES(1,I)=0
       NODES(2,I)=0
       NODES(3,I)=0
120    CONTINUE
       WRITE(4,130)
130    FORMAT(10X,'INPUT NODAL CONNECTIVITIES',//)
       DO 160 I=1,M,2
       I1=I+1
       READ(3,*) KK,NODES(1,I),NODES(2,I),
      * NODES(3,I)
       IF(I1.GT.M) GO TO 140
       READ (3,*) L,NODES(1,I1),NODES(2,I1),
      * NODES(3,I1)
140    WRITE(4,150) I,NODES(1,I),NODES(2,I),
```

```
      *  NODES(3,I),  I1,NODES(1,I1),NODES(2,I1)
      *  ,NODES(3,I1)
150      FORMAT(1X,2(I3,2X,3(I3,2X),2X))
160      CONTINUE
C*****  INITIALIZATIONS
         STORED = 0
C*****  STORED : STORES THE  NO.OF NODES NUMBERED
C*****  UPTO ANY  STAGE
         DO 170 I=1,600
         NEWNOD(I)=0
         MDIFF(I)=0
170      CONTINUE
         MMBW=0
         PROFIL=0
         NETLAB=0
         DO 340 JJ=1,N
         DO 180 I=1,N
         ORDER(I)=0
180      CONTINUE
         JJ1=N+1-JJ
         IF(JJ.EQ.1) NODE=1
         IF(JJ.EQ.1) GO TO 200
         DO 190 IL=1,N
C***    THE OLDNODE WITH NEWNODE NO.JJ/JJ1 IS FOUND
C***    OUT BY THIS LOOP TO GET STARTING NODE
         IF(NEWNOD(IL).NE.JJ1) GO TO 190
C***    USE JJ1 FOR REV.CUTHIL AND JJ FOR CUTHIL IN
C*****  THE ABOVE STATEMENT.
         NODE=IL
         GO TO 200
190      CONTINUE
200      ORDER(1)=NODE
         MAX=0
C*****  FIND AN ELEMENT HAVING STARTING NODE
         DO 210 K=1,M
         CALL FIND(NODE,K,PART,NUMBER)
         IF(.NOT.PART) GO TO 210
         ELEMNT=K
         FIRST=ELEMNT
         GO TO 220
210      CONTINUE
220      I1=1
```

```
C***** FIND OTHER TWO NODES OF THAT ELEMENT
       DO 230 I=1,3
       IF(I.EQ.NUMBER) GO TO 230
       NA(I1)=NODES(I,ELEMNT)
       I1=2
230    CONTINUE
C***** FIND THE NODE TOWARDS CCW DIRECTION
       CALL CCW(NA(1),NA(2),NODE,NCCW,NCW)
       ORDER(2)=NCW
       ORDER(3)=NCCW
C***    FIND OTHER CCW NODES CONN.TO STARTING NODE
       DO 240 KK=4,N
       I2=KK-1
       CALL SEARCH(NODE,ORDER(I2),ELEMNT,NODE3
     * ,NELEM,
     * ELEMS2)
       IF(.NOT.ELEMS2) COUNT=KK-1
       IF(.NOT.ELEMS2) GO TO 250
       IF(NODE3.EQ.ORDER(2)) COUNT=KK-1
C**    IF NODE3 EQUALS ORDER(2) THEN NODES
C**    CONNECTED TO STARTING NODE FORM A CLOSED LOOP
       IF(NODE3.EQ.ORDER(2)) GO TO 250
       ORDER(KK)=NODE3
       ELEMNT=NELEM
240    CONTINUE
       COUNT=KK-1
250    IF(NODE3.EQ.ORDER(2)) GO TO 300
       N1=ORDER(1)
       N2=ORDER(2)
C*****FIND NODES IN CW CONN. TO STARTING NODE
       DO 270 KJ=1,N
       CALL SEARCH(N1,N2,FIRST,NODE3,NELEM
     * ,ELEMS2)
       IF(ELEMS2) GO TO 260
C**MAX :STORES THE NUMBER  OF CONNECTED CW NODES
       MAX=KJ-1
       GO TO 280
260    ORDRCW(KJ)=NODE3
       FIRST=NELEM
       N2=NODE3
270    CONTINUE
C***** STORE THE CONNECTED CCW NODES
```

```
280        DO 290 IL=2,COUNT
           LL=COUNT-IL+2
           LLMAX=LL+MAX
           ORDER(LLMAX) = ORDER(LL)
290        CONTINUE
300        MAXIM=COUNT+MAX
C**        MAXIM :THE TOTAL NUMBER OF NODES CONN. TO THE
C*****     STARTING NODE IN BOTH CW AND CCW DIRNS
           IF(MAX.EQ.0.0) GO TO 320
           DO 310 IN=1,MAX
           KN=MAX+2-IN
           ORDER(KN)=ORDRCW(IN)
310        CONTINUE
320        CONTINUE
C*****     RENUMBER
           DO 330 IT=1,MAXIM
           ITK=ORDER(IT)
           IF(NEWNOD(ITK).NE.0) GO TO 330
           INDEX=STORED+1
           IND=N+1-INDEX
           NEWNOD(ITK)=IND
C*****     EQUATE NEWNOD(ITK) TO 'INDEX' FOR CUTHIL-MCKEE
C*****     ALGORITHM;EQUATE TO 'IND' FOR REV.CUTHIL-MCKEE
           STORED=STORED+1
           IF(STORED.EQ.N) GO TO 350
C***       WHEN 'STORED' EQUALS 'N' ALL NODES WOULD'VE
C***       BEEN NUMBERED. SO EXIT FROM THE LOOP.
330        CONTINUE
340        CONTINUE
C*****     OUTPUT
350        DO 360 I=1,M
           NE1=NODES(1,I)
           NE2=NODES(2,I)
           NE3=NODES(3,I)
           CALL BWIDTH(NE1,NE2,NE3,MBW,MDIFF)
           IF(MBW.GT.MMBW) MMBW=MBW
360        CONTINUE
           DO 380 IM=1,N
           PROFIL=PROFIL+MDIFF(IM)+1
           ME=MDIFF(IM)+1
           NETLAB=NETLAB+(2*ME)+(ME*(ME+1)/2)
380        CONTINUE
```

```
      WRITE(4,390)MMBW,PROFIL,NETLAB
390   FORMAT(10X,'MAX.BANDWIDTH OF THE INPUT=',
     * I3//5X, 'PROFILE=',I5, 5X,'LABOUR=', I6,
     * '(REF.TABLE 2.3)')
C*****   PRINT NEW NUMBERING
      WRITE(4,400)
400   FORMAT(1X,2(1X,'OLD',3X,'NEW',17X))
      WRITE(4,410)
410   FORMAT(1X,2(1X,'NODE',1X,'NODE',4X,'X',
     * 6X,'Y', 5X))
      DO 430 I=1,N,2
      I1=I+1
      WRITE(4,420) I,NEWNOD(I),X(I),Y(I),I1,
     * NEWNOD(I1),X(I1),Y(I1)
420   FORMAT(1X,2(1X,I3,2X,I3,2X,F6.2,1X,F6.2,3X))
430   CONTINUE
      WRITE(4,440)
440   FORMAT(1X,2('ELEMENT',2X,'NEW',3X,'NEW',3X,
     * 'NEW',3X))
      WRITE(4,450)
450   FORMAT(1X,2('NUMBER',2X,'NODE1',1X,'NODE2',
     * 1X, 'NODE3',2X))
      DO 470 I=1,M,2
      NE1=NODES(1,I)
      NE2=NODES(2,I)
      NE3=NODES(3,I)
      I1=I+1
      NF1=NODES(1,I1)
      NF2=NODES(2,I1)
      NF3=NODES(3,I1)
      NODES(1,I)=NEWNOD(NE1)
      NODES(2,I)=NEWNOD(NE2)
      NODES(3,I)=NEWNOD(NE3)
      NODES(1,I1)=NEWNOD(NF1)
      NODES(2,I1)=NEWNOD(NF2)
      NODES(3,I1)=NEWNOD(NF3)
      WRITE(4,460) I,NEWNOD(NE1),NEWNOD(NE2),
     * NEWNOD(NE3),I1,NEWNOD(NF1),NEWNOD(NF2),
     * NEWNOD(NF3)
460   FORMAT(1X,2(2X,I3,4X,I3,3X,I3,3X,I3,3X))
470   CONTINUE
C***** PRINT OUTPUT'S PROFILE,MAX,BWIDTH,LABOR
```

```
        MMBW=0
        PROFIL=0
        NETLAB=0
        DO 480 I=1,N
        MDIFF(I)=0
480     CONTINUE
        DO 490 IL=1,M
        I1=NODES(1,IL)
        I2=NODES(2,IL)
        I3=NODES(3,IL)
        CALL BWIDTH(I1,I2,I3,MBW,MDIFF)
        IF(MBW.GT.MMBW) MMBW=MBW
490     CONTINUE
        DO 500 IM=1,N
        PROFIL=PROFIL+MDIFF(IM)+1
        ME=MDIFF(IM)+1
        NETLAB=NETLAB+(2*ME)+(ME*(ME+1)/2)
500     CONTINUE
        WRITE(4,510) MMBW,PROFIL,NETLAB
510     FORMAT(3X//1X,'MAX.BWIDTH OF REARRANGED
      * MATRIX=',I3//3X,'PROFILE=',I5,2X,'TOTAL
      * LABOUR=',I6,'(REFER TABLE 2.3)')
        STOP
        END
C
        SUBROUTINE BWIDTH(I1,I2,I3,MBW,MDIFF)
C**     BWIDTH FINDS THE MAX.DIFF.AMONG I1,I2,I3.
C*****  IT ALSO FINDS THE DIFF.FROM THE MAXIMUM
C*****  (AMONG I1,I2,I3) AND STORES +FOR EACH
C*****   NODE IN MDIFF
        DIMENSION MDIFF(600)
        LARGE=I1
        IF(I2.GT.LARGE.AND.I2.GT.I3) GO TO 520
        IF(I3.GT.LARGE) GO TO 530
        MM=LARGE-I2
        NN=LARGE-I3
        GO TO 540
520     LARGE=I2
        MM=LARGE-I1
        NN=LARGE-I3
        GO TO 540
530     LARGE=I3
```

```
          MM=LARGE-I1
          NN=LARGE-I2
540       MBW=MM
          IF(MM.LT.NN) MBW=NN
          LMM=LARGE-MM
          LNN=LARGE-NN
          IF(MDIFF(LMM).LT.MM) MDIFF(LMM)=MM
          IF(MDIFF(LNN).LT.NN) MDIFF(LNN)=NN
          RETURN
          END


          SUBROUTINE CCW(N1,N2,NODE,NCCW,NCW)
C*        CCW GETS THE MORE COUNTERCLOCKWISE OF N1& N2
C*        RELATIVE TO STARTING NODE;STORES THIS IN NCCW.
          COMMON/A/X(600),Y(600)
          PI=4*ATAN(1.0)
          CALL ANGLE(N1,NODE,ANGLE1)
          CALL ANGLE(N2,NODE,ANGLE2)
          IF(ABS(ANGLE1-ANGLE2).GT.PI) GO TO 560
          IF(ANGLE1.GT.ANGLE2) GO TO 550
          NCCW=N2
          NCW=N1
          GO TO 580
550       NCCW=N1
          NCW=N2
          GO TO 580
560       IF(ANGLE1.GT.ANGLE2) GO TO 570
          NCCW=N1
          NCW=N2
          GO TO 580
570       NCCW=N2
          NCW=N1
          GO TO 580
580       RETURN
          END


          SUBROUTINE ANGLE(NODE1,NODE,ANG)
C***      ANGLE GETS THE ANGLE IN RADIANS MADE BY THE
C**       SEGMENT JOINING 'NODE1'AND'NODE'WITH +VE X-AX
          COMMON/A/X(600),Y(600)
          PI=4*ATAN(1.0)
          NM=0
```

```
      XDIFF=X(NODE1)-X(NODE)
      YDIFF=Y(NODE1)-Y(NODE)
      IF(XDIFF.EQ.0.0)GO TO 610
      IF(YDIFF.EQ.0.0)GO TO 590
      SLOPE=YDIFF/XDIFF
      RADIAN=ATAN(SLOPE)
         IF(XDIFF.GT.0.0.AND.YDIFF.GT.0.0)
    * ANG=RADIAN
         IF(XDIFF.LT.0.0.AND.YDIFF.GT.0.0)
    * ANG=PI+RADIAN
         IF(XDIFF.LT.0.0.AND.YDIFF.LT.0.0)
      *ANG=PI+RADIAN
         IF(XDIFF.GT.0.0.AND.YDIFF.LT.0.0)
      *ANG=2.0*PI+RADIAN
      GO TO 630
590   IF(XDIFF.GT.0.0)GO TO 600
      ANG=PI
      GO TO 630
600   ANG=0.0
      GO TO 630
610   IF(YDIFF.GT.0.0) GO TO 620
      ANG=3.0*PI/2.0
      GO TO 630
620   ANG=PI/2.0
630   RETURN
      END
         SUBROUTINE FIND(NODENO,NELEM, PART,
      NUMBER)
C**   'FIND' CHECKS WHETHER 'NODENO' IS A PART OF
C***  'NELEM'.IF SO,'PART' IS '.TRUE.'
      LOGICAL  PART,L,M,N
      COMMON/B/NODES(3,600)
      NUMBER=0
      L=(NODENO.EQ.NODES(1,NELEM))
      M=(NODENO.EQ.NODES(2,NELEM))
      N=(NODENO.EQ.NODES(3,NELEM))
      IF(L)NUMBER=1
      IF(M)NUMBER=2
      IF(N)NUMBER=3
      PART=L.OR.M.OR.N
      RETURN
      END
```

```
      SUBROUTINE SEARCH(N1,N2,NEL,NODE3,NELEM,
   *ELEMS2)
C***  THIS SUBROUTINE FINDS A SECOND ELEMENT
C**** 'NELEM' OTHER THAN NEL,HAVING NODES 'N1' &
C**** 'N2'.'NODE3'IS THE THIRD NODE OF THIS
C***** ELEMENT.'ELEMS2'IS TRUE.'IF SUCH AN
C***** ELEMENT EXISTS.
      LOGICAL ELEMS2,PART
      COMMON/B/NODES(3,600)/C/M
      ELEMS2=.FALSE.
      NELEM=0
      NODE3=0
      DO 650 I=1,M
      IF(I.EQ.NEL)GO TO 650
      CALL FIND(N1,I,PART,NUMBER)
      IF(.NOT.PART)GO TO 650
      CALL FIND(N2,I,PART,NUMBER)
      IF(.NOT.PART)GO TO 650
      NELEM=I
      ELEMS2=.TRUE.
      DO 640 J=1,3
      IF(NODES(J,NELEM).EQ.N1) GO TO 640
      IF(NODES(J,NELEM).EQ.N2) GO TO 640
      NODE3=NODES(J,NELEM)
640   CONTINUE
650   CONTINUE
      RETURN
      END
```

```
         INPUT NODAL CONNECTIVITIES
    1    1    9    8    2    1    2    9
    3    2    3   10    4    2   10    9
    5    3    4   11    6    3   11   10
    7    4    5   12    8    4   12   11
    9    5    6   13   10    5   13   12
   11    6    7   14   12    6   14   13
   13    8    9   16   14    8   16   15
   15    9   10   17   16    9   17   16
   17   10   11   18   18   10   18   17
   19   11   12   19   20   11   19   18
   21   12   13   20   22   12   20   19
   23   13   14   21   24   13   21   20
   25   15   16   30   26   15   30   29
```

27	16	17	31	28	16	31	30
29	17	18	32	30	17	32	31
31	18	19	33	32	18	33	32
33	19	20	34	34	19	34	33
35	20	21	35	36	20	35	34
37	21	22	36	38	21	36	35
39	22	23	37	40	22	37	36
41	23	24	38	42	23	38	37
43	24	25	39	44	24	39	38
45	25	26	40	46	25	40	39
47	26	27	41	48	26	41	40
49	27	28	42	50	27	42	41
51	29	30	44	52	29	44	43
53	30	31	45	54	30	45	44
55	31	32	46	56	31	46	45
57	32	33	47	58	32	47	46
59	33	34	48	60	33	48	47
61	34	35	49	62	34	49	48
63	35	36	50	64	35	50	49
65	36	37	51	66	36	51	50
67	37	38	52	68	37	52	51
69	38	39	53	70	38	53	52
71	39	40	54	72	39	54	53
73	40	41	55	74	40	55	54
75	41	42	56	76	41	56	55
77	43	44	62	78	43	62	61
79	44	45	63	80	44	63	62
81	45	46	64	82	45	64	63
83	46	47	65	84	46	65	64
85	47	48	66	86	47	66	65
87	48	49	67	88	48	67	66
89	49	50	68	90	49	68	67
91	50	51	69	92	50	69	68
93	51	52	70	94	51	70	69
95	52	53	71	96	52	71	70
97	53	54	72	98	53	72	71
99	54	55	73	100	54	73	72
101	55	56	74	102	55	74	73
103	56	57	75	104	56	75	74
105	57	58	76	106	57	76	75
107	58	59	77	108	58	77	76
109	59	60	78	110	59	78	77

111	61	62	80	112	61	80	79
113	62	63	81	114	62	81	80
115	63	64	82	116	63	82	81
117	64	65	83	118	64	83	82
119	65	66	84	120	65	84	83
121	66	67	85	122	66	85	84
123	67	68	86	124	67	86	85
125	68	69	87	126	68	87	86
127	69	70	88	128	69	88	87
129	70	71	89	130	70	89	88
131	71	72	90	132	71	90	89
133	72	73	91	134	72	91	90
135	73	74	92	136	73	92	91
137	74	75	93	138	74	93	92
139	75	76	94	140	75	94	93
141	76	77	95	142	76	95	94
143	77	78	96	144	77	96	95
145	79	80	98	146	79	98	97
147	80	81	99	148	80	99	98
149	81	82	100	150	81	100	99
151	82	83	101	152	82	101	100
153	83	84	102	154	83	102	101
155	84	85	103	156	84	103	102
157	85	86	104	158	85	104	103
159	86	87	105	160	86	105	104
161	87	88	106	162	87	106	105
163	88	89	107	164	88	107	106
165	89	90	108	166	89	108	107
167	90	91	109	168	90	109	108
169	100	101	111	170	100	111	110
171	101	102	112	172	101	112	111
173	102	103	113	174	102	113	112
175	103	104	114	176	103	114	113
177	104	105	115	178	104	115	114
179	105	106	116	180	105	116	115
181	106	107	117	182	106	117	116
183	107	108	118	184	107	118	117
185	108	109	119	186	108	119	118
187	116	117	121	188	117	118	122
189	116	121	120	190	118	122	121
191	120	124	123	192	121	122	125
193	120	121	124	194	121	125	124

195	123	124	127	196	124	125	128
197	123	127	126	198	124	128	127
199	126	127	130	200	127	128	131
201	126	130	129	202	127	131	130
203	129	130	133	204	130	131	134
205	129	133	132	206	130	134	133
207	132	133	136	208	133	134	137
209	132	136	135	210	133	137	136
211	135	136	139	212	136	137	140
213	135	139	138	214	136	140	139
215	138	139	142	216	139	140	143
217	138	142	141	218	139	143	142
219	141	142	145	220	142	143	146
221	141	145	144	222	142	146	145
223	144	145	148	224	145	146	149
225	144	148	147	226	145	149	148
227	147	148	151	228	148	149	152
229	147	151	150	230	148	152	151
231	153	135	138	232	153	138	154
233	154	138	141	234	154	141	155
235	155	141	144	236	155	144	156
237	156	144	147	238	156	147	157
239	157	147	150	240	157	150	158
241	159	156	157	242	159	157	160
243	160	157	158	244	160	158	161
245	137	162	163	246	137	163	140
247	140	163	164	248	140	164	143
249	143	164	165	250	143	165	146
251	146	165	166	252	146	166	149
253	149	166	167	254	149	167	152
255	162	168	169	256	162	169	163
257	163	169	170	258	163	170	164
259	164	170	171	260	164	171	165
261	165	171	172	262	165	172	166
263	166	172	173	264	166	173	167
265	170	174	175	266	170	175	171
267	171	175	176	268	171	176	172
269	172	176	177	270	172	177	173
271	174	178	179	272	174	179	175
273	175	179	180	274	175	180	176
275	176	180	181	276	176	181	177
277	181	180	182	278	182	184	183

279	180	184	182	280	184	185	183
281	180	179	184	282	184	186	185
283	179	186	184	284	186	187	185
285	179	178	186	286	186	188	187
287	178	188	186	288	188	189	187
289	178	190	188	290	191	189	188
291	190	191	188	292	191	192	189
293	190	193	191	294	191	194	192
295	193	194	191	296	194	195	192

MAX.BANDWIDTH OF THE INPUT= 26

PROFILE= 2214   LABOUR= 23045(REF.TABLE 2.3)

OLD NODE	NEW NODE	X	Y	OLD NODE	NEW NODE	X	Y
1	195	0.00	0.00	2	194	25.00	0.00
3	191	50.00	0.00	4	186	75.00	0.00
5	179	100.00	0.00	6	170	123.00	0.00
7	159	143.00	0.00	8	192	0.00	10.00
9	193	25.00	10.00	10	190	50.00	10.00
11	185	75.00	10.00	12	178	100.00	10.00
13	169	123.00	10.00	14	158	143.00	10.00
15	187	0.00	20.00	16	188	25.00	20.00
17	189	50.00	20.00	18	184	75.00	20.00
19	177	100.00	20.00	20	168	123.00	20.00
21	157	143.00	20.00	22	146	156.00	20.00
23	133	166.00	20.00	24	121	171.00	20.00
25	114	174.75	20.00	26	105	178.50	20.00
27	94	183.50	20.00	28	84	193.50	20.00
29	180	0.00	25.00	30	181	25.00	25.00
31	182	50.00	25.00	32	183	75.00	25.00
33	176	100.00	25.00	34	167	123.00	25.00
35	156	143.00	25.00	36	145	156.00	25.00
37	132	166.00	25.00	38	120	171.00	25.00
39	113	174.75	25.00	40	104	178.50	25.00
41	93	183.50	25.00	42	83	193.50	25.00
43	171	0.00	32.50	44	172	25.00	32.50
45	173	50.00	32.50	46	174	75.00	32.50
47	175	100.00	32.50	48	166	123.00	32.50
49	155	143.00	32.50	50	144	156.00	32.50
51	131	166.00	32.50	52	119	171.00	32.50
53	112	174.75	32.50	54	103	178.50	32.50
55	92	183.50	32.50	56	82	193.50	32.50
57	76	217.50	32.50	58	70	232.50	32.50

59	64	242.50	32.50	60	55	250.00	32.50
61	160	0.00	37.50	62	161	25.00	37.50
63	162	50.00	37.50	64	163	75.00	37.50
65	164	100.00	37.50	66	165	123.00	37.50
67	154	143.00	37.50	68	143	156.00	37.50
69	130	166.00	37.50	70	118	171.00	37.50
71	111	174.75	37.50	72	102	178.50	37.50
73	91	183.50	37.50	74	81	193.50	37.50
75	75	217.50	37.50	76	69	232.50	37.50
77	63	242.50	37.50	78	54	250.00	37.50
79	147	0.00	42.50	80	148	25.00	42.50
81	149	50.00	42.50	82	150	75.00	42.50
83	151	100.00	42.50	84	152	123.00	42.50
85	153	143.00	42.50	86	142	156.00	42.50
87	129	166.00	42.50	88	117	171.00	42.50
89	110	174.75	42.50	90	101	178.50	42.50
91	90	183.50	42.50	92	80	193.50	42.50
93	74	217.50	42.50	94	68	232.50	42.50
95	62	242.50	42.50	96	53	250.00	42.50
97	134	0.00	50.00	98	135	25.00	50.00
99	136	50.00	50.00	100	137	75.00	50.00
101	138	100.00	50.00	102	139	123.00	50.00
103	140	143.00	50.00	104	141	156.00	50.00
105	128	166.00	50.00	106	116	171.00	50.00
107	109	174.75	50.00	108	100	178.50	50.00
109	89	183.50	50.00	110	122	75.00	64.00
111	123	100.00	64.00	112	124	123.00	64.00
113	125	143.00	64.00	114	126	156.00	64.00
115	127	166.00	64.00	116	115	171.00	64.00
117	108	174.75	64.00	118	99	178.50	64.00
119	88	183.50	64.00	120	106	171.00	98.00
121	107	174.75	98.00	122	98	178.50	98.00
123	95	171.00	137.00	124	96	174.75	137.00
125	97	178.50	137.00	126	85	171.00	177.00
127	86	174.75	177.00	128	87	178.50	177.00
129	77	171.00	216.00	130	78	174.75	216.00
131	79	178.50	216.00	132	71	171.00	256.00
133	72	174.75	256.00	134	73	178.50	256.00
135	65	171.00	295.00	136	66	174.75	295.00
137	67	178.50	295.00	138	57	171.00	310.00
139	58	174.75	310.00	140	59	178.50	310.00
141	46	171.00	325.00	142	47	174.75	325.00

143	48	178.50	325.00	144	38	171.00	335.00
145	39	174.75	335.00	146	40	178.50	335.00
147	28	171.00	350.00	148	29	174.75	350.00
149	30	178.50	350.00	150	15	171.00	365.00
151	16	174.75	365.00	152	17	178.50	365.00
153	56	166.00	295.00	154	45	166.00	310.00
155	37	166.00	325.00	156	27	166.00	335.00
157	14	166.00	350.00	158	6	166.00	165.00
159	13	151.00	335.00	160	5	151.00	350.00
161	2	151.00	360.00	162	61	188.50	295.00
163	60	188.50	310.00	164	49	188.50	325.00
165	41	188.50	335.00	166	31	188.50	350.00
167	18	188.50	365.00	168	52	196.00	295.00
169	51	196.00	310.00	170	50	196.00	325.00
171	42	196.00	335.00	172	32	196.00	350.00
173	19	196.00	365.00	174	44	211.00	325.00
175	43	211.00	335.00	176	33	211.00	350.00
177	20	211.00	365.00	178	36	230.00	325.00
179	35	230.00	335.00	180	34	230.00	350.00
181	21	230.00	365.00	182	22	240.00	365.00
183	7	250.00	365.00	184	23	240.00	350.00
185	8	250.00	350.00	186	24	240.00	335.00
187	10	250.00	335.00	188	25	240.00	325.00
189	9	250.00	325.00	190	26	230.00	310.00
191	11	240.00	310.00	192	3	250.00	310.00
193	12	230.00	295.00	194	4	240.00	295.00
195	1	250.00	295.00	196	0	0.00	0.00

ELEMENT NUMBER	NEW NODE1	NEW NODE2	NEW NODE3	ELEMENT NUMBER	NEW NODE1	NEW NODE2	NEW NODE3
1	195	193	192	2	195	194	193
3	194	191	190	4	194	190	193
5	191	186	185	6	191	185	190
7	186	179	178	8	186	178	185
9	179	170	169	10	179	169	178
11	170	159	158	12	170	158	169
13	192	193	188	14	192	188	187
15	193	190	189	16	193	189	188
17	190	185	184	18	190	184	189
19	185	178	177	20	185	177	184
21	178	169	168	22	178	168	177
23	169	158	157	24	169	157	168
25	187	188	181	26	187	181	180

27	188	189	182	28	188	182	181
29	189	184	183	30	189	183	182
31	184	177	176	32	184	176	183
33	177	168	167	34	177	167	176
35	168	157	156	36	168	156	167
37	157	146	145	38	157	145	156
39	146	133	132	40	146	132	145
41	133	121	120	42	133	120	132
43	121	114	113	44	121	113	120
45	114	105	104	46	114	104	113
47	105	94	93	48	105	93	104
49	94	84	83	50	94	83	93
51	180	181	172	52	180	172	171
53	181	182	173	54	181	173	172
55	182	183	174	56	182	174	173
57	183	176	175	58	183	175	174
59	176	167	166	60	176	166	175
61	167	156	155	62	167	155	166
63	156	145	144	64	156	144	155
65	145	132	131	66	145	131	144
67	132	120	119	68	132	119	131
69	120	113	112	70	120	112	119
71	113	104	103	72	113	103	112
73	104	93	92	74	104	92	103
75	93	83	82	76	93	82	92
77	171	172	161	78	171	161	160
79	172	173	162	80	172	162	161
81	173	174	163	82	173	163	162
83	174	175	164	84	174	164	163
85	175	166	165	86	175	165	164
87	166	155	154	88	166	154	165
89	155	144	143	90	155	143	154
91	144	131	130	92	144	130	143
93	131	119	118	94	131	118	130
95	119	112	111	96	119	111	118
97	112	103	102	98	112	102	111
99	103	92	91	100	103	91	102
101	92	82	81	102	92	81	91
103	82	76	75	104	82	75	81
105	76	70	69	106	76	69	75
107	70	64	63	108	70	63	69
109	64	55	54	110	64	54	63

111	160	161	148	112	160	148	147
113	161	162	149	114	161	149	148
115	162	163	150	116	162	150	149
117	163	164	151	118	163	151	150
119	164	165	152	120	164	152	151
121	165	154	153	122	165	153	152
123	154	143	142	124	154	142	153
125	143	130	129	126	143	129	142
127	130	118	117	128	130	117	129
129	118	111	110	130	118	110	117
131	111	102	101	132	111	101	110
133	102	91	90	134	102	90	101
135	91	81	80	136	91	80	90
137	81	75	74	138	81	74	80
139	75	69	68	140	75	68	74
141	69	63	62	142	69	62	68
143	63	54	53	144	63	53	62
145	147	148	135	146	147	135	134
147	148	149	136	148	148	136	135
149	149	150	137	150	149	137	136
151	150	151	138	152	150	138	137
153	151	152	139	154	151	139	138
155	152	153	140	156	152	140	139
157	153	142	141	158	153	141	140
159	142	129	128	160	142	128	141
161	129	117	116	162	129	116	128
163	117	110	109	164	117	109	116
165	110	101	100	166	110	100	109
167	101	90	89	168	101	89	100
169	137	138	123	170	137	123	122
171	138	139	124	172	138	124	123
173	139	140	125	174	139	125	124
175	140	141	126	176	140	126	125
177	141	128	127	178	141	127	126
179	128	116	115	180	128	115	127
181	116	109	108	182	116	108	115
183	109	100	99	184	109	99	108
185	100	89	88	186	100	88	99
187	115	108	107	188	108	99	98
189	115	107	106	190	99	98	107
191	106	96	95	192	107	98	97
193	106	107	96	194	107	97	96

195	95	96	86	196	96	97	87
197	95	86	85	198	96	87	86
199	85	86	78	200	86	87	79
201	85	78	77	202	86	79	78
203	77	78	72	204	78	79	73
205	77	72	71	206	78	73	72
207	71	72	66	208	72	73	67
209	71	66	65	210	72	67	66
211	65	66	58	212	66	67	59
213	65	58	57	214	66	59	58
215	57	58	47	216	58	59	48
217	57	47	46	218	58	48	47
219	46	47	39	220	47	48	40
221	46	39	38	222	47	40	39
223	38	39	29	224	39	40	30
225	38	29	28	226	39	30	29
227	28	29	16	228	29	30	17
229	28	16	15	230	29	17	16
231	56	65	57	232	56	57	45
233	45	57	46	234	45	46	37
235	37	46	38	236	37	38	27
237	27	38	28	238	27	28	14
239	14	28	15	240	14	15	6
241	13	27	14	242	13	14	5
243	5	14	6	244	5	6	2
245	67	61	60	246	67	60	59
247	59	60	49	248	59	49	48
249	48	49	41	250	48	41	40
251	40	41	31	252	40	31	30
253	30	31	18	254	30	18	17
255	61	52	51	256	61	51	60
257	60	51	50	258	60	50	49
259	49	50	42	260	49	42	41
261	41	42	32	262	41	32	31
263	31	32	19	264	31	19	18
265	50	44	43	266	50	43	42
267	42	43	33	268	42	33	32
269	32	33	20	270	32	20	19
271	44	36	35	272	44	35	43
273	43	35	34	274	43	34	33
275	33	34	21	276	33	21	20
277	21	34	22	278	22	23	7

279	34	23	22	280	23	8	7
281	34	35	23	282	23	24	8
283	35	24	23	284	24	10	8
285	35	36	24	286	24	25	10
287	36	25	24	288	25	9	10
289	36	26	25	290	11	9	25
291	26	11	25	292	11	3	9
293	26	12	11	294	11	4	3
295	12	4	11	296	4	1	3

MAX.BWIDTH OF REARRANGED MATRIX= 16

PROFILE=2138;TOTAL LABOUR=17970(REFER TABLE 2.3)

```
C****** PROGRAM : 3.1
C****** FINDING THE FREQUENCY RANGES OF EXAMPLE
C****** 3.14 BY STURM SEQUENCE.
        IMPLICIT REAL*8(A-H,O-Z)
        DIMENSION AK1(7,4),AM1(7,4),AK(7,4)
        DATA AK1/10.0D0,20.0D0,10.0D0,10.0D0,20.0D0,
      1 30.0D0,20.0D0,10.0D0,-10.0D0,0.D0,
      2 4*-10.0D0,9*0.0D0,-10.0D0,5*0.0D0/
        DATA AM1/28*0.0D0/
        N = 7
        KBAND = 4
        DO 25 I = 1,4
   25   AM1(I,1) = 1.0
        AM1(5,1) = 2.0
        AM1(6,1) = 5.0
        AM1(7,1) = 5.0
        WRITE(6,150)
        WRITE(6,50)((AK1(I,J),J=1,KBAND),I=1,N)
   50   FORMAT(4F10.2)
        WRITE(6,200)
        WRITE(6,50)((AM1(I,J),J=1,KBAND),I=1,N)
  150   FORMAT(//5X,'STIFFNES MATRIX AS','SUPPLIED'//)
  200   FORMAT(//5X,'MASS MATRIX AS SUPPLIED'//)
        CALL STURM(AK1,AM1,AK,N,KBAND)
        STOP
        END
        SUBROUTINE STURM(AK1,AM1,AK,N,KBAND)
C******* PURPOSE AND METHOD:-
C*******    TO FIND THE FREQUENCY RANGE USING
C*******    STURM SEQUENCE. THIS OPERATES ON THE
C*******    PRINCIPLE THAT THE NUMBER OF NEGATIVE
C*******    ELEMENTS IN THE 'D' MATRIX OF 'LDLT'
C*******    DECOMPOSITION OF (AK - FREQ*AM1)
C*******    REPRESENTS THE FREQUENCIES BELOW THE
C*******    PARTICULAR VALUE OF 'FREQ'.SINCE THE
C*******    'D' MATRIX CORRESPONDS TO THE DIAGONAL
C*******    ELEMENTS OF THE UPPER TRIANGULAR
C*******    MATRIX OBTAINED BY GAUSSIAN
C*******    ELIMINATION ONLY THIS SCHEME IS USED.
C*******    DESCRIPTION OF PARAMETERS:-
C*******    AK1    STIFFNESS MATRIX IN BANDED
C*******           FORM (N,KBAND).
```

```
C*******   AM1    MASS MATRIX IN BANDED FORM.
C*******          (N,KBAND).
C*******   AK1 & AM1 ARE NOT DESTROYED.
C*******   N      NUMBER OF EQUATIONS.
C*******   KBAND  BANDWIDTH.
C*******   OMEGA  FREQUENCY VALUE . READ IN THE
C*******          SUBROUTINE ITSELF.THIS IS DONE
C*******          TO FACILITATE THE STURM-SEQUENCE
C*******          CHECK BY CALLING THE SUBROUTINE
C*******          ONLY ONCE.THE SCHEME ENDS WHEN
C*******          A NEGATIVE VALUE IS SUPPLIED.
C*******   AK   WORKING MATRIX OF SIZE (N,KBAND)
C*******SUBROUTINE &FUNCTION SUBPROGRAM REQUIRED
C*******     NONE.
C******* NOTE:-
C*******     THE PROGRAM OPERATES IN DOUBLE
C*******     PRECISION ARITHMETIC.
C*******     SUPPLY AK1 & AM1 IN BANDED FORM.
C*******     THAT IS,THE COLUMN NUMBER STARTS
C*******     FROM THE DIAGONAL ELEMENT OF
C*******     THE RESPECTIVE ROW.
       IMPLICIT REAL*8(A-H,O-Z)
       DIMENSION AK1(N,KBAND),AM1(N,KBAND),
     1 AK(N,KBAND)
       WRITE(6,550)
 50    READ 100,OMEGA
100    FORMAT(F10.5)
       DO 150 I = 1,N
       DO 150 J = 1,KBAND
150    AK(I,J) = AK1(I,J)
       FREQ = OMEGA**2
       IF (OMEGA)1000,200,200
200    DO 300 I = 1,N
       DO 300 J = 1,KBAND
       AK(I,J) = AK(I,J) - FREQ * AM1(I,J)
       IF (AK(I,1)) 300,900,300
300    CONTINUE
       N1 = N - 1
       NBH = KBAND - 1
       DO 475 I = 1,N1
       II = I + 1
       IF (AK(I,1)) 320,900,320
```

```
320      RATIO = 1./AK(I,1)
         IHR = I + NBH
         IF (IHR - N) 360,360,340
340      IHR = N
360      DO 460 J = II,IHR
         L = J - I + 1
         DO 450 K = 1,NBH
         JK = J + K - 1
         IF (JK - N) 380,380,450
380      M = J - I + K
         IF (M-KBAND) 400,400,450
400      TERM = AK(I,L)*AK(I,M)*RATIO
         AK(J,K) = AK(J,K) - TERM
         IF ((DABS(AK(J,K)) - 1D-8)) 420,420,450
420      AK(J,K) = 0.0
450      CONTINUE
460      CONTINUE
475      CONTINUE
         KOUNT = 0
         DO 500 I = 1,N
         IF (AK(I,1).LT.0.0) KOUNT = KOUNT + 1
500      CONTINUE
550      FORMAT(/3X,'OMEGA',10X,'OMEGASQR',6X,
        1 'FREQUENCIES'/35X,'BELOW'/)
         PRINT 600,OMEGA,FREQ,KOUNT
600      FORMAT(/,F10.5,1X,F15.5,7X,I4)
         GO TO 50
900      PRINT 950,OMEGA
950      FORMAT(//,10X,' ****** THE SCHEME DOES',
        1 ' NOT WORK'/17X,'FOR OMEGA = ',F10.5/
        2   8X,'PLEASE TRY SOME OTHER VALUE OF','OMEGA')
         GO TO 50
1000     RETURN
         END

    STIFFNES MATRIX AS SUPPLIED
       10.00     -10.00      0.00      0.00
       20.00       0.00      0.00    -10.00
       10.00     -10.00      0.00      0.00
       20.00     -10.00      0.00      0.00
       30.00     -10.00      0.00      0.00
       20.00     -10.00      0.00      0.00
       10.00       0.00      0.00      0.00
```

MASS MATRIX AS SUPPLIED

1.00	0.00	0.00	0.00
1.00	0.00	0.00	0.00
1.00	0.00	0.00	0.00
1.00	0.00	0.00	0.00
2.00	0.00	0.00	0.00
5.00	0.00	0.00	0.00
5.00	0.00	0.00	0.00

OMEGA	OMEGASQR	FREQUENCIES BELOW
5.75000	33.06250	7
5.50000	30.25000	6
5.25000	27.56250	6
5.10000	26.01000	5
3.75000	14.06250	5
3.50000	12.25000	4
2.25000	5.06250	4
2.21250	4.89516	3

```
****** THE SCHEME DOES NOT WORK
       FOR OMEGA =    2.00000
 PLEASE TRY SOME OTHER VALUE OF OMEGA
```

1.90000	3.61000	2
1.50000	2.25000	2
1.00000	1.00000	1
0.00000	0.00000	0

COMPILE TIME= 0.39 SEC, EXECUTION TIME= 0.48 SEC

```
C****** PROGRAM : 3.2
C****** FINDING EIGEN PAIRS OF EXAMPLE 3.14 BY
C       SECANT ITERATION SCHEME.
        IMPLICIT REAL*8(A-H,O-Z)
        DIMENSION AK1(7,4),AM1(7,4),AK(7,4),
     1  FREQ(6,2),OMEGA(6)
        DATA AK1/10.0D0,20.0D0,10.0D0,20.0D0,
     1  30.0D0,20.0D0,10.0D0,-10.0D0,0.D0,
     2  4*-10.0D0,9*0.0D0,-10.0D0,5*0.0D0/
        DATA AM1/28*0.0D0/
        N = 7
        KBAND = 4
        DO 25 I = 1,4
25      AM1(I,1) = 1.0
        AM1(5,1) = 2.0
        AM1(6,1) = 5.0
        AM1(7,1) = 5.0
        FREQ(1,1) = 2.25
        FREQ(1,2) = 1.00
        FREQ(2,1) = 3.99001
        FREQ(2,2) = 3.61
        FREQ(3,1) = 5.0625
        FREQ(3,2) = 4.9284
        FREQ(4,1) = 14.0625
        FREQ(4,2) = 12.25
        FREQ(5,1) = 27.5625
        FREQ(5,2) = 26.01
        FREQ(6,1) = 33.0625
        FREQ(6,2) = 30.25
        WRITE(6,150)
        WRITE(6,50)((AK1(I,J),J=1,KBAND),I=1,N)
50      FORMAT(4F10.2)
        WRITE(6,200)
        WRITE(6,50)((AM1(I,J),J=1,KBAND),I=1,N)
150     FORMAT(//5X,'STIFFNES MATRIX AS',
     1  ' SUPPLIED'//)
200     FORMAT(//5X,'MASS MATRIX AS SUPPLIED'//)
250     FORMAT(/6X,'OMEGA',9X,'OMEGASQR'/)
        NEIG = 6
        NITE = 15
        IDECIM = 3
        IPRINT = 1
```

```
        CALL SECANT(AK1,AM1,AK,N,KBAND,FREQ,
     1  OMEGA,NEIG,NITE,IDECIM,IPRINT)
        WRITE(6,250)
        DO 300 I = 1,6
        FREQ(I,1) = OMEGA(I)**2
        WRITE(6,400)OMEGA(I),FREQ(I,1)
300     CONTINUE
400     FORMAT(2X,F10.5,F15.5)
        STOP
        END
        SUBROUTINE SECANT(AK1,AM1,AK,N,KBAND,
     1  FREQ,OMEGA,NEIG,NITE,IDECIM,IPRINT)
C******* METHOD AND PURPOSE:-
C*******    DETERMINES THE NATURAL FREQUENCY BY
C*******    THE SECANT ITERATION SCHEME. THIS
C*******    REQUIRES THE APPROXIMATE RANGE OF THE
C*******    PARTICULAR FREQUENCY TO BE SUPPLIED
C*******    AS AN ARRAY FREQ. THE RANGE CAN BE
C*******    OBTAINED BY PERFORMING STURM-SEQUENCE
C*******     CHECK OF THE SYSTEM USING THE
C*******      SUBROUTINE 'STURM'.
C******* DESCRIPTION OF PARAMETERS:-
C*******    AK1   STIFFNESS MATRIX IN BANDED
C*******          FORM (N,KBAND).
C*******    AM1   MASS MATRIX IN BANDED FORM
C*******          (N,KBAND)
C*******          AK1 & AM1 ARE NOT DESTROYED.
C*******    N     NUMBER OF EQUATIONS.
C*******    KBAND BANDWIDTH.
C*******    FREQ  ARRAY OF FREQUENCY RANGE.
C*******          (NEIG,2).
C*******          (NEIG,1) -- UPPER LIMIT.
C*******          (NEIG,2) -- LOWER LIMIT.
C*******    OMEGA STORES AND RETURNS THE VALUE
C*******          OF 'OMEGA' FOR LATER USE.
C*******    NEIG  NUMBER OF FREQUENCIES.
C*******    NITE  NUMBER OF SECANT ITERATIONS
C*******          TO BE PERFORMED.
C*******    IDECIM NUMBER OF DIGITS ACCURACY
C*******          REQUIRED IN 'OMEGASQR' VALUE.
C*******    IPRINT PRINT OPTION FOR INTERMEDIATE
C*******          VALUES.
```

```
C*******          1 -- PRINTING IS DONE.
C*******          0 -- SKIPS PRINTING.
C*******SUBROUTINE & FUNCTION SUBPROGRAM
C*******REQUIRED :          DETERM.
C******* NOTE:-
C*******    THE PROGRAM OPERATES IN DOUBLE
C*******    PRECISION ARITHMETIC.
C*******    SUPPLY AK1 AND AM1 IN BANDED FORM.
C*******    THAT IS, THE COLUMN NUMBER STARTS
C*******    FROM THE DIAGONAL ELEMENT OF THE
C*******    RESPECTIVE ROW.
C*******
      IMPLICIT REAL*8(A-H,O-Z)
      DIMENSION AK1(N,KBAND),AM1(N,KBAND),
    1 AK(N,KBAND),FREQ(NEIG,2),OMEGA(NEIG)
      ABS(X) = DABS(X)
      DO 2000 I = 1,NEIG
      CHEK = 0.0
      OMCUR = FREQ(I,1)
      OMOLD = FREQ(I,2)
      CALL DETERM(AK1,AM1,AK,N,KBAND,OMCUR,
    1 ANR,IWARN)
      CALL DETERM(AK1,AM1,AK,N,KBAND,OMOLD,
    1 DR1,IWARN)
      KOUNT = 0
50    DR = ANR - DR1
      KOUNT = KOUNT + 1
      OMNEW = OMCUR - ANR/DR*(OMCUR - OMOLD)
      CALL DETERM(AK1,AM1,AK,N,KBAND,OMNEW,
    1 DETNEW,IWARN)
      IF (IWARN.EQ.1) GO TO 900
      DIFF = ABS(OMNEW - CHEK)
      IF (DIFF.LE.(1./10**IDECIM)) GO TO 1000
      IF (KOUNT.EQ.NITE) GO TO 1200
      IF (ANR.LT.0) GO TO 100
      GO TO 500
100   IF (DETNEW) 150,1000,200
150   OMCUR = OMNEW
      ANR = DETNEW
      CHEK = OMNEW
      GO TO 50
200   OMOLD = OMNEW
```

```
              DR1 = DETNEW
              CHEK = OMNEW
              GO TO 50
       500    IF (DETNEW) 650,1000,700
       650    OMOLD = OMNEW
              DR1 = DETNEW
              CHEK = OMNEW
              GO TO 50
       700    OMCUR = OMNEW
              ANR = DETNEW
              CHEK = OMNEW
              GO TO 50
       900    IF (IPRINT.EQ.0) GO TO 975
              WRITE(6,950)I
              WRITE(6,1820)I,FREQ(I,1),I,FREQ(I,2)
       950    FORMAT(/2X,'PROGRAM REQUIRES DIFFERENT',
             1 ' STARTING VALUE',/2X,'FOR FREQUENCY',I5/ )
       975    OMEGA(I) = 0.0
              GO TO 2000
      1000    IF (IPRINT.EQ.1) WRITE(6,1050)I,KOUNT
      1050    FORMAT(/2X,'SUCCESSFUL CONVERGENCE',
             1 ' FOR FREQUENCY',I2,' KOUNT=',I3)
              GO TO 1800
      1200    IF (IPRINT.EQ.1) WRITE(6,1250)I
      1250    FORMAT(/2X,'***NO CONVERGENCE FOR',
             1 ' FREQUENCY',I5)
              GO TO 1800
      1800    OMEGA(I) = DSQRT(OMNEW)
              IF (IPRINT.EQ.0) GO TO 2000
              WRITE(6,1820)I,FREQ(I,1),I,FREQ(I,2)
      1820    FORMAT(2X,'FREQ(',I2,',1)=',F10.5,' FREQ
             1 (',I2,',2)=',F10.5)
              WRITE(6,1850)I,OMEGA(I),OMNEW
      1850    FORMAT(/2X,'OMEGA(',I2,')=',F10.5,'
             1 OMEGA','SQR=',F10.5,/)
      2000    CONTINUE
              RETURN
              END
              SUBROUTINE DETERM(AK1,AM1,AK,N,KBAND,
             1 FREQ,PROD,IWARN)
C******* PURPOSE AND METHOD:-
C*******    THIS FINDS THE DETERMINANT OF THE
```

```
C*******    MATRIX (AK1- FREQ*AM1). THE MATRIX IS
C*******    CONVERTED AS AN UPPER TRIANGULAR ONE
C*******    BY GAUSSIAN ELIMINATION. THE PRODUCT
C*******    OF THE DIAGONAL ELEMENTS GIVES THE
C*******    VALUE OF THE DETERMINANT.
C******* DESCRIPTION OF PARAMETERS:-
C*******    AK1     STIFFNESS MATRIX IN BANDED FORM
C*******            (N,KBAND).
C*******            NOT DESTROYED N COMPUTATION.
C*******    AM1     MASS MATRIX IN BANDED FORM
C*******            (N,KBAND)
C*******            NOT DESTROYED IN COMPUTATION.
C*******    N       NUMBER OF EQUATIONS.
C*******    KBAND   BANDWIDTH
C*******    FREQ    FREQUENCY SQUARE. THIS IS
C*******            SLIGHTLY ALTERED IN THE
C*******            ROUTINE IF ONE OF THE DIAGONAL
C*******            ELEMENTS OF 'AK' BECOMES ZERO.
C*******            THE DETERMINANT IS RECOMPUTED,
C*******            AND THE NEW 'FREQ' IS RETURNED.
C*******            ALTERATION OF 'FREQ' IS DONE
C*******            FOR A MAXIMUM OF TEN TIMES TO
C*******            AVOID ENTERING INTO INFINITE
C*******            LOOPING.
C*******    AK      WORKING VARIABLE OF SIZE
C*******            (N,KBAND)
C******* SUBROUTINE & FUNCTION SUBPROGRAM
C******* REQUIRED:   NONE.
C******* NOTE:-
C*******     THE PROGRAM OPERATES IN DOUBLE
C*******     PRECISION ARITHMETIC.
C*******     THE BANDED FORM OF THE MATRIX IS
C*******     PRESERVED WHILE PERFORMING
C*******     GAUSSIAN ELIMINATION.
C*******
        IMPLICIT REAL*8(A-H,O-Z)
        DIMENSION AK1(N,KBAND),AM1(N,KBAND),
      1 AK(N,KBAND)
        ABS(X) = DABS(X)
        KOUNT = 0
  100   IWARN = 0
        DO 150 I = 1,N
```

```
          DO 150 J = 1,KBAND
150       AK(I,J) = AK1(I,J)
200       DO 300 I = 1,N
          DO 300 J = 1,KBAND
          AK(I,J) = AK(I,J) - FREQ * AM1(I,J)
          IF (AK(I,1)) 300,900,300
300       CONTINUE
          N1 = N - 1
          NBH = KBAND - 1
          DO 475 I = 1,N1
          II = I + 1
          IF (AK(I,1)) 320,900,320
320       RATIO = 1./AK(I,1)
          IHR = I + NBH
          IF (IHR - N) 360,360,340
340       IHR = N
360       DO 460 J = II,IHR
          L = J - I + 1
          DO 450 K = 1,NBH
          JK = J + K - 1
          IF (JK - N) 380,380,450
380       M = J - I + K
          IF (M-KBAND) 400,400,450
400       TERM = AK(I,L)*AK(I,M)*RATIO
          AK(J,K) = AK(J,K) - TERM
          IF ((ABS(AK(J,K)) - 1D-8)) 420,420,450
420       AK(J,K) = 0.0
450       CONTINUE
460       CONTINUE
475       CONTINUE
          PROD = AK(1,1)
          DO 500 I = 2,N
          PROD = PROD*AK(I,1)
500       CONTINUE
          GO TO 1000
900       KOUNT = KOUNT + 1
          FREQ = FREQ*1.001
          IF (KOUNT.LE.10) GO TO 100
          FREQ = FREQ/1.001
          IWARN=1
1000      RETURN
          END
```

STIFFNES MATRIX AS SUPPLIED

10.00	-10.00	0.00	0.00
20.00	0.00	0.00	-10.00
10.00	-10.00	0.00	0.00
20.00	-10.00	0.00	0.00
30.00	-10.00	0.00	0.00
20.00	-10.00	0.00	0.00
10.00	0.00	0.00	0.00

MASS MATRIX AS SUPPLIED

1.00	0.00	0.00	0.00
1.00	0.00	0.00	0.00
1.00	0.00	0.00	0.00
1.00	0.00	0.00	0.00
2.00	0.00	0.00	0.00
5.00	0.00	0.00	0.00
5.00	0.00	0.00	0.00

SUCCESSFUL CONVERGENCE FOR FREQUENCY 1 KOUNT= 4
  FREQ(1,1)=   2.25000 FREQ(1,2)=   1.00000

  OMEGA(1)=   1.24106 OMEGASQR=   1.54023

SUCCESSFUL CONVERGENCE FOR FREQUENCY 2 KOUNT= 3
  FREQ(2,1)=   3.99001 FREQ(2,2)=   3.61000

  OMEGA(2)=   1.95440 OMEGASQR=   3.81969

SUCCESSFUL CONVERGENCE FOR FREQUENCY 3 KOUNT= 3
  FREQ(3,1)=   5.06250 FREQ(3,2)=   4.92840

  OMEGA(3)=   2.23606 OMEGASQR=   4.99998

SUCCESSFUL CONVERGENCE FOR FREQUENCY 4 KOUNT= 5
  FREQ(4,1)= 14.06250 FREQ(4,2)= 12.25000

  OMEGA(4)=   3.65534 OMEGASQR= 13.36150

SUCCESSFUL CONVERGENCE FOR FREQUENCY 5 KOUNT= 4

```
FREQ( 5,1)=  27.56250 FREQ( 5,2)=  26.00999

OMEGA( 5)=   5.11923 OMEGASQR=  26.20653

SUCCESSFUL CONVERGENCE FOR FREQUENCY 6 KOUNT= 11
FREQ( 6,1)=  33.06250 FREQ( 6,2)=  30.25000

OMEGA( 6)=   5.57651 OMEGASQR=  31.09751


    OMEGA        OMEGASQR

   1.24106       1.54023
   1.95440       3.81969
   2.23606       4.99998
   3.65534      13.36150
   5.11923      26.20653
   5.57651      31.09751
COMPILE TIME= 0.84 SEC, EXECUTION TIME= 1.47 SEC
```

```
C****** PROGRAM : 3.3
C****** FINDING THE FIRST FEW EIGEN PAIRS OF
C        EXAMPLE 3.13 BY INVERSE ITERATION SCHEME
C        COUPLED WITH GRAM-SCHMIDT DEFLATION
C        TECHNIQUE.
      IMPLICIT REAL*8 (A-H,O-Z)
      DIMENSION STIF(100),DSTIF(100),AMASS(10),
     1 EIGEN(4,4),X(10),XD(10),
     2 ADD(9),PROD(10),OMEGA(10)
      READ 20,NSYS,NEIGEN,NPROB,NITER,KOD
      PRINT 25,NSYS,NEIGEN,NPROB,NITER,KOD
      READ 30,(STIF(I),I=1,16),(X(I),AMASS(I),
     1 I=1,4)
      PRINT 40,(STIF(I),I=1,16),(AMASS(I),
     1I=1,4),(X(I),I=1,4)
      NEIG=NEIGEN-1
      DO 10 I=1,NPROB
      CALL INVERS (STIF,AMASS,X,EIGEN,XD,ADD,
     1 NSYS,NEIGEN,NEIG,NITER,PROD,DSTIF, KOD,
     1 OMEGA)
10    PRINT 50,(OMEGA(N),(EIGEN(M,N),M=1,4),
     1 N=1,4)
20    FORMAT (5I2)
25    FORMAT(1X,'NSYS = ',I2,' NEIGEN = ',I2,
     1' NPROB = ',I2,' NITER = ',I2,' KOD =',I2)
30    FORMAT (12F6.0)
40    FORMAT (//3X,'STIFFNESS MATRIX'//4(2X,
     14F10.2/),//3X,'MASS MATRIX'//3X,4F6.1
     2//3X,'STARTING VECTOR'//3X,4F6.1//)
50    FORMAT (1X,'EIGEN FREQUENCY',5X,'EIGEN ',
     1'VECTORS'//(1X,F10.4,2X,4F8.4))
      STOP
      END
      SUBROUTINE INVERS (STIF,AMASS,X,EIGEN,XD,
     1 ADD,NSYS,NEIGEN,NEIG,NITER,PROD,DSTIF,
     2 KOD,OMEGA)
C**** METHOD:-
C****       INVERSE ITERATION SCHEME COUPLED WITH
C****       GRAM-SCHMIDT DEFLATION TECHNIQUE.
C**** DESCRIPTION OF PARAMETERS :-
C**** STIF     STIFFNESS MATRIX (NSYS,NSYS)
C**** AMASS    MASS MATRIX (NSYS)
```

```
C**** EIGEN      EIGEN VECTORS STORED COLUMNWISE
C****            (NEIGEN,NSYS).
C**** NSYS       NUMBER OF EQUATIONS
C**** NEIGEN     NUMBER OF EIGEN VECTORS TO BE
C****EVALUATED
C**** X          ASSUMED TRIAL VALUES OF EIGEN VECTOR
C****            (NSYS)
C**** NPROB      NUMBER OF PROBLEMS TO BE SOLVED
C**** NITER      NUMBER OF ITERATIONS REQUIRED
C****            FOR EACH EIGEN VECTOR.
C**** XD     DUMMY VECTOR USED IN 'INVERS' (NSYS)
C**** ADD    DUMMY VECTOR USED IN 'INVERS' (NSYS)
C**** DSTIF DUMMY MATRIX USED IN 'ITER'
C****            (NSYS, NSYS)
C**** PROD   DUMMY VECTOR USED IN 'ITER' (NSYS)
C**** KOD    IF KOD=0 NO INTERMEDIATE PRINT OUTS
C****            IN  SUBROUTINES.
C**** OMEGA      EIGEN FREQUENCIES (NSYS)
C**** REMARKS :-
C****              THE PROGRAM OPERATES IN DOUBLE
C****              PRECISION ARITHMETIC.
C****              APPLICABLE FOR DIAGONAL MASS
C****              MATRIX ONLY.
C**** SUBROUTINES & FUNCTION SUBPROGRAM REQUIRED
C****              'ITER','NORMAL' AND 'GAUSS'.
C****
      IMPLICIT REAL*8 (A-H,O-Z)
      DIMENSION STIF(NSYS,NSYS),DSTIF(NSYS,NSYS)
      DIMENSION EIGEN(NEIGEN,NSYS),X(NSYS),
     1 ADD(NEIG),XD(NSYS),OMEGA(NSYS),
     2 MASS(NSYS) ,PROD(NSYS)
      DO 10 I=1,NSYS
10    XD(I)=X(I)
      DO 20 JP=1,NSYS
20    EIGEN(1,JP)=0
      DO 100 IX=1,NEIGEN
      NFREQ=IX-1
      IF(NFREQ.EQ.0)NFREQ=1
      IF(KOD.EQ.0)GO TO 40
      PRINT 110,IX
40    CONTINUE
C**** GRAM-SCHMIDT ORTHOGONALISATION
```

```
      DO 80 JX=1,NITER
      DO 60 I=1,NFREQ
      ADD(I)=0
      DO 60 J=1,NSYS
60    ADD(I)=ADD(I)+EIGEN(I,J)*AMASS(J)*X(J)
      DO 70 J=1,NSYS
      DO 70 I=1,NFREQ
70    X(J)=X(J)-ADD(I)*EIGEN(I,J)
80    CALL ITER (X,AMASS,JX,STIF,PROD,DSTIF,
     1 NSYS,OMSQR,KOD)
      OMEGA(IX)=OMSQR
      DO 90 I=1,NSYS
      EIGEN(IX,I)=X(I)
90    X(I)=XD(I)
100   CONTINUE
110   FORMAT(///25X,'EIGEN VALUE NUMBER', I2,'IS' ,//)
      RETURN
      END
      SUBROUTINE ITER (PHI,AM,IJK,AK,PROD,AKDM,
     1NS,OMEGA,KOD)
C**** CARRIES OUT INVERSE ITERATION
      IMPLICIT REAL*8(A-H,O-Z)
      DIMENSION PHI(NS),AM(NS),PROD(NS),
     1AKDM(NS,NS),AK(NS,NS)
      DO 10 I=1,NS
      PROD(I)=AM(I)*PHI(I)
      DO 10 J=1,NS
10    AKDM(I,J)=AK(I,J)
      CALL GAUSS(AKDM,NS,PROD,1,PHI)
      CALL NORMAL (PHI,AM,NS)
      DO 20 I=1,NS
      PROD(I)=0
      DO 20 J=1,NS
20    PROD(I)=PROD(I)+AK(I,J)*PHI(J)
      OMEGA=0
      DO 30 I=1,NS
30    OMEGA=OMEGA+PHI(I)*PROD(I)
      IF(KOD.EQ.0)GO TO 50
      PRINT 40,IJK,OMEGA,(PHI(I),I=1,NS)
40    FORMAT(//5X,'EIGEN FREQUENCY AFTER
     1 ITERATION',I2,' = ',F14.4/
     2/10X,'THE CORRESPONDING EIGEN VECTOR IS '
```

```
       3/10X,4E15.8/)
50     CONTINUE
       RETURN
       END
       SUBROUTINE NORMAL (PHI,AM,NS)
C****  NORMALISES MASS MATRIX
       IMPLICIT REAL*8(A-H,O-Z)
       DIMENSION PHI(NS),AM(NS)
       ADD=0
       DO 13 I=1,NS
13     ADD=ADD+PHI(I)*AM(I)*PHI(I)
       ANORM=DSQRT(ADD)
       DO 14 I=1,4
14     PHI(I)=PHI(I)/ANORM
       RETURN
       END
NSYS = 4 NEIGEN = 4 NPROB = 1 NITER = 10 KOD=  0

   STIFFNESS MATRIX
        15.00    -10.00      0.00      0.00
       -10.00     25.00    -15.00      0.00
         0.00    -15.00     35.00    -20.00
         0.00      0.00    -20.00     20.00


   MASS MATRIX

     10.0  20.0  30.0  40.0


   STARTING VECTOR

      1.0    0.0    0.0    0.0
EIGEN FREQUENCY      EIGEN VECTORS

     0.0304    0.0601   0.1582   0.1924   0.1838
     0.5993    0.0883   0.1425  -0.0022  -0.1480
     1.5002    0.1035   0.0181  -0.1253   0.0821
     2.2866    0.1103  -0.0914   0.0629  -0.0236
COMPILE TIME= 0.80 SEC, EXECUTION TIME= 1.48 SEC
```

```
C******  PROGRAM :3.4
C******  SOLVING THE EXAMPLE PROBLEM 3.14
C        BY SIMULTANEOUS ITERATION METHOD.
C        (NON-STANDARD EIGEN VALUE PROBLEM IS
C        TRANSFORMED INTO STANDARD
C        EIGEN VALUE PROBLEM.)
         IMPLICIT REAL*8(A-H,O-Z)
         COMMON ALFA
         DIMENSION AK(7,4),AM(7,4)
         DATA AK/10.0D0,20.0D0,10.0D0,20.0D0,
       * 30.0D0,20.0D0,10.0D0,-10.0D0,0.D0,
       * 4*-10.0D0,9*0.0D0,-10.0D0,5*0.0D0/
         DATA AM/4*1.0D0,2.0D0,5.0D0,5.0D0,
       * 21*0.0D0/
         WRITE(6,100)
         NEQ=7
         KB=4
         ALFA=2.0
         WRITE(6,110)
         DO 60 I=1,NEQ
60       WRITE(6,130)(AK(I,J),J=1,KB)
         WRITE(6,100)
         WRITE(6,120)
         DO 70 I=1,NEQ
70       WRITE(6,130)(AM(I,J),J=1,KB)
100      FORMAT(//)
110      FORMAT(5X,'STIFFNESS MATRIX IN BANDED
       1 FORM')
120      FORMAT(5X,'MASS MATRIX IN BANDED FORM')
130      FORMAT(5X,10F8.3)
         CALL STFRQN(AK,AM,7,4,4,12,0)
         STOP
         END
         SUBROUTINE STFRQN(ST,SM,NEQ,KB,NEIG,
       * NIT,KPRNT)
C*******  METHOD AND PURPOSE:-
C*******    TO SOLVE FOR THE FIRST FEW
C*******   EIGEN VALUES AND EIGEN VECTORS IN THE
C*******   GENERALISED EIGEN VALUE PROBLEM USING
C*******   THE SIMULTANEOUS ITERATION METHOD
C*******   THIS PROGRAM CONVERTS THE GENERALISED
C*******   EIGEN VALUE PROBLEM INTO STANDARD
```

```
C*******    EIGEN VALUE PROBLEM BEFORE EXTRACTING
C*******    EIGEN VALUES AND EIGEN VECTORS
C*******    DESCRIPTION OF PARAMETERS:
C******* ST(NEQ,KB)   STIFFNESS MATRIX IN
C*******               BANDED FORM
C******* SM(NEQ,KB)   MASS MATRIX IN BANDED FORM
C******* NEQ           NUMBER OF EQUATIONS
C******* KB            BANDWIDTH OF ST AND SM
C******* NEIG          NUMBER OF REQUIRED
C*******               EIGEN PAIRS
C******* NIT           NUMBER OF ITERATIONS
C******* KPRNT         FLAG FOR PRINTING DURING
C*******               ITERATION
C*******               EQ.0. NO PRINTING
C*******               EQ.1. PRINT
C******* BE(I,I)       EIGEN FREQUENCIES
C******* U(NEQ,NEIG    EIGEN VECTORS
C******* ALFA          ZERO FOR DEFINITE SYSTEMS
C*******               NON-ZERO FOR SEMI-DEFINITE
C*******               SYSTEMS
C*******          WORKING VARIABLES:
C******* (TO BE DIMENSIONED APPROPRIATELY)
C*******    Q(NEQ)     U(NEQ,NEIG)    V(NEQ,NEIG)
C******* AU(NEQ,NEIG)   D(NEIG,NEIG)  BE(NEIG,NEIG)
C******* TE(NEIG,NEIG)   S(NEIG,NEIG)  D(NEIG,NEIG)
C*******             SUBROUTINES USED:
C******* 1. UPTRI FOR UPPER TRIANGULARISATION
C******* 2. SOLUT FOR FORWARD SUBSTITUTION
C******* 3. SOLU FOR BACKWARD SUBSTITUTION
C******* NOTE:-THE PROGRAM OPERATES IN
C******* DOUBLE PRECISION ARITHMETIC.
        IMPLICIT REAL*8 (A-H,O-Z)
        COMMON ALFA
        DIMENSION ST(NEQ,KB),SM(NEQ,KB),Q(7)
        DIMENSION V(7,4),AU(7,4),BE(4,4),
     *  TE(4,7),S(4,7),D(4,4),U(7,4)
        ABS(X)=DABS(X)
        SQRT(X)=DSQRT(X)
        IN=NEIG
        NDF=NEQ
        DO 5 I=1,NEQ
        DO 5 J=1,KB
```

```
5         ST(I,J)=ST(I,J)+ALFA*SM(I,J)
C*******UPPER TRIANGULARISATION OF ST
          CALL UPTRI (ST,NEQ,KB)
C*******ESTABLISH STARTING ITERATION VECTORS
          DO 30 J=1,NEIG
          DO 10 I=J,NEQ
          ENI=NEQ
          EI=I
10        U(I,J)=1.*(ENI+1.-EI)/ENI
          DO 20 I=1,J
          EJ=J
          EI=I
20        U(I,J)=EI/EJ
30        CONTINUE
          DO 40 I=1,NEIG
40        U(I,I)=1.
          DO 50 I=1,NEQ
          DO 50 J=1,NEIG
50        AU(I,J)=U(I,J)
C*******START OF ITERATION LOOP
          DO 450 II=1,NIT
          DO 80 J=1,NEIG
          DO 60 I=1,NEQ
60        Q(I)=U(I,J)
C*******BACKWARD SUBSTITUTION
          CALL SOLU (ST,NEQ,KB,Q)
          DO 70 I=1,NEQ
70        U(I,J)=Q(I)
80        CONTINUE
          DO 130 J=1,NEIG
          DO 120 I=1,NEQ
          V(I,J)=0.
          IF (I.EQ.1) GO TO 100
          IM=I-1
          DO 90 K1=1,IM
          I1=I-K1+1
          IF (I1.GT.KB) GO TO 90
          V(I,J)=V(I,J)+SM(K1,I1)*U(K1,J)
90        CONTINUE
100       DO 110 K2=I,NEQ
          I2=K2-I+1
          IF (I2.GT.KB) GO TO 110
```

```
          V(I,J)=V(I,J)+SM(I,I2)*U(K2,J)
110       CONTINUE
120       CONTINUE
130       CONTINUE
          DO160 J=1,NEIG
          DO 140 I=1,NEQ
140       Q(I)=V(I,J)
C*******FORWARD SUBSTITUTION
          CALL SOLUT (ST,NEQ,KB,Q)
          DO 150 I=1,NEQ
150       V(I,J)=Q(I)
160       CONTINUE
C*******INTERACTION ANALYSIS
          DO 170 I=1,IN
          DO 170 J=1,IN
          BE(I,J)=0.
          DO 170 L=1,NDF
170       BE(I,J)=BE(I,J)+AU(L,I)*V(L,J)
          IF(KPRNT.EQ.0)GO TO 190
          IF(II.EQ.1)WRITE(6,1060)
          WRITE(6,1070)II
          DO 180 I=1,IN
180       WRITE(6,1010)(BE(I,J),J=1,IN)
C*******PREDICTION OF LATENT VECTORS
190       DO 270 I=1,IN
          DO 270 J=1,IN
          IF (I-J) 200,260,200
200       P=-2.*BE(I,J)
          T=BE(I,I)-BE(J,J)
          IF (T) 220,210,220
210       SI=1.
          GO TO 230
220       SI=T/ABS(T)
230       SQ=SQRT(T**2+ 4.*P**2)*SI
          IF (P) 250,240,250
240       TE(I,J)=0.
          GO TO 270
250       TE(I,J)=P/(T+SQ)
          GO TO 270
260       TE(I,J)=1.
270       CONTINUE
          DO 280 I=1,NDF
```

```
          DO 280 J=1,IN
          U(I,J)=0.
          DO 280 L=1,IN
280       U(I,J)=U(I,J)+V(I,L)*TE(L,J)
C*******ORTHONORMALISE THE RESULT VECTORS
290       DO 300 I=1,IN
          DO 300 J=1,IN
          D(I,J)=0.
          DO 300 L=1,NDF
300       D(I,J)=D(I,J)+U(L,I)*U(L,J)
          K=1
310       DO 390 J=1,IN
          DO 390 I=1,IN
          IF (J-K) 390,320,360
320       IF (I-K) 330,340,350
330       DD=ABS(D(K,K))
          S(I,J)=D(I,J)/SQRT(DD)
          GO TO 390
340       DD=ABS(D(K,K))
          S(I,J)=1./SQRT(DD)
          GO TO 390
350       S(I,J)=0.
          GO TO 390
360       IF(I-K)370,380,370
370       S(I,J)=D(I,J)-D(I,K)*D(K,J)/D(K,K)
          GO TO 390
380       S(I,J)=-D(K,J)/D(K,K)
390       CONTINUE
          DO 400 I=1,IN
          DO 400 J=1,IN
400       D(I,J)=S(I,J)
          K=K+1
          IF (K-IN) 310,310,410
410       DO 420 I=1,NDF
          DO 420 J=1,IN
          V(I,J)=0.
          DO 420 L=1,IN
420       V(I,J)=V(I,J)+U(I,L)*D(L,J)
          DO 440 I=1,NDF
          DO 440 J=1,IN
430       U(I,J)=V(I,J)
440       AU(I,J)=U(I,J)
```

```
450       CONTINUE
C*******END OF ITERATION LOOP
C*******CALCULATE NATURAL FREQUENCIES
          DO 460 I=1,NEIG
          XX=ABS(1.0/BE(I,I)-ALFA)
460       BE(I,I)=SQRT(XX)
C*******PRINT NATURAL FREQUENCIES
          WRITE(6,1000)
          WRITE(6,1010)(BE(I,I),I=1,NEIG)
C*******PRINT EIGEN VECTORS
          WRITE(6,1020)
          DO 490 J=1,NEIG
          DO 470 I=1,NEQ
470       Q(I)=U(I,J)
C*******OBTAIN ACTUAL EIGEN VECTORS
          CALL SOLU(ST,NEQ,KB,Q)
          DO 480 I=1,NEQ
480       U(I,J)=Q(I)
490       CONTINUE
          WRITE(6,1030)(I,I=1,NEIG)
          DO 500 I=1,NEQ
500       WRITE(6,1040)(U(I,J),J=1,NEIG)
1000      FORMAT(//17X,'NATURAL FREQUENCIES ARE'//)
1010      FORMAT(/4(3X,F9.3))
1020      FORMAT(/5X,' EIGEN VECTORS ARE AS FOLLOWS'/)
1030      FORMAT(10(1X,'MODE',I2))
1040      FORMAT(/7(2X,F5.2))
1060      FORMAT(//20X,'ITERATION RESULTS'//)
1070      FORMAT(//24X,'ITERATION',I2)
          RETURN
          END


          STIFFNESS MATRIX IN BANDED FORM
             10.000 -10.000   0.000   0.000
             20.000   0.000   0.000 -10.000
             10.000 -10.000   0.000   0.000
             20.000 -10.000   0.000   0.000
             30.000 -10.000   0.000   0.000
             20.000 -10.000   0.000   0.000
             10.000   0.000   0.000   0.000
```

MASS MATRIX IN BANDED FORM
```
1.000   0.000   0.000   0.000
1.000   0.000   0.000   0.000
1.000   0.000   0.000   0.000
1.000   0.000   0.000   0.000
2.000   0.000   0.000   0.000
5.000   0.000   0.000   0.000
5.000   0.000   0.000   0.000
```

NATURAL FREQUENCIES ARE

0.000   1.241   1.954   2.236

EIGEN VECTORS ARE AS FOLLOWS

MODE 1 MODE 2 MODE 3 MODE 4

```
0.18  -0.20   0.25   0.09
0.18  -0.17   0.15   0.04
0.18  -0.20  -0.25   0.09
0.18  -0.17  -0.15   0.04
0.18  -0.11   0.00  -0.02
0.18   0.04   0.00  -0.13
0.18   0.15  -0.00   0.09
```

COMPILE TIME= 1.22 SEC,EXECUTION TIME=3.27 SEC

```
C****** PROGRAM :3.5
C****** SOLVING THE EXAMPLE PROBLEM 3.14 BY
C****** SIMULTANEOUS ITERATION METHOD.
C****** THE NON-STANDARD EIGEN VALUE PROBLEM IS
C****** RETAINED AS NON-STANDARD EIGEN VALUE
C****** PROBLEM.
        IMPLICIT REAL*8(A-H,O-Z)
        DIMENSION AK(7,4),AM(7,4)
        DATA AK/10.0D0,20.0D0,10.0D0,20.0D0,
     *  30.0D0,20.0D0,10.0D0,-10.0D0,0.D0,
     *  4*-10.0D0,9*0.0D0,-10.0D0,5*0.0D0/
        DATA AM/4*1.0D0,2.0D0,5.0D0,5.0D0,
     *  21*0.0D0/
        WRITE(6,100)
        NEQ=7
        KB=4
        WRITE(6,110)
        DO 60 I=1,NEQ
60      WRITE(6,130)(AK(I,J),J=1,KB)
        WRITE(6,100)
        WRITE(6,120)
        DO 70 I=1,NEQ
70      WRITE(6,130)(AM(I,J),J=1,KB)
100     FORMAT(//)
110     FORMAT(5X,'STIFFNESS MATRIX IN BANDED FORM')
120     FORMAT(5X,'MASS MATRIX IN BANDED FORM')
130     FORMAT(5X,10F8.3)
        CALL GFRQN(AK,AM,7,4,4,12,0)
        STOP
        END
        SUBROUTINE GFRQN(ST,SM,NEQ,KB,NEIG,
     *  NIT,KPRNT)
C*******  METHOD AND PURPOSE:-
C*******     TO SOLVE FOR THE FIRST FEW
C*******     EIGEN VALUES AND EIGEN VECTORS IN THE
C*******     GENERALISED EIGEN VALUE PROBLEM USING
C*******     THE SIMULTANEOUS ITERATION METHOD
C*******        DESCRIPTION OF PARAMETERS:
C******* ST(NEQ,KB)    STIFFNESS MATRIX IN
C*******                BANDED FORM
C******* SM(NEQ,KB)    MASS MATRIX IN BANDED FORM
C******* NEQ           NUMBER OF EQUATIONS
```

```
C******* KB              BANDWIDTH OF ST AND SM
C******* NEIG            NUMBER OF REQUIRED
C*******                 EIGEN PAIRS
C******* NIT             NUMBER OF ITERATIONS
C******* KPRNT           FLAG FOR PRINTING DURING
C*******                 ITERATION
C*******                 EQ.0. NO PRINTING
C*******                 EQ.1. PRINT
C******* BE(I,I)         EIGEN FREQUENCIES
C******* U(NEQ,NEIG)     EIGEN VECTORS
C******* ALFA            ZERO FOR DEFINITE SYSTEMS
C*******                 NON-ZERO FOR SEMI-DEFINITE
C*******                 SYSTEMS
C*******        WORKING VARIABLES:
C******* (TO BE DIMENSIONED APPROPRIATELY)
C******* Q(NEQ)  U(NEQ,NEIG)    V(NEQ,NEIG)
C******* AU(NEQ,NEIG) D(NEIG,NEIG) BE(NEIG,NEIG)
C******* AV(NEIG,NEIG)   S(NEIG,NEIG)
C*******            SUBROUTINES USED:
C*******    1. UPTRI FOR UPPER TRIANGULARISATION
C*******    2. SOLUT FOR FORWARD SUBSTITUTION
C*******    3. SOLU FOR BACKWARD SUBSTITUTION
C*******    4. BANMUL FOR MULTIPLICATION OF TWO
C*******       MATRICES OF WHICH THE FIRST ONE IS
C*******       IN BANDED FORM
C*******    5. DECOMP FOR UPPER TRIANGULARISATION
C*******       OF A FULLY POPULATED MATRIX
C******* NOTE:-THE PROGRAM OPERATES IN
C******* DOUBLE PRECISION ARITHMETIC.
        IMPLICIT REAL*8 (A-H,O-Z)
        DIMENSION ST(NEQ,KB),SM(NEQ,KB),Q(7)
        DIMENSION V(7,4),AU(7,4),BE(4,4),
     *  AV(7,4),S(4,4),D(4,4),U(7,4)
        ABS(X)=DABS(X)
        SQRT(X)=DSQRT(X)
        IN=NEIG
        NDF=NEQ
        ALFA=2.0
        DO 5 I=1,NEQ
        DO 5 J=1,KB
5       ST(I,J)=ST(I,J)+ALFA*SM(I,J)
C*******UPPER TRIANGULARISATION OF ST
```

```
        CALL UPTRI (ST,NEQ,KB)
C*******ESTABLISH STARTING ITERATION VECTORS
        DO 30 J=1,NEIG
        DO 10 I=J,NEQ
        ENI=NEQ
        EI=I
10      U(I,J)=1.*(ENI+1.-EI)/ENI
        DO 20 I=1,J
        EJ=J
        EI=I
20      U(I,J)=EI/EJ
30      CONTINUE
        DO 40 I=1,NEIG
40      U(I,I)=1.
C*******START OF ITERATION LOOP
        DO 450 II=1,NIT
        CALL BANMUL(SM,U,V,NEQ,KB,NEIG)
        DO 50 I=1,NEQ
        DO 50 J=1,NEIG
50      AU(I,J)=V(I,J)
        DO 55 J=1,NEIG
        DO 51 I=1,NEQ
51      Q(I)=V(I,J)
C*******FORWARD SUBSTITUTION
        CALL SOLUT (ST,NEQ,KB,Q)
        DO 52 I=1,NEQ
52      V(I,J)=Q(I)
55      CONTINUE
        DO 80 J=1,NEIG
        DO 60 I=1,NEQ
60      Q(I)=V(I,J)
C*******BACKWARD SUBSTITUTION
        CALL SOLU(ST,NEQ,KB,Q)
        DO 70 I=1,NEQ
70      V(I,J)=Q(I)
80      CONTINUE
C********ORTHONORMALISATION OF RESULT VECTORS
200      CALL BANMUL(SM,V,AV,NEQ,KB,NEIG)
        DO 300 I=1,IN
        DO 300 J=1,IN
        D(I,J)=0.
        DO 300 L=1,NDF
```

```
300       D(I,J)=D(I,J)+V(L,I)*AV(L,J)
C****** ORTHONORMALISATION IS PERFORMED BY
C****** DECOMPOSITION OF MATRIX D INTO L*LT USING
C****** THE SUBROUTINE DECOMP
          CALL DECOMP(D,IN,IN)
          DO 310 I=1,NEQ
          DO 310 J=1,NEIG
          J1=J-1
          IF(J.EQ.1)U(I,J)=V(I,J)*D(J,J)
          IF(J.EQ.1)GOTO310
          X=0.0
          DO 305 K=1,J1
305       X=X+U(I,K)*D(K,J)
          X=V(I,J)-X
          U(I,J)=X*D(J,J)
310       CONTINUE
C*******INTERACTION ANALYSIS
          DO 377 I=1,NEQ
          DO 376 J=1,NEIG
          AU(I,J)=0.0
          DO 375 L=I,NEQ
          L1=L-I+1
          IF(L1.GT.KB)GOTO 376
375       AU(I,J)=AU(I,J)+ST(I,L1)*U(L,J)
376       CONTINUE
377       CONTINUE
          DO 380 I=1,IN
          DO 380 J=1,IN
          BE(I,J)=0.
          DO 380 L=1,NDF
380       BE(I,J)=BE(I,J)+AU(L,I)*AU(L,J)
          IF(KPRNT.EQ.0)GO TO 450
          IF(II.EQ.1)WRITE(6,1060)
          WRITE(6,1070)II
          DO 390 I=1,IN
390       WRITE(6,1010)(BE(I,J),J=1,IN)
450       CONTINUE
C*******END OF ITERATION LOOP
C*******CALCULATE NATURAL FREQUENCIES
          DO 460 I=1,NEIG
          XX=ABS(BE(I,I)-ALFA)
460       BE(I,I)=SQRT(XX)
```

```
C*******PRINT NATURAL FREQUENCIES
        WRITE(6,1000)
        WRITE(6,1010)(BE(I,I),I=1,NEIG)
C*******PRINT EIGEN VECTORS
        WRITE(6,1020)
        WRITE(6,1030)(I,I=1,NEIG)
        DO 500 I=1,NEQ
500     WRITE(6,1040)(U(I,J),J=1,NEIG)
1000    FORMAT(//17X,'NATURAL FREQUENCIES ARE'//)
1010    FORMAT(/4(3X,E11.4))
1020    FORMAT(/5X,' EIGEN VECTORS ARE AS FOLLOWS'/)
1030    FORMAT(10(8X,'MODE',I2))
1040    FORMAT(//7(3X,E11.4))
1060    FORMAT(//20X,'ITERATION RESULTS'//)
1070    FORMAT(//24X,'ITERATION',I2)
        RETURN
        END


        SUBROUTINE BANMUL(SM,U,V,NEQ,KBAND,NEIG)
        IMPLICIT REAL*8(A-H,O-Z)
        DIMENSION SM(NEQ,KBAND),U(NEQ,NEIG),
      1 V(NEQ,NEIG)
        DO 130 J=1,NEIG
        DO 120 I=1,NEQ
        V(I,J)=0.
        IF (I.EQ.1) GO TO 100
        IM=I-1
        DO 90 K1=1,IM
        I1=I-K1+1
        IF (I1.GT.KBAND) GO TO 90
        V(I,J)=V(I,J)+SM(K1,I1)*U(K1,J)
90      CONTINUE
100     DO 110 K2=I,NEQ
        I2=K2-I+1
        IF (I2.GT.KBAND) GO TO 110
        V(I,J)=V(I,J)+SM(I,I2)*U(K2,J)
110     CONTINUE
120     CONTINUE
130     CONTINUE
        RETURN
        END
```

```
      SUBROUTINE DECOMP(B,N,NS)
      IMPLICIT REAL*8(A-H,O-Z)
      DIMENSION B(NS,NS)
      DO 10 I=1,N
      IN=I-1
      DO 10 J=1,N
      SUM=B(I,J)
      IF(I-1)6,6,4
4     DO 5 K=1,IN
5     SUM=SUM-B(K,I)*B(K,J)
6     IF(J-I)7,8,7
7     B(I,J)=SUM*TEMP
      GO TO 10
8     TEMP=1.0/DSQRT(SUM)
      B(I,I)=TEMP
10    CONTINUE
      RETURN
      END
```

```
STIFFNESS MATRIX IN BANDED FORM
    10.000 -10.000   0.000   0.000
    20.000   0.000   0.000 -10.000
    10.000 -10.000   0.000   0.000
    20.000 -10.000   0.000   0.000
    30.000 -10.000   0.000   0.000
    20.000 -10.000   0.000   0.000
    10.000   0.000   0.000   0.000
```

```
MASS MATRIX IN BANDED FORM
     1.000   0.000   0.000   0.000
     1.000   0.000   0.000   0.000
     1.000   0.000   0.000   0.000
     1.000   0.000   0.000   0.000
     2.000   0.000   0.000   0.000
     5.000   0.000   0.000   0.000
     5.000   0.000   0.000   0.000
```

```
          NATURAL FREQUENCIES ARE

).1089D-02   0.1241D 01   0.1954D 01   0.2236D 01
```

```
     EIGEN VECTORS ARE AS FOLLOWS

     MODE 1       MODE 2       MODE 3       MODE 4

  0.2502D 00  -0.3745D 00   -0.5970D 00   -0.2352D 00

  0.2502D 00  -0.3158D 00   -0.3686D 00   -0.1179D 00

  0.2502D 00  -0.3660D 00    0.6059D 00   -0.2254D 00

  0.2502D 00  -0.3106D 00    0.3749D 00   -0.1119D 00

  0.2501D 00  -0.2078D 00    0.9871D-03    0.5792D-01

  0.2500D 00   0.6693D-01   -0.3280D-02    0.3452D 00

  0.2498D 00   0.2901D 00   -0.1611D-03   -0.2303D 00

COMPILE TIME=1.00 SEC,EXECUTION TIME=2.86 SEC
```

```
C*****   PROGRAM :3.6
C*****   SOLVING THE EXAMPLE 3.14
C        BY SUBSPACE ITERATION METHOD.
         IMPLICIT REAL*8(A-H,O-Z)
         DIMENSION AK(7,4),AM(7,4)
         DATA AK/10.0D0,20.0D0,10.0D0,20.0D0,
     *   30.0D0,20.0D0,10.0D0,-10.0D0,0.D0,
     *   4*-10.0D0,9*0.0D0,-10.0D0,5*0.0D0/
         DATA AM/4*1.0D0,2.0D0,5.0D0,5.0D0,
     *   21*0.0D0/
         WRITE(6,100)
         NEQ=7
         KB=4
         WRITE(6,110)
         DO 60 I=1,NEQ
60       WRITE(6,130)(AK(I,J),J=1,KB)
         WRITE(6,100)
         WRITE(6,120)
         DO 70 I=1,NEQ
70       WRITE(6,130)(AM(I,J),J=1,KB)
100      FORMAT(//)
110      FORMAT(5X,'STIFFNESS MATRIX IN BANDED
        1 FORM')
120      FORMAT(5X,'MASS MATRIX IN BANDED FORM')
130      FORMAT(5X,10F8.3)
         TOL=1.0D-04
         CALL SSPACE(AK,AM,7,4,4,25,TOL,0)
         STOP
         END
         SUBROUTINE SSPACE(ST,SM,NEQ,KB,NEIG,
     *   NIT,TOL,KPRNT)
C*******   METHOD AND PURPOSE:-
C*******      TO SOLVE FOR THE FIRST FEW
C*******      EIGEN VALUES AND CORRESPONDING
C*******      EIGEN VECTORS IN THE GENERALISED
C*******      EIGEN VALUE PROBLEM USING THE
C*******      SUBSPACE ITERATION METHOD
C*******         DESCRIPTION OF PARAMETERS:
C******* ST(NEQ,KB)   STIFFNESS MATRIX IN
C*******              BANDED FORM
C*******SM(NEQ,KB)    MASS MATRIX IN BANDED FORM
C******* NEQ          NUMBER OF EQUATIONS
```

```
C******* KB              BANDWIDTH OF ST AND SM
C******* NEIG            NUMBER OF REQUIRED
C*******                 EIGEN PAIRS
C******* NIT             NUMBER OF ITERATIONS
C******* KPRNT           FLAG FOR PRINTING DURING
C*******                 ITERATION
C*******                 EQ.0. NO PRINTING
C*******                 EQ.1. PRINT
C******* Q1(I)           EIGEN FREQUENCIES
C******* U(NEQ,NEIG)     EIGEN VECTORS
C******* ALFA            ZERO FOR DEFINITE SYSTEMS
C*******                 NON-ZERO FOR SEMI-DEFINITE
C*******                 SYSTEMS
C*******         WORKING VARIABLES:
C******* (TO BE DIMENSIONED APPROPRIATELY)
C******* U(NEQ,NEIG)   V(NEQ,NEIG)  AU(NEQ,NEIG)
C****** BE(NEIG,NEIG)  BH(NEIG,NEIG)
C******* Q(NEQ)  Q1(NEIG)  EIGV(NEIG)
C*******           SUBROUTINES USED:
C*******    1. JACOBI FOR SOLVING EIGEN PROBLEM
C*******       OF SUBSPACE OPERATORS
C*******    2. UPTRI FOR UPPER TRIANGULARISATION
C*******    3. SOLUT FOR FORWARD SUBSTITUTION
C*******    4. SOLU FOR BACKWARD SUBSTITUTION
C*******    5. BANMUL FOR MULTIPLICATION OF TWO
C*******       MATRICES OF WHICH THE FIRST ONE IS
C*******       IN BANDED FORM
C*******    6. DECOMP FOR UPPER TRIANGULARISATION
C*******       OF A FULLY POPULATED MATRIX
C*******    7. FRWARD FOR FORWARD SUBSTITUTION OF
C*******       A FULLY POPULATED MATRIX
C*******    8. BCWARD FOR BACKWARD SUBSTITUTION
C*******       OF A FULLY POPULATED MATRIX
C*******    9. MULT FOR MATRIX MULTIPLICATION
C*******   10. ASC FOR WRITING EIGEN VALUES IN
C*******       ASCENDING ORDER
C******* NOTE:-THE PROGRAM OPERATES IN
C******* DOUBLE PRECISION ARITHMETIC.
        IMPLICIT REAL*8 (A-H,O-Z)
        DIMENSION ST(NEQ,KB),SM(NEQ,KB),U(7,4),
     *  V(7,4),AU(7,4),BE(4,4),BH(4,4),
     *  Q(7),Q1(4),EIGV(4),S(4,4),D(4,4),EIG(4)
```

```
        ABS(X)=DABS(X)
        SQRT(X)=DSQRT(X)
        IN=NEIG
        NDF=NEQ
        ALFA=2.0
        DO 5 I=1,NEQ
        DO 5 J=1,KB
5       ST(I,J)=ST(I,J)+ALFA*SM(I,J)
C*******UPPER TRIANGULARISATION OF ST
        CALL UPTRI (ST,NEQ,KB)
C*******ESTABLISH STARTING ITERATION VECTORS,'U'
        DO 30 J=1,NEIG
        DO 10 I=J,NEQ
        ENI=NEQ
        EI=I
10      U(I,J)=1.*(ENI+1.-EI)/ENI
        DO 20 I=1,J
        EJ=J
        EI=I
20      U(I,J)=EI/EJ
30      CONTINUE
        DO 40 I=1,NEIG
        U(I,I)=1.
40      EIG(I)=0.0
C*******START OF ITERATION LOOP
        DO 310 II=1,NIT
        CALL BANMUL(SM,U,V,NEQ,KB,NEIG)
        DO 50 I=1,NEQ
        DO 50 J=1,NEIG
50      AU(I,J)=V(I,J)
        DO 80 J=1,NEIG
        DO 60 I=1,NEQ
60      Q(I)=V(I,J)
C*******FORWARD SUBSTITUTION
        CALL SOLUT (ST,NEQ,KB,Q)
        DO 70 I=1,NEQ
70      V(I,J)=Q(I)
80      CONTINUE
        DO 110 J=1,NEIG
        DO 90 I=1,NEQ
90      Q(I)=V(I,J)
C*******BACKWARD SUBSTITUTION TO GET THE RESULT
```

```
C*******VECTORS,'V'
        CALL SOLU(ST,NEQ,KB,Q)
        DO 100 I=1,NEQ
100     V(I,J)=Q(I)
110     CONTINUE
C*******CALCULATE 'BE'AND 'BH' WHICH ARE THE
C*******PROJECTIONS OF 'ST'AND 'SM' RESPECTIVELY
        DO 125 I=1,IN
        DO 125 J=I,IN
        BE(I,J)=0.
        DO 120 L=1,NDF
120     BE(I,J)=BE(I,J)+V(L,I)*AU(L,J)
125     BE(J,I)=BE(I,J)
        CALL BANMUL(SM,V,U,NEQ,KB,NEIG)
        DO 140 I=1,IN
        DO 140 J=I,IN
        BH(I,J)=0.
        DO 130 L=1,NDF
130     BH(I,J)=BH(I,J)+V(L,I)*U(L,J)
        BH(J,I)=BH(I,J)
        D(J,I)=BH(J,I)
140     D(I,J)=BH(I,J)
C*******CONVERT THE SUBSPACE OPERATORS BE AND BH
C*******INTO A STANDARD EIGEN SYSTEM 'S'TO SOLVE
C*******BY JACOBI ITERATION METHOD
        CALL DECOMP(BE,IN,IN)
        DO 170 J=1,IN
        DO 150 I=1,IN
150     Q1(I)=BH(I,J)
        CALL FRWARD(BE,IN,Q1)
        DO 160 I=1,IN
160     BH(I,J)=Q1(I)
170     CONTINUE
        DO 200 J=1,IN
        DO 180 I=1,IN
180     Q1(I)=BH(J,I)
        CALL FRWARD(BE,IN,Q1)
        DO 190 I=1,IN
190     S(I,J)=Q1(I)
200     CONTINUE
        RTOL=1.0E-12
        NSMAX=8
```

```
      IF(KPRNT.EQ.0)GOTO220
      IF(II.EQ.1)WRITE(6,1060)
      WRITE(6,1070)II
      WRITE(6,1050)
      DO 210 I=1,IN
210   WRITE(6,1010)(S(I,J),J=1,IN)
C*******SOLVE FOR EIGEN SYSTEM OF SUBSPACE OPERATOR'S'
220   CALL JACOBI(S,IN,RTOL,NSMAX,BH,EIGV)
C*******WRITE EIGEN VALUES IN INCREASING ORDER
      DO 230 I=1,IN
230   EIGV(I)=(1.0/EIGV(I)-ALFA)
      CALL ASC(EIGV,IN)
C*******PRINT CURRENT EIGEN VALUES FROM JACOBI
      WRITE(6,1080)
      WRITE(6,1010)(EIGV(I),I=1,IN)
      DO 260 J=1,IN
      DO 240 I=1,IN
240   Q1(I)=BH(I,J)
C*******OBTAIN ACTUAL EIGEN VECTORS OF SUBSPACE
C*******OPERATOR 'S'
      CALL BCWARD(BE,IN,Q1)
      DO 250 I=1,IN
250   BH(I,J)=Q1(I)
260   CONTINUE
C*******NORMALISE THE EIGEN VECTORS
      DO 275 I=1,IN
      DO 275 J=I,IN
      BE(I,J)=0.0
      DO 270 L=1,IN
      AF=0.0
      DO 265 K=1,IN
265    AF=AF+BH(K,I)*D(K,L)
270   BE(I,J)=BE(I,J)+AF*BH(L,J)
      BE(J,I)=BE(I,J)
275   CONTINUE
      DO 290 J=1,IN
      SUM=1.0/BE(J,J)
      DO 280 I=1,IN
280   BH(I,J)=BH(I,J)*SUM
290   CONTINUE
C*******OBTAIN THE EIGEN VECTORS OF ORIGINAL PROBLEM
      DO 300 I=1,NEQ
```

```
          DO 300 J=1,IN
          U(I,J)=0.0
          DO 300 L=1,IN
300       U(I,J)=U(I,J)+V(I,L)*BH(L,J)
C*******CHECK FOR CONVERGENCE
          DO 340 I=1,IN
          DIF=DABS(EIGV(I)-EIG(I))
          IF(DIF.GT.TOL*DABS(EIGV(I)))GOTO304
340       CONTINUE
          GOTO315
304       DO 305 I=1,IN
305       EIG(I)=EIGV(I)
310       CONTINUE
C*******END OF ITERATION LOOP
C*******CALCULATE NATURAL FREQUENCIES
315       DO 320 I=1,NEIG
320       EIGV(I)=SQRT(EIGV(I))
C*******PRINT NATURAL FREQUENCIES
          WRITE(6,1000)
          WRITE(6,1010)(EIGV(I),I=1,NEIG)
C*******PRINT EIGEN VECTORS
          WRITE(6,1020)
          WRITE(6,1030)(I,I=1,NEIG)
          DO 350 I=1,NEQ
350       WRITE(6,1040)(U(I,J),J=1,NEIG)
1000      FORMAT(//17X,'NATURAL FREQUENCIES ARE'//)
1010      FORMAT(/8(3X,E11.4))
1020      FORMAT(/5X,' EIGEN VECTORS ARE AS FOLLOWS'/)
1030      FORMAT(10(8X,'MODE',I2))
1040      FORMAT(//7(3X,E11.4))
1050      FORMAT(/,5X,'INTERACTION MATRIX:-')
1060      FORMAT(//20X,'ITERATION RESULTS'//)
1070      FORMAT(//24X,'ITERATION',I2)
1080      FORMAT(/,5X,'CURRENT EIGEN VALUES FROM
         1 JACOBI:-')
          RETURN
          END
          SUBROUTINE JACOBI(A,N,RTOL,NSMAX,X,EIGV)
C*******TO SOLVE THE STANDARD EIGEN PROBLEM
C******* USING THE THRESHOLD JACOBI METHOD
          IMPLICIT REAL*8(A-H,O-Z)
          DIMENSION A(N,N),X(4,4),P(4,4),EIGV(4),
```

```
      *    D(4),AP(4,4),OP(4,4)
C*******INITIALISE EIGEN VALUE AND EIGEN VECTORS
           DO 10 I=1,N
           IF(A(I,I).GT.0.)GOTO5
           WRITE(6,2020)
           STOP
5          D(I)=A(I,I)
10         EIGV(I)=D(I)
           DO 30 I=1,N
           DO 20 J=1,N
20         X(I,J)=0.0
30         X(I,I)=1.0
           IF(N.EQ.1)RETURN
           NSWEEP=0
           NR=N-1
40         NSWEEP=NSWEEP+1
C*******CHECK IF PRESENT OFF-DIAGONAL ELEMENT IS
C       LARGE ENOUGH TO REQUIRE ZEROING
           EPS=(.01**NSWEEP)**2
           DO 210 J=1,NR
           JJ=J+1
           DO 210 K=JJ,N
           EPTOL=(A(J,K)*A(J,K))/(A(J,J)*A(K,K))
           IF(EPTOL.LT.EPS)GO TO 210
C*******IF ZEROING IS REQUIRED,CALCULATE THE
C       ROTATION MATRIX ELEMENTS
           IF(A(J,J).NE.A(K,K))GOTO50
           PI=3.1415926
           THETA=PI/4.
           GOTO60
50         THETA=0.5*DATAN(2.*A(J,K)/(A(J,J)-
          1A(K,K)))
60         SN=DSIN(THETA)
           CS=DCOS(THETA)
           DO 70 IA=1,N
           DO 70 IB=1,N
           P(IA,IB)=0.0
70         P(IA,IA)=1.0
           P(J,J)=CS
           P(J,K)=-SN
           P(K,J)=SN
           P(K,K)=CS
```

```
C*******PERFORM THE TRANSFORMATION TO REDUCE
C        THE PRESENT OFF-DIAGONAL ELEMENT TO ZERO
         CALL MULT(A,P,AP,N,N,N)
         DO 90 M2=1,N
         DO 90 N2=1,N
         A(M2,N2)=0.0
         DO 90 L2=1,N
90       A(M2,N2)=A(M2,N2)+P(L2,M2)*AP(L2,N2)
         IF(NSWEEP.EQ.1.AND.K.EQ.2)GOTO101
C*******UPDATE THE EIGEN VECTOR MATRIX AFTER
C        EACH ROTATION
         CALL MULT(OP,P,X,N,N,N)
         GO TO100
101      DO 35 I4=1,N
         DO 35 J4=1,N
35       X(I4,J4)=P(I4,J4)
100      DO 200 I1=1,N
         DO 200 J1=1,N
200      OP(I1,J1)=X(I1,J1)
210      CONTINUE
C*******UPDATE EIGEN VALUES AFTER EACH ITERATION
         DO 220 I=1,N
220      EIGV(I)=A(I,I)
C*******CHECK FOR CONVERGENCE
230      DO 240 I=1,N
         TOL=RTOL*D(I)
         DIF=DABS(EIGV(I)-D(I))
         IF(DIF.GT.TOL)GOTO280
240      CONTINUE
C*******CHECK ALL OFF-DIAGONAL ELEMENTS TO SEE
C        IF ANOTHER SWEEP IS REQUIRED
         EPS=RTOL**2
         DO 250 J=1,NR
         JJ=J+1
         DO 250 K=JJ,N
         EPSA=(A(J,K)*A(J,K))/(A(J,J)*A(K,K))
         IF(EPSA.LT.EPS)GOTO250
         GO TO 280
250      CONTINUE
260      RETURN
C*******UPDATE D MATRIX AND START
C        NEW SWEEP IF ALLOWED
```

```
280      DO 290 I=1,N
290      D(I)=EIGV(I)
         DO 300 I=1,N
         DO 300 J=1,N
300      X(I,J)=OP(I,J)
         IF(NSWEEP.LT.NSMAX)GOTO40
         GOTO260
2020     FORMAT(2X,'----ERROR---- STOP----'//5X,
     *   'MATRICES ARE NOT POSITIVE DEFINITE')
         END
      SUBROUTINE FRWARD(CB,N,Q)
      IMPLICIT REAL*8(A-H,O-Z)
      DIMENSION CB(N,N),Q(N)
      DO 60 I=1,N
      IN=I-1
      SUM=Q(I)
      IF(I.EQ.1)GOTO50
      DO 40 K=1,IN
40    SUM=SUM-CB(K,I)*Q(K)
50    Q(I)=SUM/CB(I,I)
60    CONTINUE
      RETURN
      END
      SUBROUTINE BCWARD(CB,N,Q)
      IMPLICIT REAL*8(A-H,O-Z)
      DIMENSION CB(N,N),Q(N)
      DO 60 L=1,N
      I=N-L+1
      SUM=Q(I)
      II=I+1
      IF(II-N)30,30,50
30    DO 40 K=II,N
40    SUM=SUM-CB(I,K)*Q(K)
50    Q(I)=SUM/CB(I,I)
60    CONTINUE
      RETURN
      END
      SUBROUTINE MULT (A,B,C,N,MM,M)
      IMPLICIT REAL *8 (A-H,O-Z)
      DIMENSION A(N,MM),B(MM,M),C(N,M)
      DO 3 I=1,N
      DO 3 J=1,M
```

```
        C(I,J)=0.0
        DO 3 K=1,MM
        C(I,J)=C(I,J)+A(I,K)*B(K,J)
3       CONTINUE
        RETURN
        END
        SUBROUTINE ASC(EIGV,IN)
        IMPLICIT REAL*8(A-H,O-Z)
        DIMENSION EIGV(IN)
        LMT=IN-1
30      LSW=1
        DO 20 I=1,LMT
        IF(EIGV(I+1).LT.EIGV(I))GOTO10
        GOTO20
10      TEMP=EIGV(I)
        EIGV(I)=EIGV(I+1)
        EIGV(I+1)=TEMP
        LSW=I
20      CONTINUE
        IF(LSW.EQ.1)GOTO40
        LMT=LSW-1
        GOTO30
40      RETURN
        END


STIFFNESS MATRIX IN BANDED FORM
   10.000 -10.000   0.000   0.000
   20.000   0.000   0.000 -10.000
   10.000 -10.000   0.000   0.000
   20.000 -10.000   0.000   0.000
   30.000 -10.000   0.000   0.000
   20.000 -10.000   0.000   0.000
   10.000   0.000   0.000   0.000

MASS MATRIX IN BANDED FORM
    1.000   0.000   0.000   0.000
    1.000   0.000   0.000   0.000
    1.000   0.000   0.000   0.000
    1.000   0.000   0.000   0.000
    2.000   0.000   0.000   0.000
    5.000   0.000   0.000   0.000
    5.000   0.000   0.000   0.000
```

```
        CURRENT EIGEN VALUES FROM JACOBI:-
0.1464D 00   0.3727D 01    0.1138D 02    0.2523D 02

        CURRENT EIGEN VALUES FROM JACOBI:-
0.2177D-01   0.2440D 01    0.3938D 01    0.1996D 02

        CURRENT EIGEN VALUES FROM JACOBI:-
0.7739D-03   0.1611D 01    0.3827D 01    0.1057D 02

        CURRENT EIGEN VALUES FROM JACOBI:-
0.1259D-04   0.1544D 01    0.3820D 01    0.6079D 01

        CURRENT EIGEN VALUES FROM JACOBI:-
0.1572D-06   0.1540D 01    0.3820D 01    0.5197D 01

        CURRENT EIGEN VALUES FROM JACOBI:-
0.2229D-08   0.1540D 01    0.3820D 01    0.5039D 01

        CURRENT EIGEN VALUES FROM JACOBI:-
0.3531D-10   0.1540D 01    0.3820D 01    0.5008D 01

        CURRENT EIGEN VALUES FROM JACOBI:-
0.5860D-12   0.1540D 01    0.3820D 01    0.5002D 01

        CURRENT EIGEN VALUES FROM JACOBI:-
0.1066D-13   0.1540D 01    0.3820D 01    0.5000D 01

        CURRENT EIGEN VALUES FROM JACOBI:-
0.4441D-15   0.1540D 01    0.3820D 01    0.5000D 01

        CURRENT EIGEN VALUES FROM JACOBI:-
0.4441D-15   0.1540D 01    0.3820D 01    0.5000D 01


              NATURAL FREQUENCIES ARE

0.2107D-07   0.1241D 01   0.1954D 01   0.2236D 01


        EIGEN VECTORS ARE AS FOLLOWS
     MODE 1      MODE 2         MODE 3          MODE 4

0.3501D 00   -0.8620D 00   0.1032D 01   -0.5986D 00
0.3513D 00   -0.4414D 00   0.7953D 00   -0.3097D 00
0.3555D 00    0.1598D 01   0.3246D 00   -0.7290D 00
0.3547D 00    0.1079D 01   0.3581D 00   -0.3903D 00
0.3533D 00    0.2237D 00   0.3879D 00    0.1220D 00
0.3539D 00   -0.4106D-01  -0.1098D 00    0.9187D 00
0.3538D 00   -0.3371D 00  -0.5402D 00   -0.5656D 00
COMPILE TIME=1.88 SEC,EXECUTION TIME=11.50 SEC
```

```
C*****PROGRAM NO 3.7
C*****SOLVING THE EXAMPLE PROBLEM 3.12 TO OBTAIN
C*****FIRST FEW EIGEN PAIRS OF THE UNSYMMETRIC
C*****EIGEN PROBLEM OF TYPE A*X=LAMBDA*B*X
      IMPLICIT REAL*8(A-H,O-Z)
      DIMENSION A(3,3),B(3,3),AT(3,3),BE(3,3)
     1,X1(3,3),X2(3,3),XT1(3,3),XT2(3,3)
     2,BT(3,3),YX1(3,3),YXT1(3,3)
     3,D(3,3),DLT(3,3),DU(3,3),RT(3)
      DATA A/9*0.0D0/,B/9*0.0D0/,AT/9*0.0D0/
     1,X1/9*0.0D0/,XT1/9*0.0D0/
     2,YX1/9*0.0D0/,YXT1/9*0.0D0/
      READ (5,5) N,NEIG,NIT,IMAS
      WRITE(6,15) N,NEIG,NIT,IMAS
      READ (5,10) ((A(I,J),J=1,N),I=1,N)
      IF(IMAS.GT.0)READ(5,10)((B(I,J),J=1,N)
     1,I=1,N)
      IF(IMAS.EQ.0)READ(5,10)(B(I,I),I=1,N)
      WRITE(6,20)
      WRITE(6,30)((A(I,J),J=1,N),I=1,N)
      IF(IMAS.GT.0)WRITE(6,25)
      IF(IMAS.EQ.0)WRITE(6,35)
      WRITE(6,30)((B(I,J),J=1,N),I=1,N)
    5 FORMAT(16I5)
   15 FORMAT(/,10X,'NEQ=',I5,10X,'NEIG=',I5,10X,
     1'NIT=',I3,5X,'IMAS=',I2,/)
   25 FORMAT(/,10X,'CONSISTENT MASS MATRIX',/)
   35 FORMAT(/,10X,'LUMPED MASS MATRIX',/)
   10 FORMAT(3F10.3)
   20 FORMAT(/,10X,'STIFFNESS MATRIX',/)
   30 FORMAT(1X,3D13.4,/)
      CALL UNSYM(N,NEIG,NIT,IMAS,A,B,AT,BT,X1
     1 ,X2,XT1,XT2,YX1,YXT1,D,DU,DLT,BE,RT)
      STOP
      END

      SUBROUTINE UNSYM(N,IC,NIT,IMAS,A,B,AT,BT,
     1X1,X2,XT1,XT2,YX1,YXT1,D,DU,DLT,BE,RT)
C***** METHOD:-
C***** EIGEN VALUE PROBLEM FOR UNSYMMETRIC
C***** MATRICES OF THE TYPE A*X=LAMBDA*B*X
C****** DESCRIPTION OF PARAMETERS:-
```

```
C*****    N      ORDER OF THE MATRIX A
C*****    IC     NO. OF EIGEN VALUES REQUIRED
C*****    NIT    NO. OF ITERATIONS
C*****    IMAS   EQUAL TO
C*****                    0 FOR LUMPED MASS
C*****                    1 FOR CONSISTENT MASS
C*****    A      INPUT STIFFNESS MATRIX ;DESTROYED
C*****    AT     TRANSPOSE OF A
C*****    B      INPUT MASS MATRIX
C*****    BT     TRANSPOSE OF B
C*****    X1,X2 RIGHT AND LEFT EIGEN VECTORS
C*****YX1,YXT1,XT1,XT2,D,DLT ARE WORKING VECTORS
C*****         INTERACTION MATRIX
C*****         SUBROUTINES REQUIRED:-
C*****                          SOLVE
C*****                          SUBS
C*****                          DECOMP
C*****                          DMAT
C*****                          DLLT
      IMPLICIT REAL*8(A-H,O-Z)
      DIMENSION A(N,N),B(N,N),AT(N,N),BT(N,N)
     1,X1(N,IC),X2(N,IC),XT1(N,IC),XT2(N,IC),
     2 YX1(N,IC),YXT1(N,IC),RT(N),D(IC,IC),
     3 DU(IC,IC),DLT(IC,IC),BE(IC,IC)
      DO 110 I=1,N
      DO 110 J=1,N
      AT(I,J)=A(J,I)
  110 BT(I,J)=B(J,I)
C***** INITIAL TRIAL VECTORS
      DO 120 I=1,N
  120 RT(I)=B(I,I)/A(I,I)
      DO 130 J=1,IC
      BIG=0.0
      DO 140 I=1,N
      IF(DABS(RT(I))-BIG)140,140,150
  150 BIG=DABS(RT(I))
      IROW=I
  140 CONTINUE
      YX1(IROW,J)=1.0
      YXT1(IROW,J)=1.0
  130 RT(IROW)=0.0
      CALL DECOMP(A.N)
```

```
      CALL DECOMP(AT,N)
C**** ITERATION STARTS
      DO 100 II=1,NIT
      WRITE(6,7000) II
      IF(II.EQ.1)GO TO 160
      IF(IMAS.EQ.0)GO TO 170
      DO 180 I=1,N
      DO 180 J=1,IC
      X=0.0
      Y=0.0
      DO 190 K=1,N
      X=B(I,K)*X1(K,J)+X
  190 Y=BT(I,K)*X2(K,J)+Y
      YX1(I,J)=X
  180 YXT1(I,J)=Y
  170 IF(IMAS.GT.0)GO TO 160
      DO 195 I=1,N
      DO 195 J=1,IC
      YX1(I,J)=B(I,I)*X1(I,J)
  195 YXT1(I,J)=BT(I,I)*X2(I,J)
  160 CONTINUE
      CALL SUBS(A,N,YX1,XT1,IC)
      CALL SUBS(AT,N,YXT1,XT2,IC)
C***** INTERACTION MATRIX
      DO 200 I=1,IC
      DO 200 J=1,IC
      X=0.0
      DO 205 K=1,N
  205 X=XT2(K,I)*YX1(K,J)+X
  200 BE(I,J)=X
      WRITE(6,1000)
      WRITE(6,2000)((BE(I,J),J=1,IC),I=1,IC)
      CALL DMAT(XT2,B,XT1,N,IC,D,YX1,IMAS)
      CALL DLLT(D,IC)
      CALL SOLVE(D,XT1,X1,IC,N)
      DO 210 I=1,IC
      DO 210 J=I,IC
  210 D(I,J)=D(J,I)
      CALL SOLVE(D,XT2,X2,IC,N)
  100 CONTINUE
C***** ITERATIONS ARE OVER
      WRITE(6,3000)
```

```
      DO 400 I=1,IC
      X=1.0/DSQRT(BE(I,I))
  400 WRITE(6,4000)I,X
      WRITE(6,5000)
      DO 410 I=1,N
  410 WRITE(6,2000)(X1(I,J),J=1,IC)
      WRITE(6,6000)
      DO 420 I=1,N
  420 WRITE(6,2000)(X2(I,J),J=1,IC)
 1000 FORMAT(/,10X,'INTERACTION MATRIX',/)
 2000 FORMAT(3D15.5,/)
 3000 FORMAT(/,10X,'NATURAL FREQ.IN RAD/SEC',/)
 4000 FORMAT(10X,'FREQ(',I3,')=',D20.10,/)
 5000 FORMAT(/,10X,'RIGHT EIGEN VECTORS',/)
 6000 FORMAT(/,10X,'LEFT EIGEN VECTORS',/)
 7000 FORMAT(/,10X,'ITERATION NUMBER=',I3,/)
      STOP
      END
      SUBROUTINE SOLVE (D,V,U,IC,N)
C***** FIND U FROM U*D=V
C***** D-UPPER TRIANGULAR MATRIX(IC*IC)
C***** V-KNOWN VECTOR(N*IC)
C***** U-ORTHO-NORMALIZED VECTOR(N*IC)
      IMPLICIT REAL*8(A-H,O-Z)
      DIMENSION D(IC,IC),V(N,IC),U(N,IC)
      DO 10 I=1,N
      DO 10 J=1,IC
      J1=J-1
      IF(J.EQ.1) U(I,J)=V(I,J)/D(J,J)
      IF(J.EQ.1) GO TO 10
      X=0.0
      DO 20 K=1,J1
   20 X=U(I,K)*D(K,J)+X
      X=V(I,J)-X
      U(I,J)=X/D(J,J)
   10 CONTINUE
      RETURN
      END

      SUBROUTINE SUBS(A,N,X,Y,IC)
C***** FORWARD AND BACKWARD SUBSTITUTIONS
      IMPLICIT REAL*8(A-H,O-Z)
```

```
      DIMENSION A(N,N),X(N,IC),Y(N,IC)
      RT=1.0/A(1,1)
      DO 10 J=1,IC
   10 Y(1,J)=X(1,J)*RT
      DO 50 J=2,N
      RT=1.0/A(J,J)
      JM1=J-1
      JP1=J+1
      DO 40 I=1,IC
      Y(J,I)=X(J,I)
      DO 30 K=1,JM1
   30 Y(J,I)=Y(J,I)-A(J,K)*Y(K,I)
   40 Y(J,I)=Y(J,I)*RT
   50 CONTINUE
      NM1=N-1
      DO 60 K=1,IC
      DO 60 L=1,NM1
      I=N-L
      II=I+1
      DO 60 J=II,N
   60 Y(I,K)=Y(I,K)-A(I,J)*Y(J,K)
      RETURN
      END

      SUBROUTINE DECOMP(A,N)
C***** DECOMPOSE A=L*U
      IMPLICIT REAL*8(A-H,O-Z)
      DIMENSION A(N,N)
      RT=1.0/A(1,1)
      DO 10 J=2,N
   10 A(1,J)=A(1,J)*RT
      DO 50 J=2,N
      JM1=J-1
      JP1=J+1
      DO 20 I=J,N
      DO 20 K=1,JM1
   20 A(I,J)=A(I,J)-A(I,K)*A(K,J)
      RT=1.0/A(J,J)
      IF(JP1.GT.N) JP1=N
      IF(J.EQ.N) GO TO 50
      DO 40 I=JP1,N
      DO 30 K=1,JM1
```

```
   30 A(J,I)=A(J,I)-A(J,K)*A(K,I)
   40 A(J,I)=A(J,I)*RT
   50 CONTINUE
      RETURN
      END
      SUBROUTINE DMAT(XT2,B,XT1,N,IC,D,YX1,IMAS)
      IMPLICIT REAL*8(A-H,O-Z)
      DIMENSION XT2(N,IC),XT1(N,IC),B(N,N),
     1D(IC,IC),YX1(N,IC)
C***** ORTHONORMALIZATION
      IF(IMAS.EQ.0)GO TO 210
      DO 220 I=1,N
      DO 220 J=1,IC
      X=0.0
      DO 230 K=1,N
  230 X=B(I,K)*XT1(K,J)+X
  220 YX1(I,J)=X
  210 IF(IMAS.GT.0)GO TO 240
      DO 250 I=1,N
      DO 250 J=1,IC
  250 YX1(I,J)=B(I,I)*XT1(I,J)
  240 CONTINUE
      DO 260 I=1,IC
      DO 260 J=1,IC
      X=0.0
      DO 270 K=1,N
  270 X=XT2(K,I)*YX1(K,J)+X
  260 D(I,J)=X
      RETURN
      END
      SUBROUTINE DLLT(B,N)
C**** DECOMPOSE D MATRIX AS L*L(T)
      IMPLICIT REAL*8(A-H,O-Z)
      DIMENSION B(N,N)
      DO 10 I=1,N
      IN=I-1
      DO 10 J=I,N
      SUM1=B(J,I)
      SUM=B(I,J)
      IF(I-1) 6,6,4
    4 DO 5 K=1,IN
      SUM1=SUM1-B(J,K)*B(K,I)
```

```
5 SUM=SUM-B(I,K)*B(K,J)
6 IF(J-I) 7,8,7
7 B(I,J)=SUM*TEMP
  B(J,I)=SUM1*TEMP
  GO TO 10
8 B(I,I)=DABS(SUM)
  B(I,I)=DSQRT(B(I,I))
  TEMP=1.0/B(I,I)
10 CONTINUE
  RETURN
  END
```

NEQ= 3   NEIG= 3   NIT= 5   IMAS= 0

        STIFFNESS MATRIX

0.2700D 02   0.2800D 02   0.1200D 02

0.1400D 02   0.1600D 02   0.7500D 01

0.4000D 01   0.5000D 01   0.3000D 01

        LUMPED MASS MATRIX

0.1000D 01   0.0000D 00   0.0000D 00
0.0000D 00   0.1000D 01   0.0000D 00
0.0000D 00   0.0000D 00   0.1000D 01

        ITERATION NUMBER= 1
        INTERACTION MATRIX

 0.20513D 01   -0.11795D 01   0.30769D 00
-0.17692D 01   0.16923D 01   -0.61538D 00
 0.92308D 00   -0.12308D 01   0.53846D 00

        ITERATION NUMBER= 2
        INTERACTION MATRIX

 0.35543D 01   -0.40304D 00   0.35087D-02
-0.60456D 00   0.70467D 00   -0.11938D-01
 0.10526D-01   -0.23876D-01   0.23097D-01

```
                ITERATION NUMBER= 3
                INTERACTION MATRIX

     0.36349D 01    -0.70812D-01     0.22171D-04
    -0.10622D 00     0.62449D 00    -0.40425D-03
     0.66512D-04    -0.80850D-03     0.22667D-01


                ITERATION NUMBER= 4
                INTERACTION MATRIX

     0.36373D 01    -0.12119D-01     0.13822D-06
    -0.18178D-01     0.62206D 00    -0.14698D-04
     0.41465D-06    -0.29397D-04     0.22667D-01


                ITERATION NUMBER= 5
                INTERACTION MATRIX

     0.36374D 01    -0.20723D-02     0.86132D-09
    -0.31085D-02     0.62199D 00    -0.53561D-06
     0.25839D-08    -0.10712D-05     0.22667D-01


                NATURAL FREQ.IN RAD/SEC

            FREQ( 1)=    0.5243303773D 00
            FREQ( 2)=    0.1267964264D 01
            FREQ( 3)=    0.6642088446D 01


                RIGHT EIGEN VECTORS

     0.52405D 00    -0.78518D 00     0.77473D 00
    -0.79825D 00     0.47256D 00     0.41946D 00
     0.69545D 00     0.55883D 00     0.12637D 00


                LEFT EIGEN VECTORS

     0.17468D 00    -0.39259D 00     0.77473D 00
    -0.53217D 00     0.47256D 00     0.83891D 00
     0.69545D 00     0.83824D 00     0.37912D 00


 PILE TIME= 0.98 SEC  EXECUTION TIME= 0.51 SEC
```

```
C****PROGRAM 3.8
C****TO CALCULATE THE BUCKLING LOAD OF COLUMN BY
C****FINITE ELEMENT METHOD(BOTH ENDS CLAMPED)
C****I VARYING AT HALF LENGTH
          IMPLICIT REAL*8(A-H,O-Z)
          DIMENSION ID(19,2),NOD(18,2),A(34,4)
          DIMENSION B(34,4),ST(4,4),BK(4,4)
          DATA NNOD,NFREE,NNOEL,NELEM,KB,
     1 IFLAG/19,2,2,18,4,1/
          OPEN(15,FILE = 'OUT5.DAT')
          DO 111 I=1,34
          DO 111 J=1,4
          A(I,J) = 0.0
111       B(I,J) = 0.0
          E = 2.1E05
          T1 = 500.0
          T2 = 250.0
          B1 = 500.0
          H = 888.88888
          AI1 = B1*T1**3/12.0
          AI2 = B1*T2**3/12.0
          N = 34
          NEIG = 8
          NIT = 8
C         ASSIGNING THE DOF
          ISUM=0
          DO 10 M=1,NNOD
          DO 10 N=1,NFREE
10        ID(M,N) = 0
          DO 20 I=1,NFREE
          ID(1,I)=1
20        ID(NNOD,I) = 1
          DO 30 I=1,NNOD
          DO 30 J=1,NFREE
          IF(ID(I,J))40,50,40
50        ISUM=ISUM+1
          ID(I,J)=ISUM
          GOTO 30
40        ID(I,J)=0
30        CONTINUE
          DO 60 I=1,19
          DO 60 J=1,2
```

```
            WRITE(6,61) I,J,ID(I,J)
61          FORMAT(1X,3I5)
60          CONTINUE
            NEQ=ISUM
C           NODAL CONNECTIVITY
c           NNOEL = NO OF NODES PER ELEMENT
c           NELEM = NO OF ELEMENTS
            DO 70 L=1,NELEM
            DO 70 I=1,NNOEL
            J=L+I-1
70          NOD(L,I)=J
            WRITE(6,*)'NELEM,   NOD(I,J)'
            DO 80 I=1,NELEM
            WRITE(6,81)I,(NOD(I,J),J=1,2)
80          CONTINUE
81          FORMAT(I4,5X,2(I5,1X))
C           ASSEMBLING MATRICES A & B
      DO 530 LK=1,NELEM
            CALL KELEM(LK,H,E,AI1,AI2,ST,BK)
139         FORMAT(6F10.2)
            DO 540 LL=1,NNOEL
            II=NOD(LK,LL)
            IM=NFREE*(LL-1)
            DO 540 KK=1,NNOEL
            JJ=NOD(LK,KK)
            JN=NFREE*(KK-1)
            DO 551 I=1,NFREE
            MMI=ID(II,I)
            IF(MMI.EQ.0) GOTO 551
            IMI=IM+I
            DO 550 J=1,NFREE
            NJJ=ID(JJ,J)
            IF (NJJ.EQ.0) GOTO 550
            NNJ=NJJ-MMI+1
            IF(NNJ) 550,550,545
545         JNJ=JN+J
            A(MMI,NNJ)=A(MMI,NNJ)+ST(IMI,JNJ)
            B(MMI,NNJ)=B(MMI,NNJ)+BK(IMI,JNJ)
550         CONTINUE
551         CONTINUE
540         CONTINUE
530         CONTINUE
```

```
      CALL STFRQN(ID,NNOD,A,B,NEQ,KB,8,8,1)
      STOP
      END

      SUBROUTINE KELEM(LK,H,E,AI1,AI2,ST,BK)
      IMPLICIT REAL*8(A-H,O-Z)
      DIMENSION ST(4,4),BK(4,4)
      DO 100 I = 1,4
      DO 100 J = 1,4
      BK(I,J) = 0.0
100   ST(I,J) = 0.0
      IF (LK.LE.9)THEN
      AI = AI1
      ELSE
      AI = AI2
      ENDIF
      ST(1,1) = 12.0 * E*AI/H**3.0
      ST(1,2) = -6.0 * E*AI/H**2.0
      ST(1,3) = -12.0 * E*AI/H**3.0
      ST(1,4) = -6.0 * E*AI/H**2.0
      ST(2,1) = -6.0 * E*AI/H**2.0
      ST(2,2) = 4.0 * E*AI/H
      ST(2,3) = 6.0 * E*AI/H**2.0
      ST(2,4) = 2.0 * E*AI/H
      ST(3,1) = -12.0 * E*AI/H**3.0
      ST(3,2) = 6.0 * E*AI/H**2.0
      ST(3,3) = 12.0 * E*AI/H**3.0
      ST(3,4) = 6.0 * E*AI/H**2.0
      ST(4,1) = -6.0 * E*AI/H**2.0
      st(4,2) = 2.0 * e*ai/h
      ST(4,3) = 6.0 * E*AI/H**2.0
      ST(4,4) = 4.0 * E*AI/H
      BK(1,1) = 36.0/(30.0*H)
      BK(1,2) = -1.0/10.0
      BK(1,3) = -36.0/(30.0*H)
      BK(1,4) = -1.0/10.0
      BK(2,1) = -1.0/10.0
      BK(2,2) = (4.0 * H)/30.0
      BK(2,3) = 1.0/10.0
      BK(2,4) = H/30.0
      BK(3,1) = -36.0/(30.0*H)
      BK(3,2) = 1.0/10.0
```

```
BK(3,3) = 36.0/(30.0*H)
BK(3,4) = 1.0/10.0
BK(4,1) = -1.0/10.0
BK(4,2) = H/30.0
BK(4,3) = 1.0/10.0
BK(4,4) = (4.0 * H)/30.0
RETURN
END
```

OUTPUT

BUCKLING LOAD IS
0.370E+08

```
C**** PROGRAM : 3.9
C**** SOLVING THE EXAMPLE 3.14 BY LANCZOS METHOD
      IMPLICIT REAL*8(A-H,O-Z)
      DIMENSION ST(7,4),SM(7,4)
      COMMON DELTA
      DATA ST/10.0D0,20.0D0,10.0D0,20.0D0,
     +30.0D0,20.0D0,10.0D0,-10.0D0,0.0D0,
     +4*-10.0D0,9*0.0D0,-10.0D0,5*0.0D0/
      DATA SM/4*1.0D0,2.0D0,5.0D0,5.0D0,
     +21*0.0D0/
      OPEN(UNIT=16,FILE='RESUL7')
      WRITE(16,100)
      DELTA = 2.0
      NEQ = 7
      KB = 4
      NEIG = 4
      WRITE(16,110)
      DO 60 I = 1,NEQ
  60  WRITE(16,130) (ST(I,J),J=1,KB)
      WRITE(16,100)
      WRITE(16,120)
      DO 70 I = 1,NEQ
  70  WRITE(16,130)(SM(I,J),J=1,KB)
 100  FORMAT(//)
 110  FORMAT(5X,'STIFFNESS MATRIX IN BANDED
     1FORM')
 120  FORMAT(5X,'MASS MATRIX IN BANDED FORM')
 130  FORMAT(5X,10F8.0)
C     DO 140 M = 4,7
      M = 5
      WRITE(16,*)'SIZE OF TRIDIAG. MATRIX M= ',M
 140  CALL LANZS1(NEQ,KB,M,NEIG,ST,SM)
      STOP
      END


      SUBROUTINE TRAMA(LK,T)
      IMPLICIT REAL*8(A-H,O-Z)
      DIMENSION T(6,6)
      IF(LK.LT.7) THEN
      X=0.00
      Y=1.00
      ELSEIF(LK.LT.16)THEN
```

```
          X=1.00
          Y=0.00
          ELSE
          X=0.00
          Y=-1.00
          END IF
          DO70 I=1,6
          DO 80 J=1,6
          IF((I.EQ.1).OR.(I.EQ.4)) THEN
          K=I+1
          T(I,K)=Y
          T(K,I)=-T(I,K)
          T(I,I)=X
          T(K,K)=T(I,I)
          ENDIF
          IF((I.LE.3).AND.(J.GE.3))THEN
          T(I,J)=0.0
          ELSE IF((I.GE.4).AND.(J.LE.3)) THEN
          T(I,J)=0.0
          ELSE
          T(3,3)=1.0
          T(6,6)=1.0
          T(4,6)=0.0
          T(5,6)=0.0
          ENDIF
    80    CONTINUE
    70    CONTINUE
          RETURN
          END


          SUBROUTINE VECTOR(ST,SM,N,NBAND,NIT,U)
          IMPLICIT REAL*8(A-H,O-Z)
          DIMENSION ST(7,4),SM(7,4),U(7),V(7)
C**** ESTABLISHING STARTING VECTOR
C**** NIT = NUMBER OF TIMES STARTING VECTOR Y1 C****
IS IMPROVED
          DO 10 I = 1,N
          ENI = N
          EI = I
    10    U(I) = 1.*(ENI+1.-EI)/ENI
          U(1)=1.0
C*****ORTHONORMALAISING THE VECTOR
```

```
          SUM = 0.0
          DO 20 I = 1,N
   20     SUM = SUM + U(I)*U(I)
          SUM = DSQRT(SUM)
          DO 30 I = 1,N
   30     U(I) = U(I)/SUM
          DO 100 K = 1,NIT
C*****BACK SUBSTITUTION
          CALL SOLU(ST,N,NBAND,U)
C*****MULTIPLING BY MASS MATRIX
          CALL ULT(SM,U,V,N,NBAND)
C*****FORWARD SUBSTITUTION
          CALL SOLUT(ST,N,NBAND,V)
C*****ORTHONORMALISE THE RESULTANT VECTOR
          DO 40 I = 1,N
   40     U(I) = V(I)
          SUM = 0.0
          DO 50 I = 1,N
   50     SUM = SUM+U(I)*U(I)
          SUM = DSQRT(SUM)
          DO 60 I =1,N
   60     U(I) = U(I)/SUM
  100     CONTINUE
          RETURN
          END
          SUBROUTINE ULT(A,V,U,N,NBAND)
          IMPLICIT REAL*8(A-H,O-Z)
          DIMENSION A(7,4),V(7),U(7)
          DO 100 I=1,N
          U(I)=0.0
          DO 100 L=1,N
          IF(L.LT.I) GO TO 200
          L1=(L-I)+1
          IF(L1.GT.NBAND) GO TO 100
          U(I)=U(I)+A(I,L1)*V(L)
          GO TO 100
  200     I1=(I-L)+1
          IF (I1.GT.NBAND) GO TO 100
          U(I)=U(I)+A(L,I1)*V(L)
  100     CONTINUE
          RETURN
          END
```

```
      SUBROUTINE LANZS1(N,NBAND,M,NEIG,ST,SM)
      IMPLICIT REAL*8(A-H,O-Z)
      COMMON DELTA
      DIMENSION ST(7,4),SM(7,4),EI(7,4),W(7),
     *U(7),V(7),ST1(7,2),ALFA(7),BETA(7) ,
     *Y(7,7),U2(7,4),EA(20,30)
      NN1 = N-1
C*****GIVING SHIFT TO THE STIFFNESS MATRIX
      DO 5 I = 1,N
      DO 5 J = 1,NBAND
  5   ST(I,J) = ST(I,J) + DELTA * SM(I,J)
C*****CHOLESKY DECOMPOSITION OF STIFFNESS MATRIX
      CALL UPTRI(ST,N,NBAND)
      CALL VECTOR(ST,SM,N,NBAND,12,U)
      DO 50 I = 1,N
  50  Y(I,1) = U(I)
C ****COMPUTE ALFA(1)
      CALL SOLU(ST,N,NBAND,U)
      CALL ULT(SM,U,V,N,NBAND)
      CALL SOLUT(ST,N,NBAND,V)
      SUM = 0.0
      DO 120 L = 1,N
 120  SUM = SUM + Y(L,1)*V(L)
      ALFA(1) = SUM
      DO 130 L = 1,N
 130  V(L) = V(L)-ALFA(1)*Y(L,1)
C ****COMPUTING BETA VALUES ------------
      DO 1300 I = 2,M
      IM1 = I-1
      BETA(M) = 0.0
      SUM = 0.0
      DO 140 L = 1,N
 140  SUM = SUM + V(L)*V(L)
C**** CHECK FOR COMMAND CDSQRT
      BETA(IM1) = DSQRT(SUM)
      IF (SUM.EQ.0.0) GO TO 400
      DO 150 J = 1,N
 150  Y(J,I) = V(J)/BETA(IM1)
      GO TO 600
 400  DO 500 J = 1,N
 500  Y(J,I) = 0.0
      Y(I,I) = 1.0
```

```
C ****MINIMISATION OF ERROR IN MATRIX Y ---
C ****ORTHONORMALISATION OF RESULT VECTORS
  600  DO 800 J = 1,IM1
       SUM = 0.0
       DO 700 K = 1,N
       CON = Y(K,I)
  700  SUM = SUM + CON*Y(K,J)
       DO 800 K = 1,N
  800  Y(K,I) = Y(K,I)-SUM*Y(K,J)
       DO 850 K = 1,N
  850  V(K) = Y(K,I)
       SUM = 0.0
       DO 900 J = 1,N
       CON = Y(J,I)
  900  SUM = SUM + CON*V(J)
       SUM = DSQRT(SUM)
       DO 1000 J = 1,N
 1000  Y(J,I) = Y(J,I)/SUM
C ****COMPUTING SUBSEQUENT ALFAS ------
       DO 1050 K = 1,N
 1050  V(K) =Y(K,I)
       CALL SOLU(ST,N,NBAND,V)
       CALL ULT(SM,V,W,N,NBAND)
       CALL SOLUT(ST,N,NBAND,W)
       DO 250 L = 1,N
  250  W(L) = W(L) - BETA(IM1)*Y(L,IM1)
       SUM = 0.0
       DO 160 L = 1,N
  160  SUM = SUM+W(L)*Y(L,I)
       ALFA(I) = SUM
       IF ( I.EQ.M ) GO TO 2001
C****INITIALISING STEP FOR SUBSEQENT BETA VALUES
       DO 170 L = 1,N
  170  V(L) = W(L)-ALFA(I)*Y(L,I)
 1300 CONTINUE
C**** FORMING TRIDIAGONAL MATRIX IN BANDED FORM
 2001 DO 333 I = 1,M
       ST1(I,1) = ALFA(I)
  333  ST1(I,2) = BETA(I)
       WRITE(16,*)'XXXX TRANSFORMED TRIDIOGANAL
      1MATRIX XXXX'
       WRITE (16,444)((ST1(I,J),J=1,2),I=1,M)
```

```
 444  FORMAT(2(2X,D12.5))
      NIT = 20
C ****CALLING FREQNC TO OBTAIN FREQUENCIES OF
C*****TRIDIGONAL MATRIX
      CALL FREQNC(ST1,M,2,NEIG,NIT,1,U2)
      DO 777 I = 1,N
      DO 666 J = 1,NEIG
      EI(I,J) = 0.0
      DO 555 K = 1,M
 555  EI(I,J)=EI(I,J)+Y(I,K)*U2(K,J)
 666  CONTINUE
 777  CONTINUE
C*****TO GET THE ACTUAL EIGEN VECTORS OF
C*****ORIGINAL PROBLEM BY BACK SUBSTITUTION
      DO 778 I = 1,M
      DO 779 J = 1,N
 779  U(J) = EI(J,I)
      CALL SOLU(ST,N,NBAND,U)
      DO 780 L=1,N
 780  EI(L,I)=U(L)
 778  CONTINUE
      DO 3002 KN=1,10
      DO 3002 I=1,20
      DO 3003 J=1,3
      JN=(KN-1)*3+J
      IN=(I-1)*3+J
3003  EA(I,JN)=EI(IN,KN)
3002  CONTINUE
      WRITE (16,8820)
8820  FORMAT(//,25X,'-REAL EIGEN VECTORS-',3X)
      WRITE(16,*) 'REAL EIGEN VECTORS '
      WRITE(16,888) ((EI(I,J),J=1,NEIG),I=1,N)
888   FORMAT(4(2X,D12.5))
C     DO 3110 I=1,5
C     JN=I*2
C     JM=JN-1
C     WRITE(16,3660)JM,JN
C     WRITE (16,3661)
C     MN=(I-1)*6+1
C     MM=MN+5
C     DO 3121 K=1,20
C     K1=K+1
```

```
      RETURN
      END

      SUBROUTINE FREQNC(ST,NEQ,KB,NEIG,NIT
     1,KPRNT,U)
      IMPLICIT REAL*8(A-H,O-Z)
      COMMON DELTA
      DIMENSION ST(NEQ,KB),V(7,4),AU(7,4),
     *BE(4,4),TE(4,4),S(4,4),D(4,4),U(7,4)
      ABS(X)=DABS(X)
      SQRT(X)=DSQRT(X)
      IN=NEIG
      NDF=NEQ
      DO 5 I=1,NEQ
    5 ST(I,1)=ST(I,1)
C*****ESTABLISH STARTING ITERATION VECTORS
      DO 30 J=1,NEIG
      DO 10 I=J,NEQ
      ENI=NEQ
      EI=I
   10 U(I,J)=1.*(ENI+1.-EI)/ENI
      DO 20 I=1,J
      EJ=J
      EI=I
   20 U(I,J)=EI/EJ
   30 CONTINUE
      DO 40 I=1,NEIG
   40 U(I,I)=1.
      DO 50 I=1,NEQ
      DO 50 J=1,NEIG
   50 AU(I,J)=U(I,J)
C*****START OF ITERATION LOOP
      DO 450 II=1,NIT
      CALL BANMUL(ST,U,V,NEQ,KB,NEIG)
C ****INTERACTION ANALYSIS
      DO 170 I=1,IN
      DO 170 J=1,IN
      BE(I,J)=0.
      DO 170 L=1,NDF
  170 BE(I,J)=BE(I,J)+AU(L,I)*V(L,J)
      IF(KPRNT.EQ.0)GO TO 190
      IF(II.EQ.1)WRITE(16,1060)
```

```
        WRITE(16,1070)II
        DO 180 I=1,IN
180     WRITE(16,1010)(BE(I,J),J=1,IN)
C*****PREDICTION OF LATENT VECTORS
190     DO 270 I=1,IN
        DO 270 J=1,IN
        IF (I-J) 200,260,200
200     P=-2.*BE(I,J)
        T=BE(I,I)-BE(J,J)
        IF (T) 220,210,220
210     SI=1.
        GO TO 230
220     SI=T/ABS(T)
230     SQ=SQRT(T**2+ 4.*P**2)*SI
        IF (P) 250,240,250
240     TE(I,J)=0.
        GO TO 270
250     TE(I,J)=P/(T+SQ)
        GO TO 270
260     TE(I,J)=1.
270     CONTINUE
        DO 280 I=1,NDF
        DO 280 J=1,IN
        U(I,J)=0.
        DO 280 L=1,IN
280     U(I,J)=U(I,J)+V(I,L)*TE(L,J)
C*****ORTHONORMALISE THE RESULT VECTORS
290     DO 300 I=1,IN
        DO 300 J=1,IN
        D(I,J)=0.
        DO 300 L=1,NDF
300     D(I,J)=D(I,J)+U(L,I)*U(L,J)
        K=1
310     DO 390 J=1,IN
        DO 390 I=1,IN
        IF (J-K) 390,320,360
320     IF (I-K) 330,340,350
330     DD=ABS(D(K,K))
        S(I,J)=D(I,J)/SQRT(DD)
        GO TO 390
340     DD=ABS(D(K,K))
        S(I,J)=1./SQRT(DD)
```

```
          GO TO 390
350    S(I,J)=0.
          GO TO 390
360    IF(I-K)370,380,370
370    S(I,J)=D(I,J)-D(I,K)*D(K,J)/D(K,K)
          GO TO 390
380    S(I,J)=-D(K,J)/D(K,K)
390    CONTINUE
          DO 400 I=1,IN
          DO 400 J=1,IN
400    D(I,J)=S(I,J)
          K=K+1
          IF (K-IN) 310,310,410
410    DO 420 I=1,NDF
          DO 420 J=1,IN
          V(I,J)=0.
          DO 420 L=1,IN
420    V(I,J)=V(I,J)+U(I,L)*D(L,J)
          DO 440 I=1,NDF
          DO 440 J=1,IN
          U(I,J) = V(I,J)
440    AU(I,J)=U(I,J)
450       CONTINUE
C*****END OF ITERATION LOOP
C*****CALCULATE NATURAL FREQUENCIES
          DO 460 I=1,NEIG
          XX=ABS(1.0/BE(I,I)-DELTA)
460    BE(I,I)=SQRT(XX)
C*****PRINT NATURAL FREQUENCIES
          WRITE(16,1000)
          WRITE(16,1010)(BE(I,I),I=1,NEIG)
100    FORMAT(//40X,'NATURAL FREQUENCIES
      1(RAD/S) ARE'/)
1010   FORMAT(5X,10(1X,D11.4))
1060   FORMAT(//40X,'ITERATION RESULTS',3X)
1070   FORMAT(//40X,'ITERATION',I2)
          RETURN
          END
          SUBROUTINE BANMUL(SM,U,V,NEQ,KBAND,NEIG)
          IMPLICIT REAL*8(A-H,O-Z)
          DIMENSION SM(7,2),U(7,4),V(7,4)
          DO 130 J=1,NEIG
```

```
      DO 120 I=1,NEQ
      V(I,J)=0.
      IF (I.EQ.1) GO TO 100
      IM=I-1
      DO 90 K1=1,IM
      I1=I-K1+1
      IF (I1.GT.KBAND) GO TO 90
      V(I,J)=V(I,J)+SM(K1,I1)*U(K1,J)
90    CONTINUE
100   DO 110 K2=I,NEQ
      I2=K2-I+1
      IF (I2.GT.KBAND) GO TO 110
      V(I,J)=V(I,J)+SM(I,I2)*U(K2,J)
110   CONTINUE
120   CONTINUE
130   CONTINUE
      RETURN
      END
```

```
STIFFNESS MATRIX IN BANDED FORM
          10.     -10.      0.      0.
          20.       0.      0.    -10.
          10.     -10.      0.      0.
          20.     -10.      0.      0.
          30.     -10.      0.      0.
          20.     -10.      0.      0.
          10.       0.      0.      0.
```

```
MASS MATRIX IN BANDED FORM
           1.       0.      0.      0.
           1.       0.      0.      0.
           1.       0.      0.      0.
           1.       0.      0.      0.
           2.       0.      0.      0.
           5.       0.      0.      0.
           5.       0.      0.      0.
```

```
SIZE OF TRIDIAG. MATRIX M=            5
%%%% TRANSFORMED TRIDIOGANAL MATRIX %%%%
    .50000D+00     .27159D-03
    .28247D+00     .12677D-03
    .16982D+00     .73647D-02
```

```
    .14487D+00    .79046D-04
    .65097D-01    .00000D+00
```

NATURAL FREQUENCIES (RAD/S) ARE

```
2107D-07    .1241D+01    .1954D+01    .2236D+01
```

REAL EIGEN VECTORS
```
.17678D+00   -.19684D+00    .24934D+00   -.86998D-01
.17678D+00   -.16652D+00    .15410D+00   -.43499D-01
.17678D+00   -.19684D+00   -.24934D+00   -.86998D-01
.17678D+00   -.16652D+00   -.15410D+00   -.43499D-01
.17678D+00   -.11056D+00    .39556D-10    .21749D-01
.17678D+00    .35433D-01    .50400D-11    .13050D+00
.17678D+00    .15414D+00   -.16315D-11   -.86998D-01
```

```
C******* PROGRAM:5.1
C**** SOLVING THE EXAMPLE 5.1 BY
C     WILSON THETA METHOD
      IMPLICIT REAL*8(A-H,O-Z)
      DIMENSION AKK(2,2),AMM(2,2),RD(2),
     *RCAP(2,2),AKCAP(2,2),AKCAPS(2,2),DD(2)
     * ,DV(2),DA(2)
         DATA DD/2*0.0D0/,DV/2*0.0D0/,
         1DA/2*0.0D0/
      READ 101,DELT,THETA
      READ 102,NEQ,KBAND,NSTEP
      READ 101,((AKK(I,J),I=1,NEQ),J=1,KBAND)
      READ 101,((AMM(I,J),I=1,NEQ),J=1,KBAND)
      READ 101,(RD(I),I=1,NEQ)
C***  DATA INITIALISATION  ***
      DD(1)=1.0
      DA(1)=-2.0
      DA(2)=-2.0
C***  ALL OTHER VALUES REMAIN UNCHANGED  ***
      PRINT 205
      PRINT 201,DELT,NEQ,KBAND,NSTEP,THETA
      PRINT 202
      PRINT 203,((AKK(I,J),J=1,KBAND),I=1,NEQ)
      PRINT 204
      PRINT 203,((AMM(I,J),J=1,KBAND),I=1,NEQ)
      PRINT 206
      PRINT 207
      DO 10 I=1,NEQ
      PRINT 208,DD(I),DV(I),DA(I)
10    CONTINUE
      PRINT 209
      CALL WILTHE(DELT,NEQ,KBAND,NSTEP,THETA,AKK
     *,AMM,RD,AKCAP,AKCAPS,RCAP,DD,DV,DA)
101   FORMAT(F6.2)
102   FORMAT(I2)
201   FORMAT(1X,'DELT=',F5.2,3X,'NEQ=',I3,3X,
     *'KBAND=',I2,/,1X,'NSTEP=',I2,3X,
     *'THETA=',F5.2)
202   FORMAT(2X,'THE STIFFNESS MATRIX IS:')
203   FORMAT(2(5X,F5.3))
204   FORMAT(2X,'THE MASS MATRIX IS:')
205   FORMAT(2X,' INPUT INFORMATION:')
```

```
206    FORMAT(/,2X,'INITIAL CONDITIONS')
207    FORMAT(5X,'DISP',5X,'VEL',5X,'ACCN')
208    FORMAT(3(4X,F5.2))
209    FORMAT(2X,'INPUT INFORMATION ENDS')
       STOP
       END
       SUBROUTINE WILTHE(DELT,NEQ,KBAND,NSTEP,
      *THETA,AKK,AMM,RD,AKCAP,AKCAPS,RCAP,
      *DD,DV,DA)
C***   TIME INTEGRATION-WILSON THETA METHOD
C***   DESCRIPTION OF PARAMETERS
C***   AKK-INPUT STIFFNESS MATRIX IN BANDED FORM
C***   AMM-INPUT MASS MATRIX IN BANDED FORM
C***   NEQ-NUMBER OF EQUATIONS
C***   KBAND-BANDWIDTH
C***   RD-FORCE VECTOR OF SIZE NEQ;DESTROYED IN
C***   COMPUTATION AND REPLACED BY DISPLACEMENT
C***   IN THIS SAMPLE LISTING RD IS TIME INVARIANT
C***   DELT-TIME STEP
C***   NSTEP-DESIRED NUMBER OF STEPS
C***   DD-DISPLACEMENT VECTOR
C***   DV-VELOCITY VECTOR
C***   DA-ACCELERATION VECTOR
       IMPLICIT REAL*8(A-H,O-Z)
       DIMENSION AMM(NEQ,KBAND),AKK(NEQ,KBAND),
       RD(NEQ),AKCAP(NEQ,KBAND),AKCAPS(NEQ,KBAND),
      *DD(NEQ),DV(NEQ),DA(NEQ),RCAP(NEQ)
       THETA=1.4
        A0=6./((THETA*DELT)**2)
        A1=3./(THETA*DELT)
        A2=A1*2.
        A3=THETA*DELT/2.
        A4=A0/THETA
        A5=-A2/THETA
        A6=1.-3./THETA
        A7=DELT/2.
        A8=DELT**2/6.
        PRINT 221
221    FORMAT(2X,'THE CONSTANTS ARE:')
        PRINT 222,A0,A1,A2,A3,A4,A5,A6,A7
222    FORMAT(2X,'A0=',F8.3,
      *5X,'A1=',F8.3,/,2X,'A2=',F8.3,5X,'A3=',
```

```
      *F8.3,5X,'A4=',F8.3,/,2X,'A5=',F8.3,5X,
      *'A6=',F8.3,5X,'A7=',F8.3)
       DO 10 I=1,NEQ
       DV(I)=0.
       DD(I)=0.
   10  CONTINUE
       DD(1)=1.0
       DA(1)=-2.0
       DA(2)=-2.0
       DO 20 I=1,NEQ
       DO 20 J=1,KBAND
       AKCAPS(I,J)=AKK(I,J)+A0*AMM(I,J)
   20  CONTINUE
C***   TIME LOOP     ***
       T=0.
       DO 80 I=1,NSTEP
       PRINT 111,T
       PRINT 112
       DO 40 J=1,NEQ
       PRINT 113,DD(J),DV(J),DA(J)
   40  CONTINUE
       DO 50 MI=1,2
       DO 50 MJ=1,2
   50  AKCAP(MI,MJ)=AKCAPS(MI,MJ)
       DO 60 J=1,NEQ
       RCAP(J)=RD(J)+AMM(J,1)*(A0*DD(J)+A2*DV(J)+
      1 2.*DA(J))
   60  CONTINUE
       CALL UPTRI(AKCAP,NEQ,KBAND)
       CALL SOLUT(AKCAP,NEQ,KBAND,RCAP)
       CALL SOLU(AKCAP,NEQ,KBAND,RCAP)
       DO 70 L=1,NEQ
       DAA=A4*(RCAP(L)-DD(L))+A5*DV(L)+A6*DA(L)
       DVV=DV(L)+A7*DA(L)+A7*DAA
       DD(L)=DD(L)+DELT*DV(L)+A8*(DAA+2.*DA(L))
       DA(L)=DAA
       DV(L)=DVV
   70  CONTINUE
       T=T+DELT
   80  CONTINUE
  111  FORMAT(/5X,'T= ',F10.6/)
  112  FORMAT(5X,'DISP',6X,'VEL',7X,'ACCN')
```

```
113    FORMAT(3(2X,F8.3))
       RETURN
       END
```

```
 INPUT INFORMATION:
DELT= 0.25   NEQ=  2   KBAND= 2
NSTEP=12   THETA= 0.50
 THE STIFFNESS MATRIX IS:
     2.000      2.000
     5.000      0.000
 THE MASS MATRIX IS:
     1.000      0.000
     1.000      0.000
```

```
INITIAL CONDITIONS
   DISP     VEL      ACCN
   1.00     0.00     -2.00
   0.00     0.00     -2.00
INPUT INFORMATION ENDS
THE CONSTANTS ARE:
A0=  48.980    A1=   8.571
A2=  17.143    A3=   0.175    A4=  34.985
A5= -12.245    A6=  -1.143    A7=   0.125
```

```
     T=   0.000000
     DISP      VEL       ACCN
     1.000     0.000     -2.000
     0.000     0.000     -2.000

     T=   0.250000
     DISP      VEL       ACCN
     0.941    -0.461     -1.685
    -0.057    -0.432     -1.456

     T=   0.500000
     DISP      VEL       ACCN
     0.779    -0.807     -1.089
    -0.200    -0.670     -0.448

     T=   0.750000
     DISP      VEL       ACCN
     0.551    -0.987     -0.345
```

```
   -0.369    -0.632    0.754


   T=   1.000000
   DISP      VEL      ACCN
   0.301    -0.987    0.345
  -0.493    -0.317    1.762


   T=   1.250000
   DISP      VEL      ACCN
   0.070    -0.843    0.802
  -0.512     0.183    2.240


   T=   1.500000
   DISP      VEL      ACCN
  -0.115    -0.627    0.929
  -0.398     0.715    2.019


   T=   1.750000
   DISP      VEL      ACCN
  -0.244    -0.418    0.739
  -0.166     1.112    1.156


   T=   2.000000
   DISP      VEL      ACCN
  -0.330    -0.282    0.351
   0.136     1.246   -0.086


   T=   2.250000
   DISP      VEL      ACCN
  -0.394    -0.246   -0.060
   0.432     1.070   -1.323


   T=   2.500000
   DISP      VEL      ACCN
  -0.460    -0.293   -0.317
   0.649     0.632   -2.179


   T=   2.750000
   DISP      VEL      ACCN
  -0.543    -0.371   -0.303
   0.736     0.060   -2.400
COMPILE TIME=0.75 SEC,EXECUTION TIME=0.35 SEC
```

```
C*****PROGRAM :5.2
C*****SOLVING THE EXAMPLE 5.1 BY NEWMARK METHOD
      IMPLICIT REAL*8(A-H,O-Z)
      DIMENSION AKK(2,2), AMM(2,2),RD(2),
     *RCAP(2, 2),AKCAP(2,2),AKCAPS(2,2),DD(2)
     *,DV(2),DA(2)
      DATA DD/2*0.0D0/,DV/2*0.0D0/,DA/2*0.0D0/
      READ 101,DELT,DELTA
      READ 102,NEQ,KBAND,NSTEP
      READ 101,((AKK(I,J),I=1,NEQ),J=1,KBAND)
      READ 101,((AMM(I,J),I=1,NEQ),J=1,KBAND)
      READ 101,(RD(I),I=1,NEQ)
C*****DATA INITIALISATION
      DD(1)=1.0
      DA(1)=-2.0
      DA(2)=-2.0
C*****ALL OTHER VALUES REMAIN UNCHANGED
      PRINT 205
      PRINT 201,DELT,DELTA,KBAND,NSTEP,NEQ
      PRINT 202
      PRINT 203,((AKK(I,J),J=1,KBAND),I=1,NEQ)
      PRINT 204
      PRINT 203,((AMM(I,J),J=1,KBAND),I=1,NEQ)
      PRINT 206
      PRINT 207
      DO 10 I=1,NEQ
      PRINT 208,DD(I),DV(I),DA(I)
10    CONTINUE
      PRINT 209
      CALL NUMARK(DELT,NEQ,KBAND,NSTEP,DELTA,
     *AKK,AMM,RD,AKCAP,AKCAPS,RCAP,DD,DV,DA)
101   FORMAT(F6.2)
102   FORMAT(I2)
201   FORMAT(/,2X,'DELT=',F5.2,2X,'DELTA=',F5.2,
     *6X,'KBAND=',I2,/,10X,'NSTEP=',I2,6X,
     *'NEQ=',I2)
202   FORMAT(6X,'THE STIFFNESS MATRIX IS:')
203   FORMAT(2(8X,F5.3))
204   FORMAT(8X,'THE MASS MATRIX IS:')
205   FORMAT(8X,' INPUT INFORMATION:')
206   FORMAT(8X,'INITIAL CONDITIONS')
207   FORMAT(8X,'DISP',4X,'VEL',5X,'ACCN')
```

```
208    FORMAT(7X,F5.2,2F8.2)
209    FORMAT(/8X,'INPUT INFORMATION ENDS')
       STOP
       END
       SUBROUTINE NUMARK(DELT,NEQ,KBAND,NSTEP,
      * DELTA,AKK,AMM,RD,AKCAP,AKCAPS,RCAP,
      * DD,DV,DA)
C*****
C***** TIME INTEGRATION SCHEME-NEWMARK'S METHOD
C***** DESCRIPTION OF PARAMETERS
C***** AKK    INPUT STIFFNESS MATRIX
C*****              IN BANDED FORM
C***** AMM    INPUT MASS MATRIX
C*****          IN BANDED FORM
C***** NEQ    NUMBER OF EQUATIONS
C***** KBAND  BANDWIDTH
C***** RD     FORCE VECTOR OF SIZE NEQ;DESTROYED
C*****          IN COMPUTATION AND REPLACED BY THE
C*****          DISPLACEMENT VECTOR.
C*****          IN THIS SAMPLE LISTING RD IS
C*****          TIME INVARIANT.
C*****          DELT  TIME STEP
C*****          NSTEP DESIRED NUMBER OF STEPS
C***** DD     DISPLACEMENT VECTOR
C***** DV     VELOCITY VECTOR
C***** DA     ACCELERATION VECTOR
C***** NOTE:- DD,DV,DA ARE WORKING VECTORS
       IMPLICIT REAL*8(A-H,O-Z)
       DIMENSION AMM(NEQ,KBAND),AKK(NEQ,KBAND),
      *RD(NEQ),AKCAP(NEQ,KBAND), AKCAPS
      *(NEQ,KBAND),DD(NEQ),DV(NEQ),DA(NEQ),
      *RCAP(NEQ)
       ALFA=0.25*(0.5+DELTA)**2
       A0=1./(ALFA*(DELT)**2)
       A1=DELTA/(ALFA*DELT)
       A2=1./(ALFA*DELT)
       A3=1./(2*ALFA)-1.
       A4=DELTA/ALFA-1.
       A5=(DELT/2.)*((DELTA/ALFA)-2.)
       A6=DELT*(1.-DELTA)
       A7=DELTA*DELT
       PRINT 221
```

```
221    FORMAT(2X,'THE CONSTANTS ARE:')
       PRINT 222,ALFA,A0,A1,A2,A3,A4,A5,A6,A7
222    FORMAT(2X,'ALFA=',F5.3,6X,'A0=',F8.3,
      *5X,'A1=',F8.3,/,2X,'A2=',F8.3,5X,'A3=',
      *F8.3,5X,'A4=',F8.3,/,2X,'A5=',F8.3,5X,
      *'A6=',F8.3,5X,'A7=',F8.3)
       DO 30 I=1,NEQ
       DO 30 J=1,KBAND
       AKCAPS(I,J)=AKK(I,J)+A0*AMM(I,J)
 30    CONTINUE
C**** TIME LOOP BEGINS
       T=0.
       DO 80 I=1,NSTEP
       PRINT 223,T
       PRINT 224
       DO 40 J=1,NEQ
       PRINT 225,DD(J),DV(J),DA(J)
 40    CONTINUE
       DO 50 MI=1,2
       DO 50 MJ=1,2
 50    AKCAP(MI,MJ)=AKCAPS(MI,MJ)
       DO 60 J=1,NEQ
       RCAP(J)=RD(J)+AMM(J,1)*(A0*DD(J)+A2*DV(J)+
      1 A3*DA(J))
 60    CONTINUE
       CALL UPTRI(AKCAP,NEQ,KBAND)
       CALL SOLUT(AKCAP,NEQ,KBAND,RCAP)
       CALL SOLU(AKCAP,NEQ,KBAND,RCAP)
       DO 70 L=1,NEQ
       DAA=A0*(RCAP(L)-DD(L))-A2*DV(L)-A3*DA(L)
       DVV=DV(L)+A6*DA(L)+A7*DAA
       DA(L)=DAA
       DV(L)=DVV
       DD(L)=RCAP(L)
 70    CONTINUE
       T=T+DELT
 80    CONTINUE
223    FORMAT(/5X,'T= ',F10.6/)
224    FORMAT(5X,'DISP',6X,'VEL',7X,'ACCN')
225    FORMAT(3(2X,F8.3))
       RETURN
       END
```

INPUT INFORMATION:

DELT= 0.25 DELTA= 0.50   KBAND= 2
        NSTEP=12   NEQ= 2
    THE STIFFNESS MATRIX IS:
        2.000   2.000
        5.000   0.000
    THE MASS MATRIX IS:
        1.000   0.000
        1.000   0.000
    INITIAL CONDITIONS
    DISP   VEL   ACCN
    1.00   0.00   -2.00
    0.00   0.00   -2.00

    INPUT INFORMATION ENDS

THE CONSTANTS ARE:
ALFA=0.250   A0= 64.000   A1= 8.000
A2= 16.000   A3= 1.000   A4= 1.000
A5=   0.000   A6= 0.125   A7= 0.125

    T=   0.000000
    DISP   VEL   ACCN
    1.000   0.000   -2.000
    0.000   0.000   -2.000

    T=   0.250000
    DISP   VEL   ACCN
    0.941   -0.471   -1.770
    0.056   -0.450   -1.601

    T=   0.500000
    DISP   VEL   ACCN
    0.778   -0.836   -1.151
   -0.202   -0.718   -0.544

    T=   0.750000
    DISP   VEL   ACCN
    0.545   -1.022   -0.335
   -0.378   -0.686   0.799

```
T=   1.000000
DISP       VEL       ACCN
0.291     -1.010     0.430
-0.507    -0.343     1.950


T=   1.250000
DISP       VEL       ACCN
0.060     -0.841     0.925
-0.523     0.213     2.494


T=   1.500000
DISP       VEL       ACCN
-0.120    -0.596     1.031
-0.396     0.802     2.219


T=   1.750000
DISP       VEL       ACCN
-0.240    -0.371     0.765
-0.142     1.228     1.192


T=   2.000000
DISP       VEL       ACCN
-0.317    -0.241     0.276
0.179      1.344    -0.262


T=   2.250000
DISP       VEL       ACCN
-0.377    -0.234    -0.217
0.485      1.102    -1.673


T=   2.500000
DISP       VEL       ACCN
-0.446    -0.323    -0.496
0.694      0.571    -2.579


T=   2.750000
DISP       VEL       ACCN
0.541     -0.439    -0.427
0.755     -0.088    -2.691
```
COMPILE TIME= 0.68 SEC,EXECUTION TIME= 0.34 SEC

```
C***** PROGRAM NO :5.3
C***** SOLVING THE EXAMPLE PROBLEM 5.1 BY
C***** MODAL SUPERPOSITION METHOD.
      IMPLICIT REAL*8(A-H,O-Z)
      DIMENSION SM(2,2),U(2,2),V(2,2),
     *DX(2),VX(2),XM(2,1),XDM(2,1),
     *XI(2,1),XDI(2,1),F(2),XX(2),XXD(2),
     *FR(2),SMM(2),ZET(2)
      ATAN(X)=DATAN(X)
      READ(5,5)NEQ,KB,NEIG,NFV,NOMEG,NINT
      WRITE(6,1)
      WRITE(6,15)NEQ,KB,NEIG,NFV,NOMEG,NINT
      READ(5,10)((SM(I,J),J=1,KB),I=1,NEQ)
      WRITE(6,30)
      WRITE(6,25)((SM(I,J),J=1,KB),I=1,NEQ)
      READ(5,10)(XI(I,1),I=1,NEQ)
      READ(5,10)(XDI(I,1),I=1,NEQ)
      READ(5,10)(ZET(I),I=1,NEIG)
      WRITE(6,35)
      WRITE(6,40)(XI(I,1),I=1,NEQ)
      WRITE(6,45)
      WRITE(6,40)(XDI(I,1),I=1,NEQ)
      WRITE(6,50)
      WRITE(6,40)(ZET(I),I=1,NEIG)
      READ(5,10)(SMM(I),I=1,NEIG)
      WRITE(6,2)
      WRITE(6,3)(I,SMM(I),I=1,NEIG)
      READ(5,10)((U(I,J),J=1,NEIG),I=1,NEQ)
      WRITE(6,70)
      WRITE(6,75)((U(I,J),J=1,NEIG),I=1,NEQ)
      TUPI=1.0
      TUPI=8.0*ATAN(TUPI)
      DELT=TUPI/SMM(NOMEG)/10.0
C*****TO COMPARE WITH INTEGRATION
C*****METHODS ASSUME DELT=0.25
      DELT=0.25
      T=DELT
      DO 200 KK=1,NINT
      IF(KK.GT.NFV)GO TO 100
      READ(5,10) (F(I),I=1,NEQ)
      WRITE(6,65)
      WRITE(6,40)(F(I),I=1,NEQ)
```

```
100    CONTINUE
       WRITE(6,60) KK,T
       CALL MODAL(NEQ,KB,NEIG,NOMEG,KK,NFV,SM,
      *SMM,U,V,XI,XDI,XM,XDM,ZET,F,XX,XXD,FR,DX,
      *VX,DELT)
       T=T+DELT
200    CONTINUE
1      FORMAT(//,10X,'MODAL SUPERPOSITION',//)
2      FORMAT(/,10X,'NATURAL FREQUENCIES IN
      *RAD/SEC',/)
3      FORMAT(10X,'OMEGA(',I2,')=',D20.10)
5      FORMAT(16I5)
10     FORMAT(2F10.4)
15     FORMAT(/,10X,'NEQ=',I4,5X,'KB=',I3,5X,
      *'NEIG=', I2,/,10X,'NFV=',I2,5X,'NOMEG=',I2
      *,5X,'NINT=',I3,/)
25     FORMAT(2D20.10)
30     FORMAT(/,10X,'MASS MATRIX',/)
35     FORMAT(/,10X,'INITIAL DISPLACEMENT
      *VECTOR',/)
40     FORMAT(/,5X,2D13.5,/)
45     FORMAT(/,10X,'INITIAL VELOCITY VECTOR',/)
50     FORMAT(/,10X,'MODAL DAMPING PARAMETER',/)
60     FORMAT(/,10X,'RESPONSE(',I3,') AT TIME=',
      *F15.7,'SEC.',/)
65     FORMAT(/,10X,'FORCE VECTOR',/)
70     FORMAT(/,10X,'EIGEN VECTOR',/)
75     FORMAT(10X,2D13.5)
       RETURN
       END
       SUBROUTINE MODAL(NEQ,KB,NEIG,NOMEG,KK,NFV,
      *SM,SMM,U,V,XI,XDI,XM,XDM,ZET,F,XX,XXD, FR,
      *DX,VX,T)
C***** METHOD:-
C***** RESPONSE BY MODAL SUPERPOSITION
C***** DESCRIPTION OF PARAMETERS:-
C***** INPUT VARIABLES:-
C***** DELT    TIME INCREMENT IN SECS
C***** NINT    NO.OF TIME INTERVALS
C***** NEQ     SIZE OF MASS MATRIX
C***** KB      BANDWIDTH
C***** NEIG    NO.OF FREQUENCIES SUPPLIED
```

```
C***** NOMEG  NO.OF FREQUENCIES USED IN
C*****        MODAL ANALYSIS
C***** NFV    NO.OF INTERVALS FOR WHICH FORCE
C*****        IS PRESENT
C***** SM     MASS MATRIX OF SIZE(NEQ,KB)
C***** SMM    NATURAL FREQUENCY VECTOR OF
C*****        SIZE (NEIG)
C***** U      ORTHO NORMALISED EIGEN VECTOR OF
C*****        SIZE (NEQ,NEIG)
C***** XI     INITIAL DISPLACEMENT VECTOR
C*****        OF SIZE (NEQ,1)
C***** XDI    INITIAL VELOCITY VECTOR
C*****        OF SIZE (NEQ,1)
C***** ZET    DAMPING COEFFICIENT VECTOR
C*****        OF SIZE (NEIG)
C***** F      FORCING VECTOR OF SIZE (NEQ)
C*****     WORKING VARIABLES:-
C***** XM,XDM ARE WORKING VECTORS
C*****            OF SIZE (NEQ,1)
C***** XX,XXD,FR,DX,VX ARE WORKING VECTORS
C*****            OF SIZE (NEIG)
C***** V      WORKING VECTOR OF SIZE(NEQ,NEIG)
      IMPLICIT REAL*8(A-H,O-Z)
      DIMENSION SM(NEQ,KB),U(NEQ,NEIG),
     *V(NEQ,NEIG),XM(NEQ,1),XDM(NEQ,1),F(NEQ),
     *XX(NEIG),XXD(NEIG),ZET(NEIG),DX(NEIG),
     *VX(NEIG),XI(NEQ,1),FR(NEIG),SMM(NEIG),
     *XDI(NEQ,1)
      SQRT(X)=DSQRT(X)
      EXP(X)=DEXP(X)
      SIN(X)=DSIN(X)
      COS(X)=DCOS(X)
      K=KK
      IF(K.GT.1)GOTO 191
C*****FINDING INITIAL DISPLACEMENT AND VELOCITY
C*****IN MODAL COORDINATES
      CALL MATMUL(NEQ,KB,1,1,SM,XI,XM)
      CALL MATMUL(NEQ,KB,1,1,SM,XDI,XDM)
      DO 190 J=1,NOMEG
      XX(J)=0.0
      XXD(J)=0.0
      DO 190 I=1,NEQ
```

```
       XX(J)=XX(J)+U(I,J)*XM(I,1)
190    XXD(J)=XXD(J)+U(I,J)*XDM(I,1)
191    CONTINUE
C***** FINDING FORCE IN MODAL CO-ORDINATES
       IF(KK.GT.NFV) GO TO 193
       DO 192 J=1,NOMEG
       FR(J)=0.0
       DO 192 I=1,NEQ
192    FR(J)=FR(J)+U(I,J)*F(I)
       GOTO 195
193    CONTINUE
       DO 194 I=1,NOMEG
194    FR(I)=0.0
195    CONTINUE
C***** CALCULATION OF MODAL RESPONSE
       DO 210 I=1,NOMEG
       Z=ZET(I)
       OM=SMM(I)
       OM2=OM*OM
       OMD=(1.0-Z*Z)
       OMD=SQRT(OMD)*OM
       C1=(XX(I)-FR(I)/OM2)
       C2=(XXD(I)+Z*OM*C1)/OMD
       DX(I)=EXP(-Z*OM*T)*(C1*COS(OMD*T)+
      *C2*SIN(OMD*T))+FR(I)/OM2
       VX(I)=OM*EXP(-Z*OM*T)*((C2-C1*Z) *
      1 COS (OMD *T)-(C1+Z*C2)*SIN(OMD*T))
210    CONTINUE
       DO 225 I=1,NEQ
       X=0.0
       DO 220 KL=1,NEIG
220    X=X+U(I,KL)*DX(KL)
225    V(I,1)=X
       WRITE(6,1100)(V(I,1),I=1,NEQ)
       IF(K.GE.NFV)GO TO 310
       DO 300 I=1,NOMEG
       XX(I)=DX(I)
300    XXD(I)=VX(I)
310    CONTINUE
1100   FORMAT(5X,5D20.10,/)
       RETURN
       END
```

```
      SUBROUTINE MATMUL(NEQ,KB,NEIG,NOMEG,
     *SM,U,V)
C*****PERFORM V=SM*U
      IMPLICIT REAL*8(A-H,O-Z)
      DIMENSION SM(NEQ,KB),U(NEQ,NEIG),
     *V(NEQ,NEIG)
      DO 170 J=1,NOMEG
      DO 160 I=1,NEQ
      V(I,J)=0.0
      IF(I.EQ.1)GOTO140
      IM=I-1
      DO 130 K1=1,IM
      I1=I-K1+1
      IF(I1.GT.KB)GOTO130
      V(I,J)=V(I,J)+SM(K1,I1)*U(K1,J)
130   CONTINUE
140   DO 150 K2=I,NEQ
      I2=K2-I+1
      IF(I2.GT.KB)GOTO150
      V(I,J)=V(I,J)+SM(I,I2)*U(K2,J)
150   CONTINUE
160   CONTINUE
170   CONTINUE
      RETURN
      END

      MODAL SUPERPOSITION
          NEQ=  2     KB=  2     NEIG=  2
          NFV= 1    NOMEG= 2     NINT= 10

          MASS MATRIX

  0.1000000000D 01     0.0000000000D 00
  0.1000000000D 01     0.0000000000D 00

          INITIAL DISPLACEMENT VECTOR

   0.10000D 01  0.00000D 00

          INITIAL VELOCITY VECTOR

   0.00000D 00  0.00000D 00
```

MODAL DAMPING PARAMETER

0.00000D 00 0.00000D 00

NATURAL FREQUENCIES IN RAD/SEC

OMEGA(1)=   0.1000000000D 01
OMEGA(2)=   0.2449500000D 01

EIGENVECTOR

   0.89440D 00 0.44720D 00
  -0.44720D 00 0.89440D 00

FORCE VECTOR

0.00000D 00 0.00000D 00

RESPONSE(1) AT TIME=   0.2500000SEC.

   0.9387869668D 00   -0.6025090979D-01

RESPONSE(2) AT TIME=   0.5000000SEC.

   0.7699022824D 00   -0.2153605590D 00

RESPONSE(3) AT TIME=   0.7500000SEC.

   0.5327128293D 00   -0.3979520792D 00

RESPONSE(4) AT TIME=   1.0000000SEC.

   0.2782593896D 00   -0.5240858325D 00

RESPONSE(5) AT TIME=   1.2500000SEC.

   0.5289304388D-01   -0.5248586370D 00

RESPONSE(6) AT TIME=   1.5000000SEC.

  -0.1157023684D 00   -0.3728791401D 00

```
RESPONSE(  7) AT TIME=      1.7500000SEC.

  -0.2252001799D 00   -0.9390824857D-01

RESPONSE(  8) AT TIME=      2.0000000SEC.

  -0.2958115045D 00    0.2406706641D 00

RESPONSE(  9) AT TIME=      2.2500000SEC.

  -0.3592090431D 00    0.5379291592D 00

RESPONSE( 10) AT TIME=      2.5000000SEC.

  -0.4434245000D 00    0.7153417800D 00

COMPILE TIME=0.67 SEC,  EXECUTION TIME= 0.35 SEC
```

Appendix 1

1.1 Table A1 shows the positions and weighting coefficients [1] for Gaussian integration of the type

$$I = \int_{-1}^{+1} f(\xi)\, d\xi = \sum_{i=1}^{n} H_i\, f(\xi_i) \qquad (A1)$$

1.2 The most obvious way of obtaining the double integral

$$I = \int_{-1}^{+1}\int_{-1}^{+1} f(\xi, \eta)\, d\xi\, d\eta \quad \text{is to express it as}$$

$$\int_{-1}^{+1}\int_{-1}^{+1} f(\xi, n)\, d\xi\, dn = \sum_{i=1}^{n} \sum_{i=1}^{n} H_i H_j f(\xi_i, \eta_j) \qquad (A2)$$

1.3 For a right-hand prism, we similarly have

$$I = \int_{-1}^{+1}\int_{-1}^{+1}\int_{-1}^{+1} f(\xi, \eta, \zeta)\, d\xi\, d\eta\, d\zeta$$

$$= \sum_{i=1}^{n} \sum_{i=1}^{n} \sum_{i=1}^{n} H_i\, H_j\, H_m\, f(\xi_i, \eta_j, \zeta_m) \qquad (A3)$$

For integrating in each direction, it is not always necessary to have the same number of points.

1.4 In case of integration over a triangular domain in terms of area co-ordinates (Fig. A1), the integral is expressed in the following form

$$I = \int_{O}^{1}\int_{O}^{(1-L_2)} f(L_1, L_2, L_3)\, dL_1\, dL_2$$

$$= \sum_{i=1}^{n} \omega_i f(L_1, L_2, L_3) \qquad (A4)$$

Table A1

**Abscissae and Weight Coefficients of the
Gaussian Quadrature Formula**

$$\int_{-1}^{1} f(x)\, dx = \sum_{j=1}^{n} H_i f(a_j),$$

±a	H
n = 2	
0.57735 02691 89626	1.00000 00000 00000
n = 3	
0.77459 66692 41483	0.55555 55555 55556
0.00000 00000 00000	0.88888 88888 88889
n = 4	
0.86113 63115 94053	0.34785 48451 37454
0.33998 10435 84856	0.65214 51548 62546
n = 5	
0.90617 98459 38664	0.23692 68850 56189
0.53846 93101 05683	0.47862 86704 99366
0.00000 00000 00000	0.56888 88888 88889
n = 6	
0.93246 95142 03152	0.17132 44923 79170
0.66120 93864 66265	0.36076 15730 48139
0.23861 91860 83197	0.46791 39345 72691
n = 7	
0.94910 79123 42759	0.12948 49661 68870
0.74153 11855 99394	0.27970 53914 89277
0.40584 51513 77397	0.38183 00505 05119
0.00000 00000 00000	0.41795 91836 73469
n = 8	
0.96028 98564 97536	0.10122 85362 90376
0.79666 64774 13627	0.22238 10344 53374
0.52553 24099 16329	0.31370 66458 77887
0.18343 46424 95650	0.36268 37833 78362
n = 9	
0.96816 02395 07626	0.08127 43883 61574
0.83603 11073 26636	0.18064 81606 94857
0.61337 14327 00590	0.26061 06964 02935
0.32425 34234 03809	0.31234 70770 40003
0.00000 00000 00000	0.33023 93550 01260

(Contd)

	±a			H	
			n = 10		
0.97390	65285	17172	0.06667	13443	08688
0.86506	33666	88985	0.14945	13491	50581
0.67940	95682	99024	0.21908	63625	15982
0.43339	53941	29247	0.26926	67193	09996
0.14887	43389	81631	0.29552	42247	14753

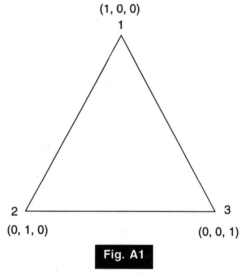

(1, 0, 0)
1

2
(0, 1, 0)

3
(0, 0, 1)

Fig. A1

Area coordinates

Table A2 shows the position and weighting coefficients which are due to Hammer et al [2].

Table A2

Numerical Integration Formulas for Triangles

Order	Fig.	Error	Points	Triangular Co-ordinates	Weights k
Linear		$R = O(h^2)$	1	$\frac{1}{3}, \frac{1}{3}, \frac{1}{3}$	1

(Contd)

Order	Fig.	Error	Points	Triangular Co-ordinates	Weights k
Quadratic		$R = O(h^3)$	1	$\frac{1}{2}, \frac{1}{2}, 0$	$\frac{1}{3}$
			2	$0, \frac{1}{2}, \frac{1}{2}$	$\frac{1}{3}$
			3	$\frac{1}{2}, 0, \frac{1}{2}$	$\frac{1}{3}$
Cubic		$R = O(h^4)$	1	$\frac{1}{3}, \frac{1}{3}, \frac{1}{3}$	$\frac{27}{48}$
			2	$\frac{11}{15}, \frac{2}{15}, \frac{2}{15}$	$\frac{25}{48}$
			3	$\frac{2}{15}, \frac{11}{15}, \frac{2}{15}$	
			4	$\frac{2}{15}, \frac{2}{15}, \frac{11}{15}$	

This formula not recommended due to negative weight and round-off error

Order	Fig.	Error	Points	Triangular Co-ordinates	Weights k
Cubic		$R = O(h^5)$	1	$\frac{1}{3}, \frac{1}{3}, \frac{1}{3}$	$\frac{27}{60}$
			2	$\frac{1}{2}, \frac{1}{2}, 0$	
			3	$0, \frac{1}{2}, \frac{1}{2}$	$\frac{8}{60}$
			4	$\frac{1}{2}, 0, \frac{1}{2}$	
			5	$1, 0, 0$	
			6	$0, 1, 0$	$\frac{3}{60}$
			7	$0, 0, 1$	
Quintic		$R = O(h^6)$	1	$\frac{1}{3}, \frac{1}{3}, \frac{1}{3}$	0.225
			2	$\alpha_1, \beta_1, \beta_1$	
			3	$\beta_1, \alpha_1, \beta_1$	0.13239415
			4	$\beta_1, \beta_1, \alpha_1$	
			5	$\alpha_2, \beta_2, \beta_2$	
			6	$\beta_2, \alpha_2, \beta_2$	0.12593918
			7	$\beta_2, \beta_2, \alpha_2$	

with $\alpha_1 = 0.05971587$
$\beta_1 = 0.47014206$
$\alpha_2 = 0.79742699$
$\beta_2 = 0.10128651$

Appendix 2

➤ **TRANSFORMATION OF CO-ORDINATES FOR THREE-DIMENSIONAL PROBLEMS**

Figure A2 shows two reference systems, the x, y, z axis representing the local system of co-ordinates and the X, Y, Z axes representing

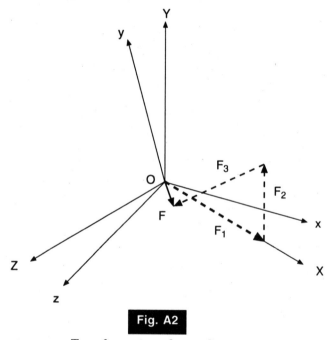

Fig. A2

Transformation of co-ordinates

the global system of co-ordinates. Also shown in this figure is a force vector F with its components F_1, F_2, F_3 along X, Y, Z. The component of the force F along the x co-ordinate is given by

$$f_1 = F_1 \cos xX + F_2 \cos xY + F_y \cos xZ \qquad (A5)$$

in which $\cos xX$ is the cosine of the angle between axes x and X and corresponding definitions for other cosines. Similarly, the y and z components of F are

$$f_2 = F_1 \cos yZ + F_2 \cos yY + F_3 \cos yZ \qquad \text{(A6)}$$

$$f_3 = F_1 \cos zX + F_2 \cos xY + F_3 \cos zZ \qquad \text{(A7)}$$

These equations are conveniently written in matrix notation as

$$\begin{Bmatrix} f_1 \\ f_2 \\ f_3 \end{Bmatrix} = \begin{bmatrix} \cos xX & \cos xY & \cos xZ \\ \cos yX & \cos yY & \cos yZ \\ \cos zX & \cos zY & \cos zZ \end{bmatrix} \begin{Bmatrix} F_1 \\ F_2 \\ F_3 \end{Bmatrix} \qquad \text{(A8)}$$

or in short notation

$$\{f\} = [T_1]\{F\} \qquad \text{(A9)}$$

In which $\{f\}$ and $\{F\}$ are respectively, the components in the local and global systems of the force F and the transformation matrix $[T_1]$ given by

$$[T_1] = \begin{bmatrix} \cos xX & \cos xY & \cos xZ \\ \cos yX & \cos yY & \cos yZ \\ \cos zX & \cos zY & \cos zZ \end{bmatrix} \qquad \text{(A10)}$$

This has application in three-dimensional beam elements and shell elements. The total number of degrees of freedom for a 3-D beam element for the 2 nodes at the ends is 12. Hence the δ_{local} and δ_{global} will each be 12×1. Hence, the relationship between the local displacement (or force) vector and the global displacement (or force) vector will be given by the following equation

$$\{\delta_{\text{local}}\}_{12\times1} = [T]\{\delta_{\text{global}}\}_{12\times1}$$

where

$$[T] = \begin{bmatrix} T_1 & 0 & 0 & 0 \\ 0 & T_1 & 0 & 0 \\ 0 & 0 & T_1 & 0 \\ 0 & 0 & 0 & T_1 \end{bmatrix}$$

REFERENCES

1. Zienkiewicz, O.C., *The Finite Element Method in Engineering Sciences*, 2nd edn, McGraw-Hill, London, 1971.
2. Hammer, P.C., O.P. Marlowe and A.H. Stroud, Numerical integration over simplexes and cones, *Math. tables aids., comput.*, vol. 10, pp 130–137, 1956.

Appendix 3

ELASTICITY MATRIX _____

Plane Stress:

$$[D_p] = \frac{E}{(1 - \mu^2)} \begin{bmatrix} 1 & \mu & 0 \\ \mu & 1 & 0 \\ 0 & 0 & \dfrac{(1 - \mu)}{2} \end{bmatrix}$$

Plate Flexure:

$$[D_f] = \frac{Eh^3}{12\,(1-\mu^2)} \begin{bmatrix} 1 & \mu & 0 \\ \mu & 1 & 0 \\ 0 & 0 & \dfrac{(1 - \mu)}{2} \end{bmatrix}$$

E — Modulus of elasticity
μ — Poisson's ratio
h — Thickness of the plate

Index